Global Analysis

Global Analysis

Differential Forms in Analysis, Geometry and Physics

Ilka Agricola
Thomas Friedrich

Translated by
Andreas Nestke

Graduate Studies
in Mathematics

Volume 52

American Mathematical Society
Providence, Rhode Island

2000 *Mathematics Subject Classification.* Primary 53–01;
Secondary 57–01, 58–01, 22-01, 74-01, 78-01, 80-01, 35-01.

This book was originally published in German by Friedr. Vieweg & Sohn Verlagsgesellschaft mbH, D-65189 Wiesbaden, Germany, as "Ilka Agricola und Thomas Friedrich: Globale Analysis. 1. Auflage (1st edition)", ©Friedr. Vieweg & Sohn Verlagsgesellschaft mbH, Braunschweig/Wiesbaden, 2001

Translated from the German by Andreas Nestke

Library of Congress Cataloging-in-Publication Data

Agricola, Ilka, 1973–
[Globale Analysis. English]
Global analysis : differential forms in analysis, geometry, and physics / Ilka Agricola, Thomas Friedrich ; translated by Andreas Nestke.
p. cm. — (Graduate studies in mathematics, ISSN 1065-7339 ; v. 52)
Includes bibliographical references and index.
ISBN 0-8218-2951-3 (alk. paper)
1. Differential forms. 2.Mathematical physics. I. Friedrich, Thomas, 1949– II. Title. III. Series.

QA381.A4713 2002
514′.74—dc21 2002027681

∞ The paper used in this book is acid-free and falls within the guidelines
established to ensure permanence and durability.
Visit the AMS home page at http://www.ams.org/

10 9 8 7 6 5 4 3 2 1 07 06 05 04 03 02

Preface

This book is intended to introduce the reader into the world of differential forms and, at the same time, to cover those topics from analysis, differential geometry, and mathematical physics to which forms are particularly relevant. It is based on several graduate courses on analysis and differential geometry given by the second author at Humboldt University in Berlin since the beginning of the eighties. From 1998 to 2000 the authors taught both courses jointly for students of mathematics and physics, and seized the opportunity to work out a self-contained exposition of the foundations of differential forms and their applications. In the classes accompanying the course, special emphasis was put on the exercices, a selection of which the reader will find at the end of each chapter. Approximately the first half of the book covers material which would be compulsary for any mathematics student finishing the first part of his/her university education in Germany. The book can either accompany a course or be used in the preparation of seminars.

We only suppose as much knowledge of mathematics as the reader would acquire in one year studying mathematics or any other natural science. From linear algebra, basic facts on multilinear forms are needed, which we briefly recall in the first chapter. The reader is supposed to have a more extensive knowledge of calculus. Here, the reader should be familiar with differential calculus for functions of several variables in euclidean space $\mathbb{R}^n$, the Riemann integral and, in particular, the transformation rule for the integral, as well as the existence and uniqueness theorem for solutions of ordinary differential equations. It is a reader with these prerequisites that we have in mind and whom we will accompany into the world of vector analysis, Pfaffian systems, the differential geometry of curves and surfaces in euclidean space, Lie groups and homogeneous spaces, symplectic geometry and mechanics, statistical mechanics and thermodynamics and, eventually, electrodynamics.

In Chapter 2 we develop the differential and integral calculus for differential forms defined on open sets in euclidean space. The central result is Stokes' formula turning the integral of the exterior derivative of a differential form over a singular chain into an integral of the form itself over the boundary of the chain. This is in fact a far-reaching generalization of the main theorem of differential and integral calculus: differentiation and integration are mutually inverse operations. At the end of a long historical development mathematicians reached the insight that a series of important integral formulas in vector analysis can be obtained by specialization from Stokes' formula. We will show this in the second chapter and derive in this way Green's first and second formula, Stokes' classical formula, and Cauchy's integral formula for complex differentiable functions. Furthermore, we will deduce Brouwer's fixed point theorem from Stokes' formula and the Weierstrass approximation theorem.

In Chapter 3 we restrict the possible integration domains to "smooth" chains. On these objects, called manifolds, a differential calculus for functions and forms can be developed. Though we only treat submanifolds of euclidean space, this section is formulated in a way to hold for every Riemannian manifold. We discuss the concept of orientation of a manifold, its volume form, the divergence of a vector field as well as the gradient and the Laplacian for functions. We then deduce from Stokes' formula the remaining classical integral formulas of Riemannian geometry (Gauss-Ostrogradski formula, Green's first and second formula) as well as the Hairy Sphere theorem, for which we decided to stick to its more vivid German name, "Hedgehog theorem". A separate section on the Lie derivative of a differential form leads us to the interpretation of the divergence of a vector field as a measure for the volume distortion of its flow. We use the integral formulas to solve the Dirichlet problem for the Laplace equation on the ball in euclidean space and to study the properties of harmonic functions on $\mathbb{R}^n$. For these we prove, among other things, the maximum principle and Liouville's theorem. Finally we discuss the Laplacian acting on forms over a Riemannian manifold, as well as the Hodge decomposition of a differential form. This is a generalization of the splitting of a vector field with compact support in $\mathbb{R}^n$ into the sum of a gradient field and a divergence-free vector field, going back to Helmholtz. In the final chapter we prove Helmholtz' theorem within the framework of electrodynamics.

Apart from Stokes' theorem, the integrability criterion of Frobenius is one of the fundamental results in the theory of differential forms. A geometric distribution (Pfaffian system) is defined by choosing a k-dimensional subspace in each tangent space of an n-dimensional manifold in a smooth way. A geometric distribution can alternatively be described as the zero set of a set of linearly independent 1-forms. What one is looking for then is an answer to the question of whether there exists a k-dimensional submanifold such that, at each point, the tangent space coincides with the value of the given geometric distribution. Frobenius' theorem gives a complete solution to this question and provides a basic tool for the integration of certain systems of first order partial differential equations. Therefore, Chapter 4 is devoted to a self-contained and purely analytical proof of this key result, which, moreover, will be needed in the sections on surfaces, symplectic geometry, and completely integrable systems.

Chapter 5 treats the differential geometry of curves and surfaces in euclidean space. We discuss the curvature and the torsion of a curve, Frenet's formulas, and prove the fundamental theorem of the theory of curves. We then turn to some special types of curves and conclude this section by a proof of Fenchel's inequality. This states that the total curvature of a closed space curve is at least 2π. Surface theory is treated in Cartan's language of moving frames. First we describe the structural equations of a surface, and then we prove the fundamental theorem of surface theory by applying Frobenius' theorem. The latter is formulated with respect to a frame adapted to the surface and the resulting 1-forms. Next we start the tensorial description of surfaces. The first and second fundamental forms of a surface as well as the relations between them as expressed in the Gauss and the Codazzi-Mainardi equations are the central concepts here. We reformulate the fundamental theorem in this tensorial description of surface theory. Numerous examples (surfaces of revolution, general graphs and, in particular, reliefs, i. e. the graph of the modulus of an analytic function, as well as the graphs of their real and imaginary parts) illustrate the differential-geometric treatment of surfaces in euclidean space. We study the normal map of a surface and are thus lead to its Gaussian curvature, which by Gauss' Theorema Egregium belongs to the inner geometry. Using Stokes' theorem, we prove the Gauss-Bonnet formula and an analogous integral formula for the mean curvature of a compact oriented surface, going back to Steiner and Minkowski. An important class of surfaces are minimal surfaces. Their normal map is always conformal, and this observation leads to the so-called Weierstrass formulas. These describe the minimal surface locally by a pair of holomorphic functions. Then we turn to the study of geodesic curves on surfaces, the integration of the geodesic flow using first integrals as well as the investigation of maps between surfaces. Chapter 5 closes with an outlook on the geometry of pseudo-Riemannian manifolds of higher dimension. In particular, we look at Einstein spaces, as well as spaces of constant curvature.

Symmetries play a fundamental role in geometry and physics. Chapter 6 contains an introduction into the theory of Lie groups and homogeneous spaces. We discuss the basic properties of a Lie group, its Lie algebra, and the exponential map. Then we concentrate on proving the fact that every closed subgroup of a Lie group is a Lie group itself, and define the structure of a manifold on the quotient space. Many known manifolds arise as homogeneous spaces in this way. With regard to later applications in mechanics, we study the adjoint representation of a Lie group.

Apart from Riemannian geometry, symplectic geometry is one of the essential pillars of differential geometry, and it is particularly relevant to the Hamiltonian formulation of mechanics. Examples of symplectic manifolds arise as cotangent bundles of arbitrary manifolds or as orbits of the coadjoint representation of a Lie group. We study this topic in Chapter 7. First we prove the Darboux theorem stating that all symplectic manifolds are locally equivalent. Then we turn to Noether's theorem and interpret it in terms of the moment map for Hamiltonian actions of Lie groups on symplectic manifolds. Completely integrable Hamiltonian systems are carefully discussed. Using Frobenius' theorem, we demonstrate an algorithm for finding the action and angle coordinates directly from the first integrals of the Hamilton function. In §7.5, we sketch the formulations of mechanics according to Newton, Lagrange, and Hamilton. In particular, we once again return to Noether's theorem within the framework of Lagrangian mechanics, which will be applied, among others, to integrate the geodesic flow of a pseudo-Riemannian manifold. Among the exercises of Chapter 7, the reader will find some of the best known mechanical systems.

In statistical mechanics, particles are described by their position probability in space. Therefore one is interested in the evolution of statistical states of a Hamiltonian system. In Chapter 8 we introduce the energy and information entropy for statistical equilibrium states. Then we characterize Gibbs states as those of maximal information entropy for fixed energy, and prove that the microcanonical ensemble realizes the maximum entropy among all states with fixed support. By means of the Gibbs states, we assign a thermodynamical system in equilibrium to a Hamiltonian system with auxiliary parameters satisfying the postulates of thermodynamics. We discuss the role of pressure and free energy. A series of examples, like the ideal gas, solid bodies, and cycles, conclude Chapter 8.

Chapter 9 is devoted to electrodynamics. Starting from the Maxwell equations, formulated both for the electromagnetic field strengths and for the dual 1-forms, we first deal with the static electromagnetic field. We prove the solution formula for the inhomogeneous Laplace equation in three-space and obtain, apart from a description of the electric and the magnetic field in the static case, at the same time a proof for Helmholtz' theorem as mentioned before. Next we turn to the vacuum electromagnetic field. Here we prove the solution formula for the Cauchy problem of the wave equation in dimensions two and three. The chapter ends with a relativistic formulation of the Maxwell equations in Minkowski space, a discussion of the Lorentz group, the Maxwell stress tensor and a thorough treatment of the Lorentz force.

We are grateful to Ms. Heike Pahlisch for her extensive work on the preparation of the manuscript and the illustrations of the German edition. We also thank the students in our courses from 1998 to 2000 for numerous comments leading to additions and improvements in the manuscript. In particular, Dipl.-Math. Uli Krähmer pointed out corrections in many chapters. Not least our thanks are due to M. A. Claudia Frank for her thourough reading and correcting of the German manuscript with regard to language.

The English version at hand does not differ by much from the original German edition. Besides small corrections and additions, we included a detailed discussion of the Lorentz force and related topics in Chapter 9. Finally, we thank Dr. Andreas Nestke for his careful translation.

Berlin, November 2000 and May 2002

Ilka Agricola
Thomas Friedrich

Contents

Chapter 1

Elements of Multilinear Algebra

Consider an n-dimensional vector space V over the field $\mathbb{K}$ of real or complex numbers. Its dual space V^* consists of all linear maps from V to $\mathbb{K}$. More generally, a multilinear and antisymmetric map,

$$\omega^k : V \times \ldots \times V \longrightarrow \mathbb{K},$$

depending on k vectors from the vector space V, is called an *exterior (multilinear) form of degree* k. The antisymmetry of ω^k means that, for all k vectors $v_1, \ldots, v_k$ from V and any permutation $\sigma \in S_k$ of the numbers $\{1, \ldots, k\}$, the following equation holds:

$$\omega^k(v_{\sigma(1)}, \ldots, v_{\sigma(k)}) = \operatorname{sgn}(\sigma) \cdot \omega^k(v_1, \ldots, v_k).$$

Here $\operatorname{sgn}(\sigma)$ denotes the sign of the permutation σ. In particular, ω^k changes sign under a transposition of the indices i and j:

$$\omega^k(v_1, \ldots, v_i, \ldots, v_j, \ldots, v_k) = -\,\omega^k(v_1, \ldots, v_j, \ldots, v_i, \ldots, v_k).$$

The vector space of all exterior k-forms will be denoted by $\bigwedge^k(V^*)$. Furthermore, we will use the conventions $\bigwedge^0(V^*) = \mathbb{K}$ and $\bigwedge^1(V^*) = V^*$.

Fixing an arbitrary basis $e_1, \ldots, e_n$ in the n-dimensional vector space V, we see that each exterior k-form ω^k is uniquely determined by its values on all k-tuples of the form $e_{i_1}, \ldots, e_{i_k}$, where the indices are always supposed to be strictly ordered, $I = (i_1 < \ldots < i_k)$. On the other hand, a k-form can be defined by arbitrarily prescribing its values on all ordered k-tuples of basis vectors and extending it to all k-tuples of vectors in an antisymmetric and multilinear way. The number of different k-tuples of n elements is equal to $\binom{n}{k} = \frac{n!}{k!(n-k)!}$. Thus we conclude

Theorem 1. *If $k > n$, $\bigwedge^k(V^*)$ consists only of the zero map. For $k \leq n$ the dimension of the vector space $\bigwedge^k(V^*)$ is equal to*

$$\dim\left(\bigwedge^k(V^*)\right) \;=\; \binom{n}{k}.$$

Exterior forms can by multiplied, and the product is again an exterior form.

Definition 1. Let $\omega^k \in \bigwedge^k(V^*)$ and $\eta^l \in \bigwedge^l(V^*)$ be two exterior forms of degrees k and l, respectively. Then the *exterior product* $\omega^k \wedge \eta^l$ is defined as a $(k+l)$-form by the formula

$$\omega^k \wedge \eta^l(v_1, \dots v_{k+l}) = \frac{1}{k!l!} \sum_{\sigma \in S_{k+l}} \operatorname{sgn}(\sigma) \omega^k(v_{\sigma(1)}, \dots v_{\sigma(k)}) \eta^l(v_{\sigma(k+1)}, \dots v_{\sigma(k+l)}).$$

Obviously, $\omega^k \wedge \eta^l$ is a multilinear and antisymmetric map acting on $(k+l)$ vectors, i. e. of degeree $k+l$. The following theorem summarizes the algebraic rules governing computations involving the exterior multiplication of forms.

Theorem 2. *The exterior product has the following properties:*

(1) $(\omega_1^k + \omega_2^k) \wedge \eta^l = \omega_1^k \wedge \eta^l + \omega_2^k \wedge \eta^l$;

(2) $\omega^k \wedge (\eta_1^l + \eta_2^l) = \omega^k \wedge \eta_1^l + \omega^k \wedge \eta_2^l$;

(3) $(a\omega^k) \wedge \eta^l = \omega^k \wedge (a\eta^l) = a(\omega^k \wedge \eta^l)$ *for any* $a \in \mathbb{K}$;

(4) $(\omega^k \wedge \eta^l) \wedge \mu^m = \omega^k \wedge (\eta^l \wedge \mu^m)$;

(5) $\omega^k \wedge \eta^l = (-1)^{kl} \eta^l \wedge \omega^k$.

Proof. Only the last two formulas require a proof, in which we shall omit some of the upper indices for better readability. First

$$\begin{aligned}&(k+l)!m!\,(\omega^k \wedge \eta^l) \wedge \mu^m(v_1, \dots, v_{k+l+m}) \\ &= \sum_{\sigma \in S_{k+l+m}} \operatorname{sgn}(\sigma)(\omega^k \wedge \eta^l)(v_{\sigma(1)}, \dots, v_{\sigma(k+l)}) \mu^m(v_{\sigma(k+l+1)}, \dots, v_{\sigma(k+l+m)}).\end{aligned}$$

Decompose the permutation group S_{k+l+m} into residue classes with respect to the subgroup $S_{k+l} \subset S_{k+l+m}$ formed by all permutations acting as the identity on the last m indices $\{k+l+1, \dots, k+l+m\}$. Each residue class R thus consists of all permutations $\sigma \in S_{k+l+m}$ with fixed values $\sigma(k+l+1), \dots, \sigma(k+l+m)$. Fixing any permutation $\sigma_0 \in R$, all the remaining elements $\sigma \in R$ are parametrized by the elements in S_{k+l}:

$$\sigma \;=\; \sigma_0 \circ \pi, \quad \pi \in S_{k+l}.$$

Hence,

$$\sum_{\sigma\in R}\operatorname{sgn}(\sigma)(\omega^k\wedge\eta^l)(v_{\sigma(1)},\ldots,v_{\sigma(k+l)})\mu^m(v_{\sigma(k+l+1)},\ldots,v_{\sigma(k+l+m)})$$

$$=\sum_{\pi\in S_{k+l}}\operatorname{sgn}(\sigma_0)\operatorname{sgn}(\pi)(\omega\wedge\eta)(v_{\sigma_0\circ\pi(1)},..v_{\sigma_0\circ\pi(k+l)})\mu(v_{\sigma_0(k+l+1)},..v_{\sigma_0(k+l+m)})$$

$$=\operatorname{sgn}(\sigma_0)(k+l)!(\omega\wedge\eta)(v_{\sigma_0(1)},\ldots,v_{\sigma_0(k+l)})\mu(v_{\sigma_0(k+l+1)},\ldots,v_{\sigma_0(k+l+m)}).$$

Using now the definition of the exterior product $\omega^k\wedge\eta^l$, this leads to the formula

$$\frac{k!\,l!}{(k+l)!}\sum_{\sigma\in R}\operatorname{sgn}(\sigma)(\omega\wedge\eta)(v_{\sigma(1)},\ldots,v_{\sigma(k+l)})\mu(v_{\sigma(k+l+1)},\ldots,v_{\sigma(k+l+m)})$$

$$=\sum_{\sigma\in R}\operatorname{sgn}(\sigma)\omega(v_{\sigma(1)},\ldots v_{\sigma(k)})\eta(v_{\sigma(k+1)},\ldots v_{\sigma(k+l)})\mu(v_{\sigma(k+l+1)},\ldots v_{\sigma(k+l+m)}).$$

In order to compute the sum over the whole group S_{k+l+m}, we sum over all the residue classes R and, simplifying the scalar factors, we obtain the equation

$$(k!\,l!\,m!)\cdot(\omega^k\wedge\eta^l)\wedge\mu^m(v_1,\ldots,v_{k+l+m}) =$$

$$\sum_{S_{k+l+m}}\operatorname{sgn}(\sigma)\omega(v_{\sigma(1)},\ldots v_{\sigma(k)})\eta(v_{\sigma(k+1)},\ldots v_{\sigma(k+l)})\mu(v_{\sigma(k+l+1)},\ldots v_{\sigma(k+l+m)}).$$

This shows the associativity of the exterior multiplication of forms. The last formula (5) is proved analogously. □

Definition 2. The *exterior algebra* $\bigwedge(V^*)$ of the vector space V is formed by the sum of all exterior forms

$$\bigwedge(V^*) = \sum_{k=0}^{n}\bigwedge^k(V^*)$$

endowed with the exterior multiplication $\wedge$ of forms as multiplication.

Next we will construct an explicit basis of the vector spaces $\bigwedge^k(V^*)$. To do so, start from any basis $e_1,\ldots,e_n$ of V and denote by $\sigma_1,\ldots,\sigma_n$ the dual basis of the dual space $V^*=\bigwedge^1(V^*)$. For an ordered k-tuple of indices (k-index for short) $I=(i_1<\ldots<i_k)$ let σ_I denote the k-form defined by the formula

$$\sigma_I := \sigma_{i_1}\wedge\ldots\wedge\sigma_{i_k}.$$

Obviously, for a fixed k-index $J=(j_1<\ldots<j_k)$,

$$\sigma_I(e_{j_1},\ldots,e_{j_k}) = \begin{cases}0 & \text{if } I\neq J,\\ 1 & \text{if } I=J.\end{cases}$$

In particular, the k-forms σ_I are linearly independent in $\bigwedge^k(V^*)$. For dimensional reasons, this immediately implies

Theorem 3. *Let $e_1, \ldots, e_n$ be a basis of the vector space V, and denote by $\sigma_1, \ldots, \sigma_n$ its dual basis in the dual space V^*. Then the forms $\sigma_I, I = (i_1 < \ldots < i_k)$, are a basis for the vector space $\bigwedge^k(V^*)$.*

Exterior forms can be pulled back under a linear map. In fact, if $L : W \longrightarrow V$ is a linear map from the vector space W to the vector space V, and $\omega^k \in \bigwedge(V^*)$ is an exterior k-form in V, then the formula

$$(L^*\omega^k)(w_1, \ldots, w_k) \ := \ \omega^k(L(w_1), \ldots, L(w_k))$$

defines an exterior k-form $(L^*\omega^k) \in \bigwedge^k(W^*)$ in the vector space W. Passing from the form ω^k to the *induced form* $L^*(\omega^k)$ is compatible with all algebraic operations. In particular, the following formula holds:

$$L^*(\omega^k \wedge \eta^l) \ = \ (L^*\omega^k) \wedge (L^*\eta^l)\,.$$

Furthermore, a vector can be inserted into an exterior form, and the result is an exterior form of one degree less. Let $\omega^k \in \bigwedge^k(V^*)$ be a k-form on V and $v_0 \in V$ any vector. Define a $(k-1)$-form $(v_0 \lrcorner\, \omega^k) \in \bigwedge^{k-1}(V^*)$ by the formula

$$(v_0 \lrcorner\, \omega^k)(v_1, \ldots, v_{k-1}) \ := \ \omega^k(v_0, v_1, \ldots, v_{k-1})\,.$$

The $(k-1)$-form $v_0 \lrcorner\, \omega^k$ is called the *inner product of the vector v_0 with the k-form ω^k*, and will also be denoted by $i_{v_0}(\omega^k)$. The antisymmetry of the k-form ω^k leads to the following relation for $(k-2)$-forms:

$$v_1 \lrcorner\, (v_0 \lrcorner\, \omega^k) \ = \ -\,v_0 \lrcorner\, (v_1 \lrcorner\, \omega^k)\,.$$

From now on, let V be a real vector space equipped with a *non-degenerate scalar product* g. This is a symmetric bilinear form,

$$g :\ V \times V \longrightarrow \mathbb{R},$$

with the property that the linear map $g^\# : V \to V^*$ from V to the dual space V^* defined by

$$g^\#(v)(w) \ := \ g(v, w)$$

is bijective. For a given basis $e_1, \ldots, e_n$ of V, the matrix

$$M(g) \ = \ (g(e_i, e_j))_{i,j=1}^n$$

is symmetric and invertible. For brevity, its entries will be denoted by $g_{ij} := g(e_i, e_j)$, the entries of the inverse matrix $(M(g))^{-1}$ by g^{ij}. Recall the following result, which goes back to *Lagrange* and *Sylvester*:

Theorem 4. *Let g be a non-degenerate scalar product on the real vector space V. Then there exists a basis $e_1, \ldots, e_n$ in V such that the matrix*

$M(g)$ is diagonal, i. e.

$$M(g) \;=\; \begin{bmatrix} 1 & & & & & \\ & \ddots & & & 0 & \\ & & 1 & & & \\ & & & -1 & & \\ & 0 & & & \ddots & \\ & & & & & -1 \end{bmatrix} .$$

The number p of $(+1)$-entries as well as the number q of (-1)-entries in this diagonal matrix are independent of the particular basis. The pair (p, q) is called the signature of the scalar product g, and the number q is called its index. □

First we extend the scalar product g to the spaces $\bigwedge^k(V^*)$ of k-forms, keeping the same symbol for its extension. This continuation is done, relative to an orthonormal basis, by means of the formula

$$g(\omega_1^k, \omega_2^k) \;=\; \sum_{i_1<\ldots<i_k} g^{i_1 i_1} \ldots g^{i_k i_k} \cdot \omega_1^k(e_{i_1}, \ldots, e_{i_k}) \cdot \omega_2^k(e_{i_1}, \ldots, e_{i_k}) .$$

It is not difficult to see that this defines a non-degenerate scalar product in all the spaces $\bigwedge^k(V^*)$ which does not depend on the choice of the orthonormal basis. The signature, however, changes. For example, if g is a scalar product with signature $(n-1, 1)$ in V, then the induced scalar product in the space $\bigwedge^k(V^*)$ of exterior k-forms has signature $\left(\binom{n-1}{k}, \binom{n-1}{k-1}\right)$. If the original basis $e_1, \ldots, e_n$ in V is orthonormal, the basis $\sigma_I = \sigma_{i_1} \wedge \ldots \wedge \sigma_{i_k}$ $(i_1 < \ldots < i_k)$ is an orthonormal basis in $\bigwedge^k(V^*)$. Here $\sigma_1, \ldots, \sigma_n$ denotes the dual basis in V^*. Moreover, the length of the k-form σ_I is computed by the formula

$$g(\sigma_I, \sigma_I) \;=\; g^{i_1 i_1} \cdot \ldots \cdot g^{i_k i_k} .$$

In particular, the scalar product in the one-dimensional space $\bigwedge^n(V^*)$ is positive or negative definite, depending on the parity of the index q of the scalar product.

Apart from a scalar product g, we fix an orientation in the real vector space V. Let us recall what this means. Consider the set $\mathcal{B}(V)$ of all ordered bases $\mathfrak{B} = (v_1, \ldots, v_n)$ in the vector space V. For two ordered bases, $\mathfrak{B} = (v_1, \ldots, v_n)$ and $\mathfrak{B}^* = (v_1^*, \ldots, v_n^*)$, there exists an $(n \times n)$ matrix $A(\mathfrak{B}, \mathfrak{B}^*) = (a_{ij})_{i,j=1}^n$ such that

$$v_i \;=\; \sum_{j=1}^n a_{ij} v_j^* .$$

We define an equivalence relation $\sim$ on the set $\mathcal{B}(V)$ by requiring

$$\mathfrak{B} \sim \mathfrak{B}^* \quad \text{if and only if} \quad \det(A(\mathfrak{B}, \mathfrak{B}^*)) > 0 .$$

Obviously, there are precisely two equivalence classes.

Definition 3. An *orientation* of the real vector space V is the choice of one of the two equivalence classes in the set $\mathcal{B}(V)$ of all bases in V.

In the plane $\mathbb{R}^2$, an orientation can be understood as a sense of rotation:

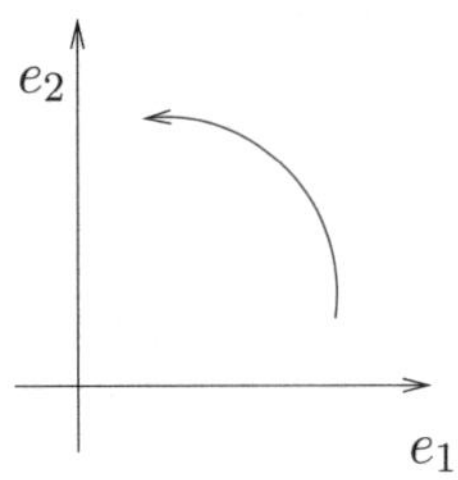

Figure 1

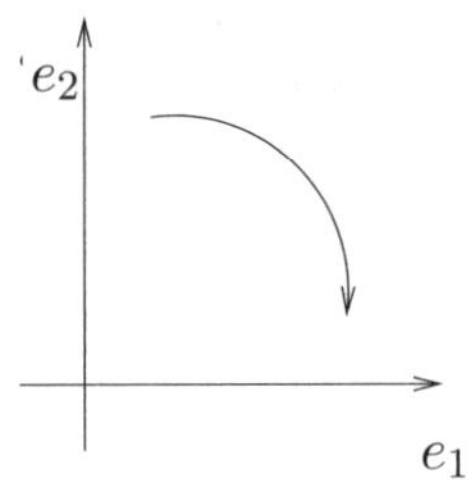

Figure 2

Figure 1 illustrates $\mathbb{R}^2$ with the orientation determined by the basis (e_1, e_2), whereas Figure 2 depicts the plane $\mathbb{R}^2$ equipped with the orientation (e_2, e_1). The transition matrix between both bases is the matrix

$$A = \begin{bmatrix} 0 & 1 \\ 1 & 0 \end{bmatrix}$$

with negative determinant, $\det(A) = -1$. Hence the bases (e_1, e_2) und (e_2, e_1) represent the two possible orientations of $\mathbb{R}^2$.

Example 1. Let (V, g) be a real vector space with fixed orientation. We choose a basis $e_1, \ldots, e_n$ in V in such a way that the matrix $M(g)$ has the diagonal format of Theorem 4, and $(e_1, \ldots, e_n)$ is positively oriented. Then the formula

$$dV(v_1, \ldots, v_n) := \det \begin{bmatrix} g(v_1, e_1), & \ldots, & g(v_n, e_1) \\ \vdots & & \vdots \\ g(v_1, e_n), & \ldots, & g(v_n, e_n) \end{bmatrix}$$

defines an n-form $dV \in \bigwedge^n(V^*)$ independent of the choice of the particular basis $e_1, \ldots, e_n$ with the stated properties. This form dV is called the *volume form* of the oriented vector space with the non-degenerate scalar product g. By means of the dual basis $\sigma_1, \ldots, \sigma_n$, the volume form is represented as

$$dV = (-1)^q \sigma_1 \wedge \ldots \wedge \sigma_n .$$

Here q denotes the index of the scalar product g. The length of the volume form is, by definition,

$$g(dV, dV) = (-1)^q .$$

The volume form of the coordinate space $\mathbb{R}^n$ with the euclidean scalar product coincides with the determinant:

$$d\mathbb{R}^n(v_1, \ldots, v_n) \;=\; \det(v_1, \ldots, v_n), \quad v_i \in \mathbb{R}^n .$$

In particular, the determinant turns out to be an n-form on the space $\mathbb{R}^n$.

Now we introduce the $*$-*operator* (*Hodge operator*) assigning a k-form to every $(n-k)$-form. Consider the given real vector space V together with a scalar product g and a fixed orientation. For each k-form $\omega^k \in \bigwedge^k(V^*)$, the rule

$$\textstyle\bigwedge^{n-k}(V^*) \ni \eta^{n-k} \longmapsto \omega^k \wedge \eta^{n-k} \in \bigwedge^n(V^*)$$

determines a linear mapping from $\bigwedge^{n-k}(V^*)$ to the 1-dimensional space $\bigwedge^n(V^*)$. The volume form dV is a basis vector in $\bigwedge^n(V^*)$, and the vector space $\bigwedge^{n-k}(V^*)$ is equipped with a non-degenerate scalar product. Hence, there exists exactly one $(n-k)$-form—to be denoted by $*\omega^k$—such that

$$\omega^k \wedge \eta^{n-k} \;=\; g(*\omega^k, \eta^{n-k})dV$$

holds for all $(n-k)$-forms η^{n-k}. Summarizing we defined a linear mapping

$$* : \textstyle\bigwedge^k(V^*) \longrightarrow \bigwedge^{n-k}(V^*) ,$$

the so-called Hodge operator.

Example 2. As above, let $e_1, \ldots, e_n$ be an orthonormal basis ($g_{ij} = \pm\delta_{ij}$) of the vector space V representing the fixed orientation, and let $\sigma_1, \ldots, \sigma_n$ be the dual basis. For an ordered k-index $I = (i_1 < \ldots < i_k)$, denote by $J = (j_1 < \ldots < j_{n-k})$ the "complementary" index, i. e. the ordered $(n-k)$-index containing the numbers $\{1, \ldots, n\}\backslash\{i_1, \ldots, i_k\}$. The equations

$$g(*\sigma_I, \sigma_J)dV \;=\; \sigma_I \wedge \sigma_J \;=\; \mathrm{sgn}\begin{pmatrix}1 \ldots n \\ I\ J\end{pmatrix}\sigma_1\wedge\ldots\wedge\sigma_n \;=\; (-1)^q\mathrm{sgn}\begin{pmatrix}1 \ldots n \\ I\ J\end{pmatrix}dV$$

and $g(\sigma_J, \sigma_J) = g_{j_1j_1}\cdots g_{j_{n-k}j_{n-k}}$ immediately imply the formula

$$*\sigma_I \;=\; (-1)^q\mathrm{sgn}\begin{pmatrix}1 \ldots n \\ I\ J\end{pmatrix}g_{j_1j_1}\cdots g_{j_{n-k}j_{n-k}}\sigma_J .$$

The following theorem assembles the main properties of the Hodge $*$-operator for a scalar product of arbitrary signature.

Theorem 5.

(1) *For each k-form ω^k, the twofold application of the Hodge operator is given by* $**\omega^k = (-1)^{k(n-k)+q}\omega^k$;

(2) *for any k-forms ω^k, η^k, the following relations hold for the exterior and the scalar product:*

$$g(*\omega^k, *\eta^k) \;=\; (-1)^q g(\omega^k, \eta^k), \quad \omega^k \wedge *\eta^k \;=\; (-1)^q g(\omega^k, \eta^k)dV .$$

Proof. To prove the first formula we first compute

$$\begin{aligned} **\sigma_I &= (-1)^q \mathrm{sgn}\begin{pmatrix}1\dots n\\ I\ \ J\end{pmatrix} g_{j_1j_1}\cdots g_{j_{n-k}j_{n-k}} * \sigma_J \\ &= (-1)^{2q}\mathrm{sgn}\begin{pmatrix}1\dots n\\ I\ \ J\end{pmatrix} g_{j_1j_1}\cdots g_{j_{n-k}j_{n-k}}\mathrm{sgn}\begin{pmatrix}1\dots n\\ I\ \ J\end{pmatrix}\cdot g_{i_1i_1}\cdots g_{i_ki_k}\sigma_I \\ &= (-1)^q(-1)^{k(n-k)}\sigma_I\,. \end{aligned}$$

The other formulas are consequences of this first one and the definition of the $*$-operator. We have

$$\begin{aligned} (*\omega^k, *\eta^k)dV &= \omega^k\wedge *\eta^k = (-1)^{k(n-k)} * \eta^k\wedge\omega^k \\ &= (-1)^{k(n-k)}g(**\eta^k,\omega^k)dV = (-1)^q g(\eta^k,\omega^k)dV\,, \end{aligned}$$

and this implies the remaining identities. □

Example 3. Let $n = 2k$ be an even number. Then the Hodge operator maps the vector space $\bigwedge^k(V^*)$ to itself,

$$*: \bigwedge^k(V^*) \longrightarrow \bigwedge^k(V^*)\,.$$

Moreover,

$$**\,\omega^k = (-1)^{k+q}\omega^k\,,$$

and hence, in case $k+q\equiv 0 \bmod 2$, the Hodge operator has the eigenvalues ± 1.

Exercises

1. Let $\sigma_1,\dots,\sigma_n$ be a basis of V^*, let $\omega^1 = \sum a_i\,\sigma_i$, $\eta^1 = \sum b_i\,\sigma_i$ be two arbitrary elements from V^*, and let $\mu^2 = \sum c_{ij}\,\sigma_i\wedge\sigma_j$ be a 2-form.

a) Compute $\omega^1\wedge\eta^1$ and, in the case $n = 3$, explain in which sense the exterior product generalizes the vector product;

b) compute $\omega^1\wedge\mu^2$ and discuss in case $n = 3$ the relation to the scalar product.

2. Prove that each 2-form $\omega^2\in\bigwedge^2(V^*)$ can be represented as

$$\omega^2 = \sigma_1\wedge\sigma_2 + \ldots + \sigma_{2r-1}\wedge\sigma_{2r}$$

for a certain basis $\sigma_1,\dots,\sigma_n$ of V^*. Prove, moreover, that the number r is independent of the choice of the basis and is characterized by the condition

$$(\omega^2)^r \neq 0,\quad (\omega^2)^{r+1} = 0\,.$$

3. Prove that k linear forms $\sigma_1, \ldots, \sigma_k$ on V^* are linearly independent if and only if

$$\sigma_1 \wedge \ldots \wedge \sigma_k \neq 0.$$

4 (Cartan's Lemma)**.** Let $\sigma_1, \ldots, \sigma_k$ be linearly independent linear forms from V^* and let $\mu_1, \ldots \mu_k$ be arbitrary elements of V^*. If $\sigma_1 \wedge \mu_1 + \ldots + \sigma_k \wedge \mu_k = 0$ holds, then the forms μ_i are linear combinations of the σ_j,

$$\mu_i = \sum_{j=1}^{k} a_{ij}\sigma_j \, .$$

Moreover, $a_{ij} = a_{ji}$.

5. Let $e_1, \ldots, e_n$ be a basis of the vector space V and $\sigma_1, \ldots, \sigma_n$ the corresponding dual basis. Then the following formula holds for every k-form ω^k:

$$\sum_{i=1}^{n} \sigma_i \wedge (e_i \lrcorner \, \omega^k) = k \cdot \omega^k \, .$$

6. For a given form $0 \neq \omega^k \in \bigwedge^k(V^*)$, define the subspace $M(\omega^k) \subset V^*$ by

$$M(\omega^k) = \left\{\eta^1 \in \textstyle\bigwedge^1(V^*) : \ \eta^1 \wedge \omega^k = 0\right\}.$$

Prove the following statements.

a) The inequality $\dim M(\omega^k) \leq k$ holds;

b) the equality $\dim M(\omega^k) = k$ holds if and only if there exist k linear forms $\sigma_1, \ldots, \sigma_k$ such that $\omega^k = \sigma_1 \wedge \ldots \wedge \sigma_k$.

Forms of the kind $\omega^k = \sigma_1 \wedge \ldots \wedge \sigma_k$ are called *decomposable.*

7. Prove the following statements.

a) A 2-form $\omega^2 = \sum_{i,j} \omega_{ij}\, \sigma_i \wedge \sigma_j$ is decomposable (see the previous exercise) if and only if

$$\omega_{ij}\, \omega_{pq} - \omega_{ip}\, \omega_{jq} + \omega_{iq}\, \omega_{jp} = 0;$$

b) a 3-form $\omega^3 = \sum_{i,j,k} \omega_{ijk}\, \sigma_i \wedge \sigma_j \wedge \sigma_k$ is decomposable if and only if

$$\omega_{ijk}\, \omega_{pqr} - \omega_{ijp}\, \omega_{kqr} + \omega_{ijq}\, \omega_{kpr} - \omega_{ijr}\, \omega_{kpq} = 0.$$

8. The Hodge operator of a 4-dimensional vector space V maps the space of 2-forms into itself,

$$* : \textstyle\bigwedge^2(V^*) \longrightarrow \bigwedge^2(V^*), \quad ** = (-1)^q \, .$$

If the index of the scalar product is even, $q = 0,\ 2$, then the Hodge operator decomposes the real vector space $\bigwedge^2(V^*)$ into the eigenspaces

$$\bigwedge\nolimits^2_{\pm}(V^*) \;=\; \big\{\omega^2 \in \bigwedge\nolimits^2(V^*) : \; *\,\omega^2 \;=\; \pm\,\omega^2\,\big\}\,.$$

In the case of an odd index, $q = 1,\ 3$, the complexification $\bigwedge^2(V^*) \otimes \mathbb{C}$ analogously decomposes into the $\pm i$-eigenspaces. Compute the dimension of the eigenspaces in both cases, and determine a basis of eigenforms.

Chapter 2

Differential Forms in $\mathbb{R}^n$

2.1. Vector Fields and Differential Forms

Vectors in euclidean space $\mathbb{R}^n$ can be understood as "free vectors" or as "place-bound vectors" at a point in $\mathbb{R}^n$. In the first case, we simply consider $\mathbb{R}^n$ as a euclidean vector space. The second point of view is based on the concept of $\mathbb{R}^n$ as a set or a metric space whose elements are called the points of space. The "place-bound vectors" then form vector spaces of their own, each one consisting of all those attached to a particular point. For example, vectors at different points cannot be added. This second concept of a vector in $\mathbb{R}^n$ leads to the notion of the tangent space at a point in $\mathbb{R}^n$.

Definition 1. Let $p \in \mathbb{R}^n$. The *tangent space* to $\mathbb{R}^n$ at the point p is defined to be the set

$$T_p\mathbb{R}^n \ := \ \big\{(p, v) : \ v \in \mathbb{R}^n\big\}.$$

An addition and a multiplication by scalars $\lambda \in \mathbb{R}$ are introduced in this set by $(p, v) + (p, w) := (p, v + w)$ and $\lambda \cdot (p, v) := (p, \lambda \cdot v)$, respectively. These operations endow each $T_p\mathbb{R}^n$ with the structure of a real n-dimensional vector space.

The ordinary differential of a smooth map f from the space $\mathbb{R}^n$ to the space $\mathbb{R}^m$ at the point p, Df_p, can now be interpreted as a linear map $f_{*,p}$ between corresponding tangent spaces.

Definition 2. Let $U \subset \mathbb{R}^n$ be open and $f : \ U \to \mathbb{R}^m$ be a differentiable map. For each point p in the set U, the linear map $f_{*,p} : \ T_p\mathbb{R}^n \to T_{f(p)}\mathbb{R}^m$

is defined by the formula

$$f_{*,p}(p, v) \;:=\; (f(p), Df_p(v)).$$

Definition 3. A *vector field* defined on an open subset U of $\mathbb{R}^n$ assigns a vector $\mathcal{V}(p) \in T_p\mathbb{R}^n$ in the corresponding tangent space to each point $p \in U$.

If $e_1, \ldots, e_n$ is the standard basis in euclidean space $\mathbb{R}^n$, the vector field determined by $p \mapsto (p, e_i)$ is usually denoted by $\partial/\partial x^i$, i.e. $(\partial/\partial x^i)(p) := (p, e_i)$. (Compare Exercise 5 for an explanation of this notation.) Obviously, every other vector field defined on U can be represented in the form

$$\mathcal{V}(p) \;=\; V^1(p)\frac{\partial}{\partial x^1}(p) + \ldots + V^n(p)\frac{\partial}{\partial x^n}(p)$$

with certain functions $V^1, \ldots, V^n$ defined on U. The vector field $\mathcal{V}$ is called differentiable of class C^k if all its component functions $V^1, \ldots, V^n$ have continuous partial derivatives up to order k. The set of all vector fields of class C^k is, on the one hand, a real vector space and, at the same time, a module over the ring $C^k(U)$ of all real-valued C^k-functions on U. Graphically, a vector field can be depicted by drawing at each point p the corresponding second component v of the vector from the tangent space. Consider, e.g., on the plane $\mathbb{R}^2$ with coordinates x, y the vector field

$$\mathcal{V}(x, y) \;=\; x\frac{\partial}{\partial y}.$$

The following figure depicts this vector field:

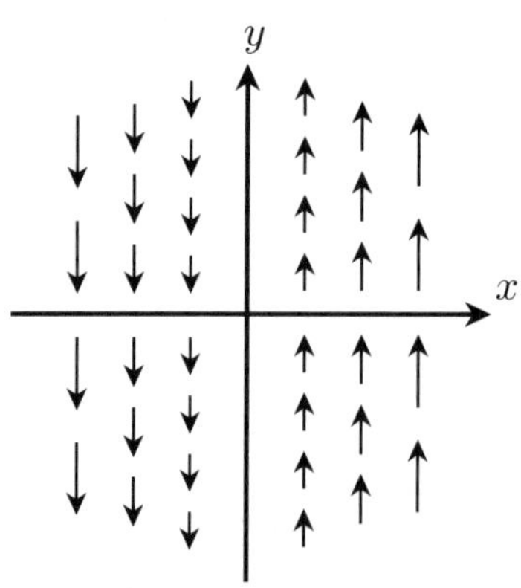

Each of the tangent spaces $T_p\mathbb{R}^n$ is a real vector space. Hence we can consider the dual space, $T_p^*\mathbb{R}^n := (T_p\mathbb{R}^n)^*$, as well as its exterior powers,

$$\Lambda^k_p(\mathbb{R}^n) \;:=\; \Lambda^k(T_p^*\mathbb{R}^n).$$

An element ω^k of the space $\bigwedge_p^k(\mathbb{R}^n)$ is thus an antisymmetric multilinear map with k arguments on the tangent space $T_p\mathbb{R}^n$:

$$\omega^k : \quad T_p\mathbb{R}^n \times \ldots \times T_p\mathbb{R}^n \longrightarrow \mathbb{R} .$$

In analogy with the notion of vector field, we will now introduce the notion of differential form.

Definition 4. A *k-form* on the open subset U of $\mathbb{R}^n$ assigns to each point $p \in U$ an element $\omega^k(p) \in \bigwedge_p^k(\mathbb{R}^n)$.

First we want to consider some examples of differential forms.

Example 1. Let $f : U \to \mathbb{R}$ be a smooth real-valued function, let $p \in U$ be a fixed point, and let $Df_p : \mathbb{R}^n \to \mathbb{R}$ be its differential at the point p. Then the formula

$$df(p)(p,v) \;:=\; Df_p(v)$$

obviously defines a 1-form df on the set U.

Example 2. A fixed basis $e_1, \ldots, e_n$ in $\mathbb{R}^n$ determines n coordinate functions $x^1, \ldots, x^n$ and hence their differentials $dx^1, \ldots, dx^n$. Thus for a tangent vector $(p,v) \in T_p\mathbb{R}^n$ we have the identity

$$dx^i(p)(p,v) \;=\; v^i ,$$

where v^i denotes the i-th component of v with respect to the basis $e_1, \ldots, e_n$. In addition, the 1-forms $dx^1(p), \ldots, dx^n(p)$ form a basis of the vector space $\bigwedge_p^1(\mathbb{R}^n) = T_p^*\mathbb{R}^n$. Arbitrary exterior products $dx^{i_1} \wedge \ldots \wedge dx^{i_k}$ as well as their linear combinations with functions as scalar coefficients lead to further examples of k-forms. Conversely, each k-form ω^k on U can be represented as

$$\omega^k \;=\; \sum_{i_1<\ldots<i_k} \omega_{i_1,\ldots,i_k}\, dx^{i_1} \wedge \ldots \wedge dx^{i_k}$$

with certain functions $\omega_{i_1,\ldots,i_k}$.

A *differential k-form* of class C^l is defined as a form ω^k whose coefficient functions $\omega_{i_1,\ldots,i_k}$ all are of class C^l. The set of all these forms is denoted by $\Omega_l^k(U)$. Obviously, this is a real vector space as well as a module over the ring $C^l(U)$. The 0-th exterior power of a vector space is its field of scalars. Therefore, the 0-forms of class C^l are taken to be the C^l-functions,

$$\Omega_l^0(U) \;:=\; C^l(U) .$$

Example 3. Let f be a real-valued function of class C^l on the open set $U \subset \mathbb{R}^n$. Then df is a 1-form of class C^{l-1}, and

$$df \;=\; \frac{\partial f}{\partial x^1}dx^1 + \ldots + \frac{\partial f}{\partial x^n}dx^n .$$

In fact, at the point $p \in U$ the following equality holds for the vector (p, v):

$$df(p)(p,v) \;=\; Df_p(v) \;=\; \sum_{i=1}^{n} \frac{\partial f}{\partial x^i}(p) v^i \,.$$

Replacing the vector components v^i by $dx^i(p)(p,v)$ and omitting the argument (p, v), we arrive at the stated formula

$$df(p) \;=\; \sum_{i=1}^{n} \frac{\partial f}{\partial x^i}(p) \cdot dx^i(p) \,.$$

The exterior product of multilinear forms can easily be extended to differential forms, defining for two forms ω^k, η^l on U a $(k+l)$-form by

$$(\omega^k \wedge \eta^l)(p) \;:=\; \omega^k(p) \wedge \eta^l(p) \,.$$

The rules known from the first chapter remain valid without change:

(1) $(\omega^k + \mu^k) \wedge \eta^l \;=\; \omega^k \wedge \eta^l + \mu^k \wedge \eta^l$;

(2) $(f \cdot \omega^k) \wedge \eta^l = f \cdot \omega^k \wedge \eta^l$;

(3) $\omega^k \wedge \eta^l = (-1)^{kl} \eta^l \wedge \omega^k$.

In particular, the last property implies that the exterior product of a form of odd degree with itself always vanishes, e.g., $dx^i \wedge dx^i = 0$. Forms of even degree do not in general have this property: A straightforward calculation shows that for the forms on $\mathbb{R}^4$ below the following relations hold:

$$(dx^1 \wedge dx^2 + dx^3 \wedge dx^4) \wedge (dx^1 \wedge dx^2 + dx^3 \wedge dx^4) \;=\; 2\, dx^1 \wedge dx^2 \wedge dx^3 \wedge dx^4 \,.$$

For conceptual as well as computational reasons, it is important that differential forms can be "pulled back" by maps.

Definition 5. Let $f:\; U_1 \to U_2$ be a differentiable map between two open subsets $U_1 \subset \mathbb{R}^n$ and $U_2 \subset \mathbb{R}^m$, and let, moreover, ω^k be a k-form on U_2. Then a k-form $f^*(\omega^k)$ (the *pullback* or *induced* form) on U_1 is defined by

$$f^*(\omega^k)((p,v_1), \ldots, (p,v_k)) \;:=\; \omega^k(f_{*,p}(p,v_1), \ldots, f_{*,p}(p,v_k)) \,.$$

The form $f^*(\omega^k)$ obeys the following rules.

Theorem 1. *Let $f:\; U_1 \to U_2$ be a differentiable map between the open sets $U_1 \subset \mathbb{R}^n$ and $U_2 \subset \mathbb{R}^m$ with component functions f^i. Then*

$$f^*(dx^i) \;=\; \sum_{j=1}^{n} \frac{\partial f^i}{\partial x^j} dx^j \,.$$

Let, moreover, differential forms on U_2 and a function $g:\; U_2 \to \mathbb{R}$ be given. Then

(1) $f^*(\omega_1^k + \omega_2^k) = f^*(\omega_1^k) + f^*(\omega_2^k)$;

(2) $f^*(g \cdot \omega^k) = (g \circ f) \cdot f^*(\omega^k)$;

(3) $f^*(\omega^k \wedge \eta^l) = f^*(\omega^k) \wedge f^*(\eta^l)$.

Furthermore, in the case $n = m$, the following additional equality holds:

$$f^*(g \cdot dx^1 \wedge \ldots \wedge dx^n) \;=\; (g \circ f) \cdot \det(Df) \cdot dx^1 \wedge \ldots \wedge dx^n .$$

Proof. The definition of f^* implies

$$f^*(dx^i)(p)(p,v) \;=\; dx^i(f(p))(f_{*,p}(p,v)) \;=\; dx^i(f(p))(f(p),\, Df_p(v)) .$$

But this is precisely the i-th component of $Df_p(v)$. Hence

$$f^*(dx^i)(p)(p,v) \;=\; \sum_{j=1}^{n} \frac{\partial f^i}{\partial x^j}(p) v^j \;=\; \sum_{j=1}^{n} \frac{\partial f^i}{\partial x^j}(p) dx^i(p)(p,v) .$$

The three rules immediately follow from the definitions and will therefore not be proved here. To derive the last equation, we use the second rule and the identity just proved:

$$\begin{aligned} f^*(g \cdot dx^1 \wedge \ldots \wedge dx^n) &= (g \circ f) f^*(dx^1) \wedge \ldots \wedge f^*(dx^n) \\ &= (g \circ f) \left[\sum_{j_1=1}^{n} \frac{\partial f^1}{\partial x^{j_1}} dx^{j_1} \right] \wedge \ldots \wedge \left[\sum_{j_n=1}^{n} \frac{\partial f^n}{\partial x^{j_n}} dx^{j_n} \right] \\ &= (g \circ f) \sum_{j_1, \ldots, j_n = 1}^{n} \frac{\partial f^1}{\partial x^{j_1}} \cdot \ldots \cdot \frac{\partial f^n}{\partial x^{j_n}} dx^{j_1} \wedge \ldots \wedge dx^{j_n} . \end{aligned}$$

Since the exterior square of a 1-form always vanishes, in this sum only those terms remain whose n-index $(j_1, \ldots, j_n)$ is a permutation of $(1, \ldots, n)$. Using the antisymmetry of the multiplication in the exterior algebra, we obtain the formula

$$f^*(g \cdot dx^1 \wedge \ldots \wedge dx^n) \;=\; (g \circ f) \sum_J \operatorname{sgn} \begin{pmatrix} 1 \ldots n \\ J \end{pmatrix} \frac{\partial f^1}{\partial x^{j_1}} \cdot \ldots \cdot \frac{\partial f^n}{\partial x^{j_n}} dx^1 \wedge \ldots \wedge dx^n$$

and thus the determinant of the differential Df. □

Recall that a C^l-function f on U can be considered as a 0-form, so the differential d is also a map turning a 0-form f into a 1-form df. Applying this map d in a suitable way to the coefficient functions of a differential form, the exterior derivative can be extended to act on k-forms in general.

Definition 6. Let ω^k be a k-form on the open set U,

$$\omega^k \;=\; \sum_{i_1 < \ldots < i_k} \omega_{i_1, \ldots, i_k} \, dx^{i_1} \wedge \ldots \wedge dx^{i_k} .$$

We define its *exterior derivative* $d\omega^k$ by the formula

$$\begin{aligned} d\omega^k &= \sum_{i_1<\ldots<i_k} d(\omega_{i_1,\ldots,i_k})\, dx^{i_1}\wedge\ldots\wedge dx^{i_k} \\ &= \sum_{i_1<\ldots<i_k}\sum_{\alpha=1}^{n} \frac{\partial\,\omega_{i_1,\ldots,i_k}}{\partial x^\alpha}\, dx^\alpha\wedge dx^{i_1}\wedge\ldots\wedge dx^{i_k}\,. \end{aligned}$$

Hence $d\omega^k$ is a $(k+1)$-form of class C^{l-1}, and d becomes a linear operator between corresponding spaces,

$$d:\quad \Omega_l^k(U)\longrightarrow \Omega_{l-1}^{k+1}(U)\,.$$

Example 4. Consider on $\mathbb{R}^2$ the 1-form

$$\omega^1 \;=\; \sin x\, dy \,+\, \sin y\, dx\,.$$

Since $d(\sin x)=\cos x\cdot dx+0\cdot dy$ and $d(\sin y)=0\cdot dx+\cos y\cdot dy$, we have

$$d\omega^1 \;=\; \cos x\, dx\wedge dy \,+\, \cos y\, dy\wedge dx \;=\; (\cos x-\cos y)\, dx\wedge dy\,.$$

Theorem 2. *The exterior derivative obeys the following rules:*

(1) $d(\omega^k+\eta^k)=d\omega^k+d\eta^k$ *for* $\omega^k,\,\eta^k\in\Omega_1^k(U)$*;*

(2) $d(\omega^k\wedge\eta^l)=(d\omega^k)\wedge\eta^l+(-1)^k\omega^k\wedge(d\eta^l)$*;*

(3) $d(d\omega^k)=0$ *for* $\omega^k\in\Omega_2^k(U)$*;*

(4) $f^*(d\omega^k)=d(f^*\omega^k)$ *for* $\omega^k\in\Omega_1^k(U_2)$ *and* $f:\;U_1\to U_2\subset\mathbb{R}^m$*.*

Proof. The first identity is trivial. For the second, we use the multi-index notation $I=(i_1<\ldots<i_k)$ and set $dx^I=dx^{i_1}\wedge\ldots\wedge dx^{i_k}$ as well as

$$\omega^k \;=\; \sum_I \omega_I\, dx^I,\quad \eta^l \;=\; \sum_J \eta_J\, dx^J\,.$$

Thus

$$\omega^k\wedge\eta^l \;=\; \sum_{I,J}\omega_I\eta_J\, dx^I\wedge dx^J\,,$$

and, applying the exterior derivative, we obtain the formula

$$d(\omega^k\wedge\eta^l) \;=\; \sum_{I,J}\sum_{\alpha=1}^{n}\left(\frac{\partial\omega_I}{\partial x^\alpha}\eta_J+\omega_I\frac{\partial\eta_J}{\partial x^\alpha}\right) dx^\alpha\wedge dx^I\wedge dx^J\,.$$

The first summand equals $d\omega^k\wedge\eta^l$, and in the case of the second precisely k transpositions lead to

$$\begin{aligned} d(\omega^k\wedge\eta^l) &= d\omega^k\wedge\eta^l+(-1)^k\sum_{I,J}\sum_{\alpha=1}^{n}\omega_I\frac{\partial\eta_J}{\partial x^\alpha}\, dx^I\wedge dx^\alpha\wedge dx^J \\ &= d\omega^k\wedge\eta^l+(-1)^k\omega^k\wedge d\eta^l. \end{aligned}$$

Now we will show that $dd = 0$. Applying the derivative twice yields the expression

$$\begin{aligned} dd\omega^k &= \sum_I \sum_{\beta=1}^{n} \sum_{\alpha=1}^{n} \frac{\partial^2 \omega_I}{\partial x^\beta x^\alpha}\, dx^\beta \wedge dx^\alpha \wedge dx^I \\ &= \sum_I \sum_{\alpha<\beta} \left(\frac{\partial^2 \omega_I}{\partial x^\beta x^\alpha} - \frac{\partial^2 \omega_I}{\partial x^\alpha x^\beta} \right) dx^\beta \wedge dx^\alpha \wedge dx^I . \end{aligned}$$

By assumption, the second partial derivatives of the functions ω_I are continuous. Hence, Schwarz' lemma implies that the expression in brackets vanishes for each multi-index I. Finally, we show that pulling back differential forms commutes with the exterior derivative. To this end, let $y^1, \ldots, y^m$ be the coordinates on V, let the form $\omega^k = \sum_I \omega_I \, dy^I$ be expressed in these, and let $f^1, \ldots, f^m$ be the components of the map $f: U_1 \to U_2$. From

$$d\omega^k = \sum_I \sum_{\alpha=1}^{m} \frac{\partial \omega_I}{\partial y^\alpha}\, dy^\alpha \wedge dy^I$$

we obtain, for the pullback of the form,

$$f^*(d\omega^k) = \sum_I \sum_{\alpha=1}^{m} \frac{\partial \omega_I}{\partial y^\alpha}(f(x))\, df^\alpha \wedge df^{i_1} \wedge \ldots \wedge df^{i_k} .$$

On the other hand,

$$f^*\omega^k = \sum_I \omega_I(f(x))\, df^{i_1} \wedge \ldots \wedge df^{i_k},$$

and its derivative is computed as follows:

$$\begin{aligned} d(f^*\omega^k) &= \sum_I \sum_{\beta=1}^{n} \frac{\partial \omega_I}{\partial x^\beta}(f(x))\, dx^\beta \wedge df^{i_1} \wedge \ldots \wedge df^{i_k} \\ &= \sum_I \sum_{\beta=1}^{n} \sum_{\alpha=1}^{m} \frac{\partial \omega_I}{\partial y^\alpha}(f(x))\, \frac{\partial f^\alpha}{\partial x^\beta}\, dx^\beta \wedge df^{i_1} \wedge \ldots \wedge df^{i_k} . \end{aligned}$$

Making use of $df^\alpha = \sum (\partial f^\alpha / \partial x^\beta)\, dx^\beta$, this can be simplified:

$$d(f^*\omega^k) = \sum_I \sum_{\alpha=1}^{m} \frac{\partial \omega_I}{\partial y^\alpha}(f(x))\, df^\alpha \wedge df^{i_1} \wedge \ldots \wedge df^{i_k} = f^*(d\omega^k) . \qquad \square$$

Example 5. The following example illustrates that, for computational purposes, the rules for f^* and d formulated in Theorems 1 and 2 often suffice. Consider, e.g., the map

$$f: \quad \mathbb{R}^2 \longrightarrow \mathbb{R}^3, \quad f(u,v) = (u^2, v^3, uv)$$

and the 1-form on $\mathbb{R}^3$ defined in the coordinates x, y, z by

$$\omega^1 = y\,dx + x\,dy + xyz\,dz\,.$$

Compute $f^*(\omega^1)$ as follows. First,

$$\begin{aligned} f^*(\omega^1) &= (y \circ f)\, f^*(dx) + (x \circ f)\, f^*(dy) + (xyz \circ f)\, f^*(dz) \\ &= v^3\, f^*(dx) + u^2\, f^*(dy) + u^3 v^4\, f^*(dz)\,. \end{aligned}$$

Then we use the fact that the exterior derivative commutes with the induced map:

$$f^*(dx) = d(f^*x) = d(x \circ f) = d(u^2) = 2u\,du\,.$$

Similarly,

$$f^*(dy) = 3\,v^2\,dv, \quad f^*(dz) = v\,du + u\,dv\,,$$

from which the result follows:

$$f^*(\omega^1) = (2\,uv^3 + u^3v^5)\,du + (3\,u^2v^2 + u^4v^4)\,dv\,.$$

2.2. Closed and Exact Differential Forms

From the theory of the Riemann integral, it is well-known that every continuous function on $\mathbb{R}$ has a primitive function. In the language of forms, this can be expressed by saying that for every 1-form $\mu^1 = g(x)\,dx$ with a continuous coefficient function $g : \mathbb{R} \to \mathbb{R}$, there exists a function f such that $df = \mu^1$. If, in addition, g is differentiable, then certainly $d\mu^1 = 0$, since each 2-form on $\mathbb{R}$ vanishes. Now we want to pose the analogous question of whether any differential form has a "primitive form":

> Let $\mu^k \in \Omega^k_1(U)$ be a k-form. What are the conditions guaranteeing the existence of a $(k-1)$-form $\eta^{k-1} \in \Omega^{k-1}_2(U)$ whose exterior derivative coincides with μ^k, $d\eta^{k-1} = \mu^k$?

The equality $dd = 0$ immediately implies that $d\mu^k = 0$ is a necessary condition, but, in general, it is not sufficient for the solvability of the equation $d\eta^{k-1} = \mu^k$.

Definition 7.

(1) A k-form $\omega^k \in \Omega^k_1(U)$ is called *closed* if $d\omega^k = 0$;

(2) a k-form $\omega^k \in \Omega^k_1(U)$ is called *exact* if there exists a $(k-1)$-form $\eta^{k-1} \in \Omega^{k-1}_2(U)$ such that $d\eta^{k-1} = \omega^k$.

The property $dd = 0$ states that each exact form is closed.

Example 6. Consider on the open set $U = \mathbb{R}^2 - \{0\}$ the *winding form*

$$\omega^1 = \frac{-y}{x^2+y^2}\,dx + \frac{x}{x^2+y^2}\,dy$$

and calculate its derivative $d\omega^1$:

$$\left[\frac{-dy}{x^2+y^2}+\frac{y(2x\,dx+2y\,dy)}{(x^2+y^2)^2}\right]\wedge dx+\left[\frac{dx}{x^2+y^2}-\frac{x(2x\,dx+2y\,dy)}{(x^2+y^2)^2}\right]\wedge dy$$
$$=\frac{y^2+x^2-2y^2+x^2+y^2-2x^2}{(x^2+y^2)^2}\,dx\wedge dy \;=\; 0\,.$$

Hence, ω^1 is closed; later we will see that ω^1 is not exact (Example 10). The integral of the winding form along a closed curve surrounding zero measures how often it "turns around" the origin (Exercises 1 and 2).

In algebraic topology, it is common to describe the difference between closed and exact forms by the so-called *de Rham cohomology*. The vector spaces of "cycles" $Z^k(U)$ and "boundaries" $B^k(U)$ are defined by

$$Z^k(U) \;:=\; \{\omega^k\in\Omega^k_\infty(U):\ \omega^k \text{ is closed}\},$$
$$B^k(U) \;:=\; \{\omega^k\in\Omega^k_\infty(U):\ \omega^k \text{ is exact}\}\,.$$

Thus $B^k(U)$ is a subspace of $Z^k(U)$, and the k-th de Rham cohomology of U is defined as the quotient space

$$H^k_{DR}(U) \;:=\; Z^k(U)\,/\,B^k(U)\,.$$

The winding form, e.g., is a non-trivial element in $H^1_{DR}(\mathbb{R}^2-\{0\})\neq 0$, and later (Exercise 14) we will show that

$$H^1_{DR}(\mathbb{R}^2-\{0\}) \;\cong\; \mathbb{R}\,.$$

The k-th de Rham cohomology only depends on the topological shape of the set U. For example, the de Rham cohomology of a convex set vanishes. For slightly more general sets this is the contents of *Poincaré's lemma*, which we are going to discuss now.

Definition 8. A subset U of $\mathbb{R}^n$ is called *star-shaped* (or a *star-region* if it is open in addition) if there exists a point $p_0\in U$ with the property that for every second point $x\in U$ the segment joining p_0 with x is completely contained in U. Obviously, star-regions are path-connected.

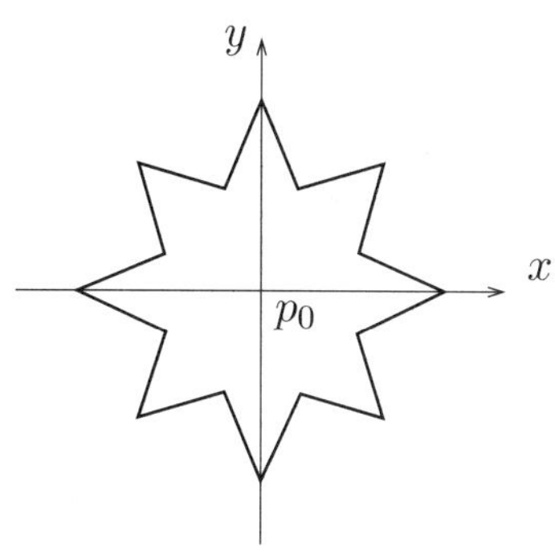

Theorem 3 (Poincaré's Lemma). *Let U be a star-shaped open set in $\mathbb{R}^n$. Then*

$$H^k_{DR}(U) \;=\; 0$$

for every $k = 1, \ldots, n$.
In other words: For each closed k-form $\omega^k \in \Omega^k_1(U)$ there exists a $(k-1)$-form $\eta^{k-1} \in \Omega^{k-1}_2$ such that $d\eta^{k-1} = \omega^k$.

Proof. In order to show this, we assign to every k-form $\omega^k = \sum \omega_I \, dx^I$ a $(k-1)$-form $P(\omega^k)$ satisfying the identity

$$\omega^k \;=\; P(d\omega^k) + dP(\omega^k)\,.$$

For closed forms the first term vanishes, which proves the assertion. The form $P(\omega^k)$ is defined as follows:

$$P(\omega^k) = \sum_{\substack{\alpha=1,\\ i_1<\ldots<i_k}}^{k} (-1)^{\alpha-1}\left[\int_0^1 t^{k-1}\omega_{i_1\ldots i_k}(tx)dt\right] x^{i_\alpha}\, dx^{i_1}\wedge\ldots\wedge\widehat{dx^{i_\alpha}}\wedge\ldots\wedge dx^{i_k}.$$

Here the notation $\widehat{dx^{i_\alpha}}$ is intended to indicate that the corresponding factor is omitted. Hence, $P(\omega^k)$ is of degree one less than ω^k, as claimed. Let us compute the exterior derivative of $P(\omega^k)$. To do so, apply the product rule to the coefficient functions: The integral has to be differentiated with respect to x implicitly, whereas the derivative of x^{i_α} each time yields the term

$$dx^{i_\alpha} \wedge dx^{i_1} \wedge \ldots \wedge \widehat{dx^{i_\alpha}} \wedge \ldots \wedge dx^{i_k}\,.$$

The sign $(-1)^{\alpha-1}$ is chosen in order to render this expression equal to $dx^{i_1} \wedge \ldots \wedge dx^{i_k}$. Altogether, we thus obtain

$$\begin{aligned} dP(\omega^k) \;&=\; k \sum_{i_1<\ldots<i_k}\left[\int_0^1 t^{k-1}\omega_{i_1\ldots i_k}(tx)dt\right] dx^{i_1}\wedge\ldots\wedge dx^{i_k} \\ &+ \sum_{\substack{j,\alpha=1,\\ i_1<\ldots<i_k}}^{n,k} (-1)^{\alpha-1}\left[\int_0^1 t^k \frac{\partial\omega_{i_1\ldots i_k}}{\partial x^j}(tx)dt\right] x^{i_\alpha} dx^j\wedge dx^{i_1}\wedge\ldots\wedge\widehat{dx^{i_\alpha}}\wedge\ldots\wedge dx^{i_k}. \end{aligned}$$

Calculating, on the other hand, $d\omega^k$ as usual leads to

$$d\omega^k \;=\; \sum_{i_1<\ldots<i_k}\sum_{j=1}^{n}\frac{\partial\omega_{i_1\ldots i_k}}{\partial x^j}\, dx^j\wedge dx^{i_1}\wedge\ldots dx^{i_k}\,.$$

We apply the operator P to this result:

$$P(d\omega^k) = \sum_{i_1<\ldots<i_k}\sum_{j=1}^{n}\left[\int_0^1 t^k \frac{\partial \omega_{i_1\ldots i_k}}{\partial x^j}(tx)dt\right] x^j\, dx^{i_1}\wedge\ldots\wedge dx^{i_k}$$

$$-\sum_{i_1<\ldots<i_k}\sum_{j,\alpha=1}^{n,k}(-1)^{\alpha-1}\left[\int_0^1 t^k \frac{\partial \omega_{i_1\ldots i_k}}{\partial x^j}(tx)dt\right] x^{i_\alpha} dx^j\wedge dx^{i_1}\wedge\ldots\wedge\widehat{dx^{i_\alpha}}\wedge\ldots\wedge dx^{i_k}.$$

Adding $P(d\omega^k)$ and $dP(\omega^k)$, two terms cancel and what remains yields

$$\begin{aligned}
P(d\omega^k)+dP(\omega^k) &= k\sum_{i_1<\ldots<i_k}\left[\int_0^1 t^{k-1}\omega_{i_1\ldots i_k}(tx)dt\right] dx^{i_1}\wedge\ldots\wedge dx^{i_k}\\
&\quad + \sum_{i_1<\ldots<i_k}\sum_{j=1}^{n}\left[\int_0^1 t^k \frac{\partial \omega_{i_1\ldots i_k}}{\partial x^j}(tx)dt\right] x^j\, dx^{i_1}\wedge\ldots\wedge dx^{i_k}\\
&= \sum_I\left[\int_0^1 \frac{d}{dt}\Big(t^k\omega_I(tx)\Big)\,dt\right] dx^I\\
&= \sum_I\left(\omega_I(x)-0\cdot\omega_I(0)\right)dx^I \;=\; \sum_I \omega_I(x)dx^I\,. \qquad \square
\end{aligned}$$

Poincaré's lemma is but an existence result for the desired form η^{k-1}, and does not claim its uniqueness. For example, if η^{k-1} is a solution of the equation $d\eta^{k-1}=\omega^k$, the sum $\eta^{k-1}+d\xi^{k-2}$ also solves the equation for any $(k-2)$-form ξ^{k-2}. Conversely, for any two forms η_1^{k-1} and η_2^{k-1} satisfying $d\eta_1^{k-1}=d\eta_2^{k-1}=\omega^k$, we have

$$d(\eta_1^{k-1}-\eta_2^{k-1}) \;=\; 0\,.$$

Then, by Poincaré's lemma, there exists a $(k-2)$-form ξ_0^{k-2} such that $\eta_1^{k-1}-\eta_2^{k-1}=d\xi_0^{k-2}$. In other words,

$$\eta_1^{k-1} \;=\; \eta_2^{k-1}+d\xi_0^{k-2}\,,$$

and we conclude that the general solution can always be expressed as the sum of a particular solution and the derivative of an arbitrary $(k-2)$-form.

Example 7. Consider on $\mathbb{R}^3$ the closed 2-form

$$\omega^2 \;=\; xy\,dx\wedge dy+2x\,dy\wedge dz+2y\,dx\wedge dz\,.$$

We will determine, in two different ways, a 1-form whose derivative coincides with ω^2. Let us first explain the *Ansatz* method to be used. The 1-form η^1 will be taken as

$$\eta^1 \;=\; f(x,y,z)\,dx+g(x,y,z)\,dy+h(x,y,z)\,dz,$$

where the functions $f, g, h : \mathbb{R}^3 \to \mathbb{R}$ still have to be determined. The exterior derivative is easily computed:

$$d\eta^1 = \left[\frac{\partial g}{\partial x} - \frac{\partial f}{\partial y}\right] dx \wedge dy + \left[\frac{\partial h}{\partial x} - \frac{\partial f}{\partial z}\right] dx \wedge dz + \left[\frac{\partial h}{\partial y} - \frac{\partial g}{\partial z}\right] dy \wedge dz\,.$$

Hence the functions have to satisfy the conditions

$$\frac{\partial g}{\partial x} - \frac{\partial f}{\partial y} = xy, \quad \frac{\partial h}{\partial x} - \frac{\partial f}{\partial z} = 2y \quad \text{and} \quad \frac{\partial h}{\partial y} - \frac{\partial g}{\partial z} = 2x\,.$$

Integrating, e.g., the first two with respect to x yields

$$g = \frac{1}{2}x^2 y + \int \frac{\partial f}{\partial y}\, dx, \quad h = 2xy + \int \frac{\partial f}{\partial z}\, dx\,.$$

Inserting the result into the last condition,

$$2x = 2x + \int \frac{\partial^2 f}{\partial y \partial z}\, dx - \int \frac{\partial^2 f}{\partial z \partial y}\, dx\,,$$

we see that it is satisfied for any function f; in particular, we may choose $f = 0$. Then $g = \frac{1}{2}x^2 y$, $h = 2xy$, and hence, using

$$\eta^1 = \frac{1}{2}x^2 y\, dy + 2xy\, dz\,,$$

we obtain a solution, as is easily checked. The *integration method* computes the "primitive form" by means of the map $P(\omega^2)$ introduced in the proof of Poincaré's lemma. In the example, this turns out to be the sum of 6 terms, namely

$$\begin{aligned} P(\omega^2) = &+ \left(\int_0^1 t(tx)(ty)dt\right) x\, dy - \left(\int_0^1 t(tx)(ty)dt\right) y\, dx \\ &+2 \left(\int_0^1 t(tx)dt\right) y\, dz - 2\left(\int_0^1 t(tx)dt\right) z\, dy \\ &+2 \left(\int_0^1 t(ty)dt\right) x\, dz - 2\left(\int_0^1 t(ty)dt\right) z\, dx\,. \end{aligned}$$

Each of the summands can easily be computed:

$$\begin{aligned} P(\omega^2) &= \frac{1}{4}x^2 y\, dy - \frac{1}{4}xy^2\, dx + \frac{2}{3}xy\, dz - \frac{2}{3}xz\, dy + \frac{2}{3}xy\, dz - \frac{2}{3}yz\, dx \\ &= -\left(\frac{1}{4}xy^2 + \frac{2}{3}yz\right) dx + \left(\frac{1}{4}x^2 y - \frac{2}{3}xz\right) dy + \frac{4}{3}xy\, dz\,. \end{aligned}$$

Which of these two methods leads to a solution more quickly depends on the particular situation; since an integration has to be carried out in any case, it might not be possible to find an explicit elementary solution (just as not every continuous function is elementary integrable).

2.3. Gradient, Divergence and Curl

Each tangent space $T_p\mathbb{R}^n$ of the coordinate space is an oriented, euclidean vector space, and hence there is the volume form

$$d\mathbb{R}^n(p) \in \textstyle\bigwedge_p^n(\mathbb{R}^n)$$

as well as the Hodge operator

$$*: \quad \textstyle\bigwedge_p^k(\mathbb{R}^n) \longrightarrow \bigwedge_p^{n-k}(\mathbb{R}^n)\,.$$

This allows us to associate with every differential form ω^k of degree k on $\mathbb{R}^n$ a corresponding $(n-k)$-form $*\omega^k$ defined by applying the $*$-operator pointwise, i.e., at each point $p \in \mathbb{R}^n$ to $\omega^k(p)$. Consider an orthonormal basis $e_1, \ldots, e_n$ of the space $\mathbb{R}^n$ and the corresponding coordinate functions $x^1, \ldots, x^n$. For a k-form $\omega^k = \sum \omega_I dx^I$ expressed in these coordinates the associated form is thus determined by the following formula:

$$*\omega^k \;=\; \sum_I \operatorname{sgn}\begin{pmatrix} 1 \ldots n \\ I, J \end{pmatrix} \omega_I \, dx^J\,.$$

Here, $J = (j_1 < \ldots < j_{n-k})$ is the complementary multi-index to $I = (i_1 < \ldots < i_k)$. The volume form $d\mathbb{R}^n$ itself is simply

$$d\mathbb{R}^n \;=\; dx^1 \wedge \ldots \wedge dx^n\,.$$

In addition, we have the possibility to pass from a vector field $\mathcal{V}$ to a 1-form $\omega^1_{\mathcal{V}}$, and vice versa. This is accomplished by

Definition 9. For a vector field $\mathcal{V}$ the *dual 1-form* $\omega^1_{\mathcal{V}}$ is defined by the equation

$$*\omega^1_{\mathcal{V}} \;:=\; \mathcal{V} \lrcorner\, d\mathbb{R}^n\,.$$

If the vector field $\mathcal{V} = \sum V^i \partial/\partial x^i$ is expressed in cartesian coordinates, then the corresponding representation for $\omega^1_{\mathcal{V}}$ is obtained as follows:

$$\begin{aligned} \omega^1_{\mathcal{V}} &= (-1)^{n-1} * (V^1 dx^2 \wedge \ldots \wedge dx^n - V^2 dx^1 \wedge dx^3 \wedge \ldots \wedge dx^n \pm \ldots) \\ &= V^1\, dx^1 + \ldots + V^n\, dx^n\,. \end{aligned}$$

This transition from vector fields to 1-forms will be used now to introduce the gradient of a function and the divergence and curl of a vector field in an invariant way.

Definition 10. Let $f : U \to \mathbb{R}$ be a C^1-function defined on an open subset $U \subset \mathbb{R}^n$. The vector field $\operatorname{grad}(f)$ associated with the 1-form df is called the *gradient* of the function f. The defining equation for the vector field $\operatorname{grad}(f)$ thus reads as follows:

$$*\omega^1_{\operatorname{grad}(f)} \;=\; \operatorname{grad}(f) \lrcorner\, d\mathbb{R}^n \;=\; *df\,.$$

In the chosen coordinates we have

$$\operatorname{grad}(f) \;=\; \frac{\partial f}{\partial x^1}\frac{\partial}{\partial x^1} + \ldots + \frac{\partial f}{\partial x^n}\frac{\partial}{\partial x^n}\,.$$

Definition 11. The *divergence* of a C^1-vector field $\mathcal{V}$ is the function determined by the equation

$$d(*\omega^1_{\mathcal{V}}) \;=\; d(\mathcal{V}\,\lrcorner\, d\mathbb{R}^n) \;:=\; \operatorname{div}(\mathcal{V})\cdot d\mathbb{R}^n\,.$$

The formula

$$\operatorname{div}(\mathcal{V}) \;=\; \sum_{i=1}^{n}\frac{\partial \mathcal{V}^i}{\partial x^i}$$

expresses the divergence of $\mathcal{V}$ through its components.

The next theorem contains a few simple properties of this operation.

Theorem 4.

(1) *Let f and g be C^1-functions. Then*

$$\operatorname{grad}(f\cdot g) \;=\; f\cdot\operatorname{grad}(g) + g\cdot\operatorname{grad}(f)\,.$$

(2) *For a function f and a vector field $\mathcal{V}$ of class C^1 the following identity holds:*

$$\operatorname{div}(f\cdot\mathcal{V}) \;=\; f\cdot\operatorname{div}(\mathcal{V}) + df(\mathcal{V})\,.$$

Proof. We only prove the second formula. By definition

$$\begin{aligned}\operatorname{div}(f\,\mathcal{V})\cdot d\mathbb{R}^n &= d(f\,(\mathcal{V}\,\lrcorner\, d\mathbb{R}^n)) \;=\; df\wedge(\mathcal{V}\,\lrcorner\, d\mathbb{R}^n) + f\,d(\mathcal{V}\,\lrcorner\, d\mathbb{R}^n)\\ &= df\wedge *\omega^1_{\mathcal{V}} + f\operatorname{div}(\mathcal{V})\,d\mathbb{R}^n \;=\; [df(\mathcal{V}) + f\operatorname{div}(\mathcal{V})]\,d\mathbb{R}^n.\end{aligned}$$

The last step makes use of the following equation, valid for every vector field $\mathcal{V}$ and any 1-form η^1:

$$\eta^1\wedge *\omega^1_{\mathcal{V}} \;=\; \eta^1(\mathcal{V})\cdot d\mathbb{R}^n\,. \qquad \square$$

Definition 12. Let f be a C^2-function defined on an open subset $U\subset\mathbb{R}^n$. The divergence of the gradient of f is called the *Laplacian* $\Delta(f)$ of the function f:

$$\Delta(f) \;=\; \operatorname{div}(\operatorname{grad}(f))\,.$$

In the chosen coordinates the Laplacian is computed by the formula

$$\Delta(f) \;=\; \sum_{i=1}^{n}\frac{\partial^2 f}{(\partial x^i)^2}\,.$$

An immediate consequence of the identities in the preceding theorem is the formula

$$\Delta(f\cdot g) \;=\; \Delta(f)\cdot g + f\cdot\Delta(g) + 2\,\langle\operatorname{grad}(f),\operatorname{grad}(g)\rangle\,.$$

In particular, there exists an additional operation acting on vector fields in dimension $n = 3$, the so-called *curl*. For a vector field $\mathcal{V}$ and the corresponding 1-form $\omega^1_{\mathcal{V}}$ in $\mathbb{R}^3$, the form $*d(\omega^1_{\mathcal{V}})$ again is a 1-form and hence in turn defines a vector field, the curl $\mathrm{curl}(\mathcal{V})$. In short, we have the following

Definition 13. The *curl* of the vector field $\mathcal{V}$ is the unique vector field determined by the following condition:

$$d\omega^1_{\mathcal{V}} \;=:\; \mathrm{curl}(\mathcal{V}) \lrcorner\, d\mathbb{R}^3 \;=\; *\omega^1_{\mathrm{curl}(\mathcal{V})}\,.$$

The definition of the exterior derivative $d\omega^1_{\mathcal{V}}$

$$d\omega^1_{\mathcal{V}} =$$

$$\left[\frac{\partial V^2}{\partial x^1} - \frac{\partial V^1}{\partial x^2}\right] dx^1 \wedge dx^2 + \left[\frac{\partial V^3}{\partial x^1} - \frac{\partial V^1}{\partial x^3}\right] dx^1 \wedge dx^3 + \left[\frac{\partial V^3}{\partial x^2} - \frac{\partial V^2}{\partial x^3}\right] dx^2 \wedge dx^3$$

leads immediately to the formula

$$\mathrm{curl}(\mathcal{V}) \;=\; \left[\frac{\partial V^3}{\partial x^2} - \frac{\partial V^2}{\partial x^3}\right] \frac{\partial}{\partial x^1} + \left[\frac{\partial V^1}{\partial x^3} - \frac{\partial V^3}{\partial x^1}\right] \frac{\partial}{\partial x^2} + \left[\frac{\partial V^2}{\partial x^1} - \frac{\partial V^1}{\partial x^2}\right] \frac{\partial}{\partial x^3}\,.$$

The properties of the curl of a vector field are summarized in the following theorem.

Theorem 5. *If the function f and the vector field $\mathcal{V}$ are of class C^2 and defined on an open subset $U \subset \mathbb{R}^3$, then*

(1) $\mathrm{div}(\mathrm{curl}(\mathcal{V})) = 0$ *and* $\mathrm{curl}(\mathrm{grad}(f)) = 0$*;*

(2) $\mathrm{curl}(f \cdot \mathcal{V}) = f \cdot \mathrm{curl}(\mathcal{V}) + \mathrm{grad}(f) \times \mathcal{V}$*, where $\times$ denotes the vector product in $\mathbb{R}^3$;*

(3) *if the curl of $\mathcal{V}$ vanishes,* $\mathrm{curl}(\mathcal{V}) = 0$*, and U is star-shaped, then there exists a function f such that* $\mathcal{V} = \mathrm{grad}(f)$*;*

(4) *if the divergence of $\mathcal{V}$ vanishes,* $\mathrm{div}(\mathcal{V}) = 0$*, and U is star-shaped, then there exists a vector field $\mathcal{W}$ such that* $\mathcal{V} = \mathrm{curl}(\mathcal{W})$.

Proof. The first equations immediately follow from the fact that the square of the exterior derivative vanishes, $dd = 0$. In fact, we obtain

$$\mathrm{div}(\mathrm{curl}(\mathcal{V})) \cdot d\mathbb{R}^3 \;=\; d(*\omega^1_{\mathrm{curl}(\mathcal{V})}) \;=\; dd\omega^1_{\mathcal{V}} \;=\; 0$$

and

$$\omega^1_{\mathrm{curl}(\mathrm{grad}(f))} \;=\; *d\omega^1_{\mathrm{grad}(f)} \;=\; *ddf \;=\; 0\,.$$

Claims (3) and (4) are special cases of Poincaré's lemma. For $\mathrm{curl}(\mathcal{V}) = 0$ the associated form $\omega^1_{\mathcal{V}}$ is closed. By Poincaré's lemma there exists a function f such that $\omega^1_{\mathcal{V}} = df$. Hence $\mathcal{V}$ is the gradient of the function f. Property (4) is proved similarly. □

2.4. Singular Cubes and Chains

We intend to develop the integral calculus for differential forms, and to do so, we need suitable sets as integration domains. First, in this chapter, we will only allow for subsets of $\mathbb{R}^n$ which are higher-dimensional analogues of parametrized curves, so-called *singular chains.* The k-dimensional unit cube will be denoted by $[0,1]^k$,

$$[0,1]^k \;=\; \{(x^1,\ldots,x^k) \in \mathbb{R}^k : \quad 0 \le x^1,\ldots,x^k \le 1\}\,.$$

Definition 14. A *singular k-cube* in $U \subset \mathbb{R}^n$ is understood to be a C^1-map

$$c^k : \quad [0,1]^k \longrightarrow U\,.$$

Once and for all we agree to call a map defined on any (not necessarily open) subset of $\mathbb{R}^k$ *smooth* if it has a smooth extension to an open neighborhood of this subset.

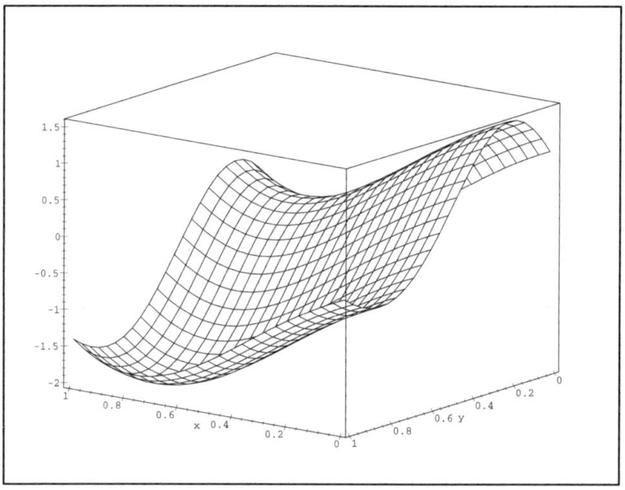

Picture of the 2-cube $[0,1]^2 \ni (x,y) \mapsto \cos(x+1)^2 + \sin(y-3)^2$

Note that not only does a singular cube consist of the image set of c^k, but the notion also involves an explicit parametrization. For example, in case $k = 0$ a singular 0-cube is a point in the set U. For $k = 1$, in general, it is a parametrized curve, but it may also degenerate to a point, which explains why these cubes are called "singular". In some situations, it can be helpful to view a singular k-cube as a cube of some dimension larger than k. Notice also that the boundary of the image set will, in general, have "corners".

Definition 15. A *singular k-chain* in U is a formal sum of singular k-cubes c_i^k in U with integer coefficients $l_i \in \mathbb{Z}$,

$$l_1 c_1^k + \ldots + l_m c_m^k \quad =: \quad s^k \, .$$

k-chains are added in the obvious way, and with addition as composition they form the abelian group of singular chains in U, which will be denoted by $\mathcal{C}_k(U)$. For example, the inverse of the k-chain s^k is its negative $-s^k$.

Definition 16. The *standard cube* in $\mathbb{R}^k$ is defined to be the identity map of the k-dimensional unit cube

$$I^k : \quad [0,1]^k \longrightarrow \mathbb{R}^k, \quad I^k(x) \; = \; x \, .$$

Now we define a boundary for every k-chain. For each index i between 1 and k, we parametrize a part of the boundary of $[0,1]^k$ by maps $I^k_{(i,0)}$, $I^k_{(i,1)} \; : \; [0,1]^{k-1} \; \to \; [0,1]^k$ which insert the value 0 or 1, respectively, at position i:

$$I^k_{(i,0)} : [0,1]^{k-1} \longrightarrow \mathbb{R}^k, \; I^k_{(i,0)}(x^1, \ldots, x^{k-1}) = (x^1, \ldots, x^{i-1}, 0, x^i, \ldots, x^{k-1})$$
$$I^k_{(i,1)} : [0,1]^{k-1} \longrightarrow \mathbb{R}^k, \; I^k_{(i,1)}(x^1, \ldots, x^{k-1}) = (x^1, \ldots, x^{i-1}, 1, x^i, \ldots, x^{k-1}).$$

The boundary of the standard cube I^k is then taken to be the $(k-1)$-chain

$$\partial I^k \; := \; \sum_{i=1}^{k} (-1)^i \left(I^k_{(i,0)} - I^k_{(i,1)} \right) .$$

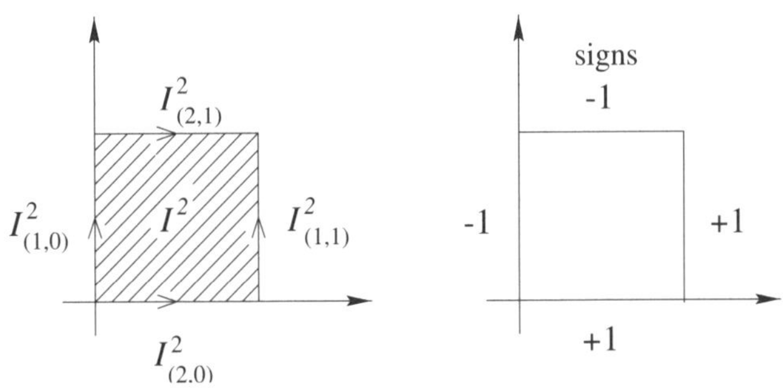

For an arbitrary singular k-cube $c^k : \; [0,1]^k \to U \subset \mathbb{R}^n$ we set

$$\partial c^k \; := \; \sum_{i=1}^{k} (-1)^i \left(c^k \circ I^k_{(i,0)} - c^k \circ I^k_{(i,1)} \right) .$$

Finally, for a singular k-chain $s^k = \sum l_j c_j^k$ we define the *boundary operator* by

$$\partial s^k \; := \; \sum_j l_j \partial c_j^k \, .$$

Example 8. We compute the boundary of the 2-dimensional standard cube I^2. To this end, we label the 4 vertices by p_1 to p_4 starting at the origin and moving counterclockwise, and the 4 edges by s_1 to s_4, again moving counterclockwise and starting at p_1. Then

$$\partial I^2 \;=\; s_1 + s_2 + s_3 + s_4,$$

and the boundary of this chain is

$$\partial\partial I^2 \;=\; (p_2 - p_1) + (p_3 - p_2) + (p_4 - p_3) + (p_1 - p_4) \;=\; 0\,.$$

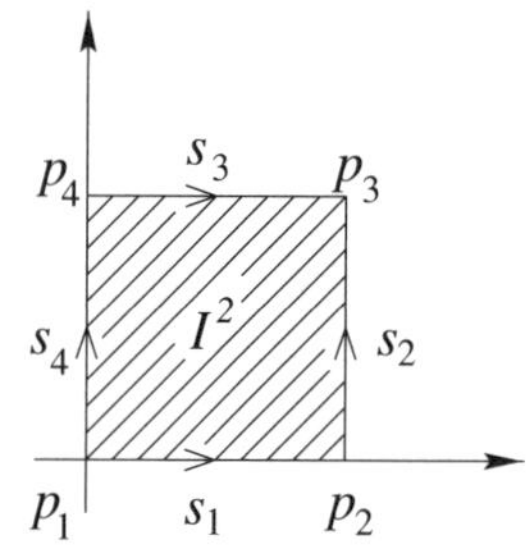

As in the case of the exterior derivative d, the square of the boundary operator vanishes.

Theorem 6.

(1) *The boundary operator* $\partial : \; \mathcal{C}_k(U) \to \mathcal{C}_{k-1}(U)$ *is a group homomorphism;*

(2) *in the sequence*

$$\ldots \xrightarrow{\partial} \mathcal{C}_{k+1}(U) \xrightarrow{\partial} \mathcal{C}_k(U) \xrightarrow{\partial} \mathcal{C}_{k-1}(U) \xrightarrow{\partial} \ldots \xrightarrow{\partial} \mathcal{C}_1(U) \xrightarrow{\partial} \mathcal{C}_0(U)$$

the equality $\partial\partial = 0$ *holds, i. e., for every* k*-chain* $s^k \in \mathcal{C}_k(U)$ *we have*

$$\partial(\partial\, s^k) \;=\; 0\,.$$

Proof. The property $\partial(s_1^k + s_2^k) = \partial s_1^k + \partial s_2^k$ immediately follows from the definition of the boundary operator. Because of linearity, this obviously suffices to prove the second statement for an arbitrary k-cube $c^k : \; [0,1]^k \to U \subset \mathbb{R}^n$. By definition the boundary is determined by

$$\partial c^k \;=\; \sum_{j=1}^{k} (-1)^j \left(c^k \circ I^k_{(j,0)} - c^k \circ I^k_{(j,1)} \right)\,.$$

Applying the boundary operator again, we obtain (omitting the composition signs in the second line)

$$\begin{aligned}\partial(\partial c^k) &= \sum_{j=1}^{k}(-1)^j \left(\partial(c^k \circ I^k_{(j,0)}) - \partial(c^k \circ I^k_{(j,1)})\right) \\ &= \sum_{j=1}^{k}\sum_{i=1}^{k-1}(-1)^{i+j} \left[c^k I^k_{(j,0)} I^{k-1}_{(i,0)} - c^k I^k_{(j,0)} I^k_{(i,1)} - c^k I^k_{(j,1)} I^{k-1}_{(i,0)} + c^k I^k_{(j,1)} I^k_{(i,1)}\right].\end{aligned}$$

In order to be able to transform this expression, we introduce two additional summation indices $\alpha, \beta = 0, 1$, and rewrite it as

$$\partial(\partial c^k) = \sum_{j=1}^{k}\sum_{i=1}^{k-1}\sum_{\substack{\alpha=0,1\\ \beta=0,1}} (-1)^{i+j+\alpha+\beta}\left[c^k \circ I^k_{(j,\alpha)} \circ I^{k-1}_{(i,\beta)}\right].$$

For $j \leq i$ and $\alpha, \beta = 0, 1$ we have

$$\begin{aligned}I^k_{(j,\alpha)} \circ I^{k-1}_{(i,\beta)}(x^1, \ldots, x^{k-2}) &= I^k_{(j,\alpha)}(x^1, \ldots, x^{i-1}, \beta, x^i, \ldots, x^{k-2}) \\ &= (x^1, \ldots, x^{j-1}, \alpha, x^j, \ldots, x^{i-1}, \beta, x^i, \ldots, x^{k-2}).\end{aligned}$$

Similarly, one rewrites $I^k_{(i+1,\beta)} \circ I^{k-1}_{(j,\alpha)}$ as

$$\begin{aligned}I^k_{(i+1,\beta)} \circ I^{k-1}_{(j,\alpha)}(x^1, \ldots, x^{k-2}) &= I^k_{(i+1,\beta)}(x^1, \ldots, x^{j-1}, \alpha, x^j, \ldots, x^{k-2}) \\ &= (x^1, \ldots, x^{j-1}, \alpha, x^j, \ldots, x^{i-1}, \beta, x^i, \ldots x^{k-2}).\end{aligned}$$

Together these yield the identity

$$I^k_{(j,\alpha)} \circ I^{k-1}_{(i,\beta)} = I^k_{(i+1,\beta)} \circ I^{k-1}_{(j,\alpha)}.$$

In order to apply this, we split the sum appearing in the expression for $\partial(\partial c^k)$ into those terms for which $j \leq i$ and the remaining ones:

$$\begin{aligned}\partial(\partial c^k) = &\sum_{1\leq j\leq i\leq k-1}\sum_{\alpha,\beta}(-1)^{i+j+\alpha+\beta} c^k \circ I^k_{(j,\alpha)} \circ I^{k-1}_{(i,\beta)} \\ &+ \sum_{i<j}\sum_{\alpha,\beta}(-1)^{i+j+\alpha+\beta} c^k \circ I^k_{(j,\alpha)} \circ I^{k-1}_{(i,\beta)}.\end{aligned}$$

In the second sum we replace i by j as well as j by $i+1$:

$$\begin{aligned}\partial(\partial c^k) = &\sum_{1\leq j\leq i\leq k-1}\sum_{\alpha,\beta}(-1)^{i+j+\alpha+\beta} c^k \circ I^k_{(j,\alpha)} \circ I^{k-1}_{(i,\beta)} \\ &+ \sum_{1\leq j<i+1}\sum_{\alpha,\beta}(-1)^{i+j+1+\alpha+\beta} c^k \circ I^k_{(i+1,\alpha)} \circ I^{k-1}_{(j,\beta)} \\ = &\ 0,\end{aligned}$$

by the identity proved before. □

In a similar way as the de Rham cohomology, we can define the k-th *cubic homology group* of a set $U \subset \mathbb{R}^n$:

$$H_k^{\text{cub}}(U) \;:=\; \ker\left(\partial : \mathcal{C}_k(U) \to \mathcal{C}_{k-1}(U)\right) / \operatorname{im}\left(\partial : \mathcal{C}_{k+1}(U) \to \mathcal{C}_k(U)\right) .$$

2.5. Integration of Differential Forms and Stokes' Theorem

Consider a singular k-cube $c^k : [0,1]^k \to U \subset \mathbb{R}^n$ of class C^1, as well as a k-form ω^k on U. Then the induced differential form $(c^k)^*\omega^k$ is defined on the unit cube $[0,1]^k$, and, as such a multiple of $dx^1 \wedge \ldots \wedge dx^n$,

$$(c^k)^*\omega^k \;=\; f(x)\, dx^1 \wedge \ldots \wedge dx^n$$

for some function $f : [0,1]^k \to \mathbb{R}$. We can thus define the integral of ω^k over the singular cube c^k as follows:

Definition 17. Set

$$\int_{c^k} \omega^k \;:=\; \int_{[0,1]^k} f(x)$$

and extend this definition linearly to any k-chain $s^k = \sum_j l_j c_j^k$ in U by

$$\int_{s^k} \omega^k \;:=\; \sum_j l_j \int_{c_j^k} \omega^k .$$

Example 9. For $k = 1$ a singular k-cube is simply a parametrized C^1-curve $c : [0,1] \to U \subset \mathbb{R}^n$. In this case, the integral of the 1-form $\omega^1 = p_1 dy^1 + \ldots + p_n dy^n$ is called the *line integral* of ω^1 along c. If $c^1, \ldots, c^n$ are the component functions of c, the pullback of the form is written as

$$c^*\omega^1(t) \;=\; p_1(c(t)) \cdot \frac{dc^1(t)}{dt}\, dt \;+\; \ldots \;+\; p_n(c(t)) \cdot \frac{dc^n(t)}{dt}\, dt \,,$$

and hence, in this situation, we obtain the following general formula for the line integral:

$$\int_c \omega^1 \;=\; \int_0^1 \left[\sum_{i=1}^n p_i(c(t)) \cdot \frac{dc^i(t)}{dt} \right] dt \,.$$

If the 1-form ω^1 is the differential of a smooth function f, then by

$$c^*(\omega^1) \;=\; c^*(df) \;=\; d(f \circ c)$$

we obtain the following value for the integral over df:

$$\int_c df \;=\; \int_0^1 \frac{d(f \circ c)(t)}{dt}\, dt \;=\; f(c(1)) - f(c(0)) \,.$$

In particular, the line integral of an exact 1-form depends only on the end points of the curve and not on its shape. It vanishes for a closed curve.

Example 10. We will use the last remark to show that the winding form,

$$\omega^1 \;=\; \frac{-y}{x^2+y^2}\,dx + \frac{x}{x^2+y^2}\,dy,$$

is not exact on $\mathbb{R}^2 - \{0\}$. To this end, consider the parametrization $c(t) = (\cos 2\pi t, \sin 2\pi t)$ of the circle by the interval $[0,1]$. Then

$$c^*\omega^1 \;=\; 2\pi\,dt \quad\text{and}\quad \int_c \omega^1 \;=\; 2\pi\,.$$

The integral of a k-form over a singular k-cube is, up to sign, independent of the parametrization of the cube. This is a consequence of a well-known transformation rule for higher-dimensional integrals: Let $\varphi:\ U \to V$ be a diffeomorphism, and let $f:\ V \to \mathbb{R}$ be integrable; then

$$\int_U (f\circ\varphi)(y)\cdot|\det(D\varphi(y))|\,dy \;=\; \int_V f(x)dx\,.$$

Two parametrizations c_1^k and c_2^k : $[0,1]^k \to U$ of one and the same point set in U differ by a diffeomorphism φ : $[0,1]^k \to [0,1]^k$ of the unit cube, $c_2^k = c_1^k \circ \varphi$. The determinant of the differential $D\varphi(x)$ has constant sign, which will be denoted by $\varepsilon(\varphi)$.

Theorem 7. *In the above situation we have*

$$\int_{c_1^k} \omega^k \;=\; \varepsilon(\varphi)\int_{c_2^k}\omega^k\,.$$

Proof. By Theorem 1 the pullback of the form

$$(c_2^k)^*\omega^k \;=\; \varphi^*(c_1^k)^*(\omega^k) \;=\; \varphi^*(f\,dx^1\wedge\ldots\wedge dx^n) \;=\; (f\circ\varphi)\cdot\varphi^*(dx^1\wedge\ldots\wedge dx^n)$$

coincides with

$$(c_2^k)^*\omega^k \;=\; (f\circ\varphi)\cdot\det(D\varphi(x))\cdot dx^1\wedge\ldots\wedge dx^n\,.$$

Thus the statement follows from the quoted transformation rule for n-dimensional integrals. □

In case $k=1$, changing the direction of a curve results in a change of sign for the line integral.

Now we prove the essential result of this chapter, *Stokes' theorem.* This is a far-reaching generalization of the Fundamental Theorem of Calculus and comprises the classical integral formulas of the 19-th century as special cases. The proof to follow will, however, reveal to the careful reader that the core of the matter is the fact that integration and differentiation are mutually inverse operations.

Theorem 8 (Stokes' Theorem). *Let ω^k be a differential form defined on the open subset $U \subset \mathbb{R}^n$, and let $s^{k+1} : [0,1]^{k+1} \to U$ be a $(k+1)$-chain. Then*

$$\int_{\partial s^{k+1}} \omega^k = \int_{s^{k+1}} d\omega^k .$$

Proof. Since the integral is additive, it suffices to prove the formula for singular $(k+1)$-cubes. First we consider the standard cube $I^{k+1} : [0,1]^{k+1} \to \mathbb{R}^{k+1}$ and represent the k-form ω^k on $\mathbb{R}^{k+1}$ as

$$\omega^k = \sum_{i=1}^{k+1} f_i \, dx^1 \wedge \ldots \wedge \widehat{dx^i} \wedge \ldots \wedge dx^{k+1} .$$

The derivative is then

$$d\omega^k = \left[\sum_{i=1}^{k+1} (-1)^{i-1} \frac{\partial f_i}{\partial x^i} \right] dx^1 \wedge \ldots \wedge dx^{k+1} .$$

Hence, by the definition of the integral,

$$\int\limits_{I^{k+1}} d\omega^k = \sum_{i=1}^{k+1} (-1)^{i-1} \int\limits_{[0,1]^k} \frac{\partial f_i}{\partial x^i} \, dx .$$

On the other hand, applying the maps $I^{k+1}_{(j,\alpha)}$ parametrizing the different parts of the boundary of the unit cube leads to the formula

$$\left(I^{k+1}_{(j,\alpha)} \right)^* \omega^k = f_j(x^1, \ldots, x^{j-1}, \alpha, x^j, \ldots, x^k) \cdot dx^1 \wedge \ldots \wedge dx^k ,$$

from which we conclude that

$$\begin{aligned} &\int\limits_{[0,1]^k} \left(I^{k+1}_{(j,1)} \right)^* \omega^k - \int\limits_{[0,1]^k} \left(I^{k+1}_{(j,0)} \right)^* \omega^k \\ &= \int\limits_{[0,1]^k} \int_0^1 \frac{\partial f_j}{dx^j} (x^1, \ldots, x^{j-1}, t, x^j, \ldots, x^k) dt \cdot dx^1 \cdot \ldots \cdot dx^k \\ &= \int\limits_{[0,1]^{k+1}} \frac{\partial f_j}{dx^j} dx . \end{aligned}$$

By the definition of the boundary of I^{k+1}, this implies that

$$\int\limits_{\partial I^{k+1}} \omega^k = \sum_{j=1}^{k+1} \sum_{\alpha=0,1} \int\limits_{I^{k+1}_{(j,\alpha)}} (-1)^{j+\alpha} \omega^k = \sum_{j=1}^{k+1} (-1)^{j+1} \int\limits_{[0,1]^{k+1}} \frac{\partial f_j}{dx^j} dx = \int\limits_{I^{k+1}} d\omega^k ,$$

and hence we have verified Stokes' formula for the standard cube. For an arbitrary singular $(k+1)$-cube $c^{k+1} : [0,1]^{k+1} \to U \subset \mathbb{R}^n$ we now use the fact that the exterior derivative commutes with the pullback of forms. We have

$$\begin{aligned}\int\limits_{c^{k+1}} d\omega^k &= \int\limits_{I^{k+1}} \left(c^{k+1}\right)^* (d\omega^k) = \int\limits_{I^{k+1}} d\left(\left(c^{k+1}\right)^* \omega^k\right) \\ &= \int\limits_{\partial I^{k+1}} \left(c^{k+1}\right)^* \omega^k = \sum_{j=1}^{k+1} \sum_{\alpha=0,1} \int\limits_{I^{k+1}_{(j,\alpha)}} \left(c^{k+1}\right)^* \omega^k \\ &= \sum_{j=1}^{k+1} \sum_{\alpha=0,1} \int\limits_{c^{k+1} \circ I^{k+1}_{(j,\alpha)}} \omega^k = \int\limits_{\partial c^{k+1}} \omega^k . \qquad \square\end{aligned}$$

We already emphasized that the line integral of an *exact* 1-form is independent of the particular shape of the curve. As a first application of Stokes' theorem, we will prove that the line integrals of a *closed* 1-form along two different curves coincide if there exists a continuously differentiable deformation of one curve into the other leaving their initial and end points fixed. This leads to the notion of homotopy, which is fundamental in topology.

Definition 18. Two C^1-curves $c_0, c_1 : [0,1] \to U \subset \mathbb{R}^n$ are called *homotopic* if there exists a C^1-map $F : [0,1] \times [0,1] \to U$ with the properties

$$\begin{aligned} F(t,0) &= c_0(t), \; F(0,s) = c_0(0) = c_1(0), \\ F(t,1) &= c_1(t), \; F(1,s) = c_0(1) = c_1(1). \end{aligned}$$

The map F is called a *homotopy* between the curves c_0 and c_1.

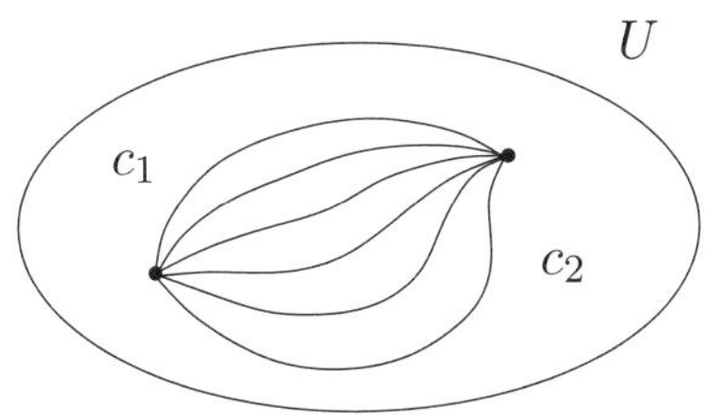

Theorem 9. *Let $c_0, c_1 : [0,1] \to U$ be two homotopic C^1-curves, and let ω^1 be a closed 1-form on U. Then the line integrals along c_0 and c_1 coincide:*

$$\int_{c_0} \omega^1 = \int_{c_1} \omega^1 .$$

Proof. Choose any homotopy $F : [0,1]^2 \to U$ between c_0 and c_1. This map F is, at the same time, a singular 2-cube, and hence the 2-form $d\omega^1$ can be integrated over F. By assumption we have $d\omega^1 = 0$, and the corresponding integral vanishes:

$$0 = \int_F d\omega^1 .$$

On the other hand, by Stokes' theorem, we have for the right-hand side

$$\int_F d\omega^1 = \int_{\partial F} \omega^1 = \int_{F(t,0)} \omega^1 + \int_{F(1,s)} \omega^1 - \int_{F(t,1)} \omega^1 - \int_{F(0,s)} \omega^1 .$$

These four integrals are computed using the homotopy property of F. First, $F(t,\alpha) = c_\alpha(t)$ immediately implies, for $\alpha = 0, 1$,

$$F|^*_{(t,\alpha)}\omega^1 = c^*_\alpha \omega^1 .$$

Moreover, $F(\alpha, s)$ is independent of s, and hence

$$F|^*_{(\alpha,s)}\omega^1 = 0.$$

Lastly, these relations combine to yield the equation

$$0 = \int_0^1 c^*_0\omega^1 - \int_0^1 c^*_1\omega^1 ,$$

and by the definition of the integral this concludes the proof. □

We will now generalize these observations concerning line integrals to the higher-dimensional case. Let $s^k = \sum_j l_j \cdot c^k_j$ be a singular k-chain with image set $A \subset \mathbb{R}^n$, and suppose that $\partial s^k = 0$. The set A can, e.g., be a k-dimensional sphere S^k in $\mathbb{R}^n$. Consider a smooth map $f : A \to A$ defined on a neighborhood of A which is homotopic to the identity Id_A. This is again supposed to mean that there is a smooth map defined on a neighborhood U of $A \times [0,1]$ in $\mathbb{R}^{n+1}$,

$$F : A \times [0,1] \longrightarrow A \quad \text{such that} \quad F(a,0) = f(a) \quad \text{and} \quad F(a,1) = a .$$

Then

$$F \circ (s^k \times \mathrm{Id}_{[0,1]}) := \sum l_j \cdot F \circ (c^k_j \times \mathrm{Id}_{[0,1]})$$

is a singular $(k+1)$-chain in $\mathbb{R}^n$, and, because $\partial s^k = 0$, the boundary is

$$\partial F \circ (s^k \times \mathrm{Id}_{[0,1]}) = s^k - f \circ s^k .$$

For a k-form ω^k defined on an open neighborhood of the set $A \subset \mathbb{R}^n$, the following holds.

Theorem 10. *If $\partial s^k = 0$ and f is homotopic to the identity, then*

$$\int_{s^k} \omega^k = \int_{f \circ s^k} \omega^k .$$

Proof. We compute the difference using Stokes' theorem:

$$\int_{s^k} \omega^k - \int_{f\circ s^k} \omega^k \;=\; \int_{\partial F\circ(s^k\times \mathrm{Id}_{[0,1]})} \omega^k \;=\; \int_{F\circ(s^k\times \mathrm{Id}_{[0,1]})} d\omega^k .$$

The $(k+1)$-form $(s^k \times \mathrm{Id}_{[0,1]})^* F^*(d\omega^k)$ vanishes. This follows from the implicit function theorem together with the assumption that A is the image of a k-dimensional chain. But this immediately implies the statement. □

2.6. The Classical Formulas of Green and Stokes

In this section, we will discuss the classical two-dimensional special cases of Stokes' formula. Let $D \subset \mathbb{R}^2$ be a subset of $\mathbb{R}^2$ which can be represented as the image of a C^1-map f defined on the standard cube $[0,1]^2$. By ∂D we denote the boundary of this set, considered as a singular 1-chain. The derivative of the 1-form $\omega^1 := (x\cdot dy - y\cdot dx)/2$ is the volume form on $\mathbb{R}^2$. Hence

$$\mathrm{vol}(D) \;=\; \int_D d\omega^1 \;=\; \frac{1}{2}\int_{\partial D} (x\cdot dy - y\cdot dx)\,.$$

The formula transforms the calculation of a two-dimensional volume into the evaluation of a line integral, and this turns out to be a special case of Green's formula to be discussed now. Consider the 1-form

$$\omega^1 \;:=\; P(x,y)\cdot dx + Q(x,y)\cdot dy\,,$$

and compute its derivative:

$$d\omega^1 \;=\; \left[\frac{\partial Q}{\partial x} - \frac{\partial P}{\partial y}\right] dx\wedge dy\,.$$

Thus we arrive at *Green's first formula.*

Theorem 11. *Let $P(x,y)$ and $Q(x,y)$ be functions of class C^1. Then*

$$\int_{\partial D} P(x,y)\cdot dx + Q(x,y)\cdot dy \;=\; \int_D \left[\frac{\partial Q}{\partial x} - \frac{\partial P}{\partial y}\right] dx\wedge dy\,.$$

Suppose now that $P(x,y)$ and $Q(x,y)$ are of class C^2, and consider the 1-forms

$$\omega^1 \;:=\; P\left[-\frac{\partial Q}{\partial y}\cdot dx + \frac{\partial Q}{\partial x}\cdot dy\right] \quad\text{and}\quad \eta^1 \;:=\; Q\left[-\frac{\partial P}{\partial y}\cdot dx + \frac{\partial P}{\partial x}\cdot dy\right].$$

The derivative of the difference $\omega^1 - \eta^1$ is easily computed to be

$$d(\omega^1 - \eta^1) \;=\; [P\cdot\Delta(Q) - Q\cdot\Delta(P)]\,dx\wedge dy\,.$$

Applying Stokes' formula leads to *Green's second formula.*

Theorem 12. *Let $P(x,y)$ and $Q(x,y)$ be functions of class C^2. Then*

$$\int_D [P\Delta(Q) - Q\Delta(P)]dx \wedge dy = \int_{\partial D} \left[Q\frac{\partial P}{\partial y} - P\frac{\partial Q}{\partial y}\right]dx + \left[P\frac{\partial Q}{\partial x} - Q\frac{\partial P}{\partial x}\right]dy.$$

Stokes' formula concerning certain surface integrals is, in a similarly simple way, a special case of the above general integral formula. In fact, let $F \subset \mathbb{R}^3$ be a subset of $\mathbb{R}^3$, which can be represented as the image of a C^1-map f defined on the standard cube $[0,1]^2$ (a surface piece). By ∂F we denote the boundary of this set considered as a singular 1-chain in $\mathbb{R}^3$. Consider a vector field $\mathcal{V}$ defined on an open neighborhood of F. Its curl is defined by the following condition:

$$d\omega^1_{\mathcal{V}} \ := \ \mathrm{curl}(\mathcal{V}) \lrcorner \, d\mathbb{R}^3 \ = \ *\omega^1_{\mathrm{curl}(\mathcal{V})} \, .$$

Integrating the 2-form $*\omega^1_{\mathrm{curl}(\mathcal{V})}$ over the surface piece F, we obtain *Stokes' theorem* in its classical form.

Theorem 13. *Let $\mathcal{V}$ be a smooth vector field on a neighborhood of F. Then*

$$\int_{\partial F} (V^1dx^1 + V^2dx^2 + V^3dx^3) \ = \ \int_F d\omega^1_{\mathcal{V}} \ =$$

$$\int_F \left[\frac{\partial V^2}{\partial x^1} - \frac{\partial V^1}{\partial x^2}\right]dx^1\wedge dx^2 + \left[\frac{\partial V^3}{\partial x^1} - \frac{\partial V^1}{\partial x^3}\right]dx^1\wedge dx^3 + \left[\frac{\partial V^3}{\partial x^2} - \frac{\partial V^2}{\partial x^3}\right]dx^2\wedge dx^3.$$

Remark. Using the volume form dF of a regular surface piece yet to be discussed, the classical Stokes formula can be written more concisely as

$$\int_{\partial F} \omega^1_{\mathcal{V}} = \int_F \langle \mathrm{curl}(\mathcal{V}), \mathcal{N} \rangle \cdot dF \, .$$

Here $\mathcal{N}$ denotes the normal vector to the surface in $\mathbb{R}^3$. We will return to this in Chapter 3.

2.7. Complex Differential Forms and Holomorphic Functions

The complexification of the vector space of all real-valued forms is called the space of *complex-valued forms* on an open subset of $\mathbb{R}^n$. Such a form can be split into its real and imaginary parts;

$$\omega^k \ = \ \omega^k_0 + i \cdot \omega^k_1 \, ,$$

and differentiation as well as integration are defined with respect to this decomposition:

$$d\omega^k \ := \ d\omega^k_0 + i \cdot d\omega^k_1, \quad \int_{\sigma^k} \omega^k \ := \ \int_{\sigma^k} \omega^k_0 + i \cdot \int_{\sigma^k} \omega^k_1 \, .$$

In a similar way we extend the exterior product to complex-valued forms:

$$\omega^k \wedge \eta^l \;:=\; (\omega_0^k \wedge \eta_0^l - \omega_1^k \wedge \eta_1^l) + i \cdot (\omega_0^k \wedge \eta_1^l + \omega_1^k \wedge \eta_0^l)\,.$$

Then the previous computational rules and Stokes' theorem still hold. Now we want to apply complex-valued forms to study holomorphic functions and, to do so, first identify the real vector space $\mathbb{R}^2$ with the complex numbers $\mathbb{C}$. From $z := x + i \cdot y$ and $\bar{z} := x - i \cdot y$ we obtain the differential forms

$$dz \;:=\; dx + i \cdot dy\,, \qquad d\bar{z} \;:=\; dx - i \cdot dy \quad \text{and} \quad dz \wedge d\bar{z} \;=\; -2i \cdot dx \wedge dy\,.$$

Let f be a complex-valued function with real and imaginary parts u and v of class C^1, $f(z) = u + i \cdot v$. Denote by u_x, u_y, v_x, v_y the partial derivatives with respect to the corresponding variables. Then $f(z) \cdot dz$ is a complex-valued differential form. We compute its differential:

$$d(f(z) \cdot dz) \;=\; (-u_y - v_x) \cdot dx \wedge dy + i \cdot (u_x - v_y) \cdot dx \wedge dy\,.$$

Now let $f : U \to \mathbb{C}$ be a complex-differentiable function defined on an open subset of $\mathbb{C}$. Elementary complex analysis starts by proving that its real and imaginary parts are smooth functions in the sense of real analysis (Goursat's theorem). Furthermore, the *Cauchy-Riemann equations* hold:

$$u_x = v_y \quad \text{and} \quad u_y = -v_x\,.$$

Theorem 14. *If $f(z)$ is a complex-differentiable function, then the 1-form $f(z) \cdot dz$ is closed,*

$$d(f(z) \cdot dz) \;=\; 0\,.$$

An immediate consequence is *Cauchy's theorem.*

Theorem 15. *Let U be an open subset of $\mathbb{C}$, and let γ be a closed curve in U, which is the boundary of a singular 2-cube. Then the integral*

$$\int_\gamma f(z) \cdot dz \;=\; 0$$

vanishes for each complex-differentiable function f.

In a similar way we derive Cauchy's integral formula. To do so, we assume that $f(z)$ is a complex-differentiable function on a neighborhood of the disc

$$K(z_0, M) \;=\; \big\{z \in \mathbb{C} : ||z - z_0|| \le M\big\}\,.$$

Fix $0 < \varepsilon \le M$. The function $f(z)/(z - z_0)$ is complex-differentiable in $K(z_0, M) \diagdown K(z_0, \varepsilon)$, and, since $d((f(z) \cdot dz)/(z - z_0)) = 0$, we have

$$\begin{aligned}\int_{\partial K(z_0,M)} \frac{f(z) \cdot dz}{z - z_0} = \int_{\partial K(z_0,\varepsilon)} \frac{f(z) \cdot dz}{z - z_0} &= \int_{\partial K(z_0,\varepsilon)} (\bar{z} - \bar{z}_0) \frac{f(z) \cdot dz}{||z - z_0||^2} \\ &= \frac{1}{\varepsilon^2} \int_{\partial K(z_0,\varepsilon)} (\bar{z} - \bar{z}_0) f(z) \cdot dz\,.\end{aligned}$$

We compute the derivative of the 1-form $(\bar{z}-\bar{z}_0)f(z)\cdot dz$:

$$d((\bar{z}-\bar{z}_0)f(z)\cdot dz) \;=\; 2i\cdot f(z)\cdot dx\wedge dy\,.$$

Thus

$$\int_{\partial K(z_0,M)} \frac{f(z)\cdot dz}{z-z_0} \;=\; \frac{2i}{\varepsilon^2}\int_{K(z_0,\varepsilon)} f(z)\cdot dx\wedge dy\,.$$

The mean-value theorem of integral calculus states that there is a number z_ε in $K(z_0,\varepsilon)$ for which

$$\int_{K(z_0,\varepsilon)} f(z)\cdot dx\wedge dy \;=\; f(z_\varepsilon)\mathrm{vol}(K(z_0,\varepsilon)) \;=\; \pi\cdot\varepsilon^2\cdot f(z_\varepsilon)\,.$$

If ε tends to zero, then $f(z_\varepsilon)$ converges to $f(z_0)$, and we arrive at *Cauchy's integral formula*:

Theorem 16.

$$f(z_0) \;=\; \frac{1}{2\pi i}\int_{\partial K(z_0,M)} \frac{f(z)\cdot dz}{z-z_0}\,.$$

This formula is fundamental for the theory of functions. It implies, e.g., that every function which is complex-differentiable in the neighborhood of a point can be expanded into a power series (i. e. is an analytic function).

2.8. Brouwer's Fixed Point Theorem

A *fixed point* of a map $f: X\to X$ from a set to itself is defined to be a point x_0 which is not moved by f, $f(x_0)=x_0$. In topology, several fixed point theorems are known. They state that certain continuous maps from a metric space to itself necessarily have at least one fixed point. If, e.g., X is a complete metric space, and $f: X\to X$ is a contracting map, then by the *Banach fixed point theorem* the map f has at least one fixed point. There are, of course, (non-contracting) maps from a complete metric space to itself without fixed points; translations in $\mathbb{R}^n$ are examples for this. A topological space X is said to have the *fixed point property* if every continuous map $f: X\to X$ from X to itself has a fixed point. This is obviously a topological property, i. e., homeomorphic spaces either all have the fixed point property or none of them has it. The unit circle $X=S^1$ is compact; nevertheless it does not have the fixed point property; rotations are continuous maps from the unit circle to itself without stationary points. *Brouwer's fixed point theorem* states that the closed n-dimensional ball of radius R,

$$D^n(R) \;=\; \{x\in\mathbb{R}^n:\ ||x||\le R\}\,,$$

is a metric space having the fixed point property. We will first prove this theorem in the case of smooth maps from the ball to itself, and afterwards extend the proof to continuous maps by means of an approximation argument.

Theorem 17. *Every C^1-map $f : D^n \to D^n$ from the n-dimensional ball to itself has at least one fixed point.*

Proof. Suppose that $f : D^n \to D^n$ does not have any fixed points. Then define the map $F : D^n \longrightarrow S^{n-1}$ from the ball to its boundary by assigning to every point $x \in D^n$ the point of intersection of the ray from $f(x)$ through x with the sphere S^{n-1}. The formula for F,

$$\begin{aligned} F(x) &= x \\ &+ \left[\sqrt{1 - ||x||^2 + \left\langle x, \frac{x - f(x)}{||x - f(x)||}\right\rangle^2} - \left\langle x, \frac{x - f(x)}{||x - f(x)||}\right\rangle\right] \frac{x - f(x)}{||x - f(x)||}, \end{aligned}$$

shows that F is smooth. Moreover, F acts on the boundary of the ball as the identity, $F(x) = x$ for all $x \in S^{n-1}$. Let $F^1, \ldots, F^n$ be the components of F. Differentiating the following relation, which is valid for all $x \in D^n$,

$$\sum_{i=1}^{n} (F^i(x))^2 = 1,$$

yields

$$2\sum_{i=1}^{n} F^i \cdot dF^i = 2\sum_{i,j=1}^{n} \left(F^i \frac{\partial F^i(x)}{\partial x^j}\right) dx^j = 0,$$

and hence for each index j

$$\sum_{i=1}^{n} F^i(x) \frac{\partial F^i(x)}{\partial x^j} = 0.$$

Therefore, the system of equations

$$\sum_{i=0}^{n} \alpha^i \frac{\partial F^i(x)}{\partial x^j} = 0$$

has a non-trivial solution $(\alpha^1, \ldots, \alpha^n) = (F^1(x), \ldots, F^n(x)) \neq (0, \ldots, 0)$. Hence the determinant of the matrix

$$\det\left(\frac{\partial F^i(x)}{\partial x^j}\right) = 0$$

vanishes. Now we apply this observation to the differential form

$$\omega^{n-1} = F^1 \wedge dF^2 \wedge \ldots \wedge dF^n$$

and conclude that its differential vanishes:

$$d\omega^{n-1} = dF^1 \wedge dF^2 \wedge \ldots \wedge dF^n = \det\left(\frac{\partial F^i(x)}{\partial x^j}\right) dx^1 \wedge \ldots \wedge dx^n = 0.$$

By Stokes' theorem the integral of the form ω^{n-1} over the boundary S^{n-1} of the singular cube D^n is equal to zero:

$$0 = \int_{D^n} d\omega^{n-1} = \int_{S^{n-1}} \omega^{n-1} .$$

On the other hand, F acts on the sphere S^{n-1} as the identity; hence

$$\omega^{n-1}|_{S^{n-1}} = x^1\, dx^2 \wedge \ldots \wedge dx^n|_{S^{n-1}} .$$

This implies

$$0 = \int_{S^{n-1}} x^1\, dx^2 \wedge \ldots \wedge dx^n = \int_{D^n} dx^1 \wedge \ldots \wedge dx^n = \operatorname{vol}(D^n) ,$$

and we arrive at a contradiction. □

Theorem 18 (Brouwer's Theorem). *Every continuous map $f : D^n \to D^n$ from the n-dimensional ball to itself has at least one fixed point.*

Proof. The proof will be reduced to the case of a C^1-map by applying the Stone-Weierstrass approximation theorem (see, e.g., [Rudin, 1998], Theorem 7.32). The ball D^n is compact. Consider the ring $C^0(D^n)$ of all real-valued continuous functions on it, as well as the subring $\mathcal{R}$ of those functions which are the restriction to D^n of a C^1-function with strictly larger, open domain of definition. Obviously, the subring $\mathcal{R}$ contains the constant functions and separates points. By the Stone-Weierstrass theorem, it is dense in $C^0(U)$. Applying this to the components $f^1, \ldots, f^n$ of f, we conclude that for each $\varepsilon > 0$ there exists a C^1-map

$$p : D^n \to \mathbb{R}^n \quad \text{such that} \quad ||f(x) - p(x)|| < \varepsilon \quad \text{for all } x \in D^n .$$

Consider the renormalized map $\tilde{p}(x) := p(x)/(1+\varepsilon)$. Since

$$||p(x)|| - ||f(x)|| \leq ||p(x) - f(x)|| < \varepsilon \quad \text{and} \quad ||f(x)|| \leq 1 ,$$

we have $||p(x)|| \leq 1 + \varepsilon$; hence $||\tilde{p}(x)|| \leq 1$. Therefore, $\tilde{p}$ is a map from the ball to itself. Moreover, $\tilde{p}$ can be estimated against f:

$$\begin{aligned} ||f(x) - \tilde{p}(x)|| &\leq ||f(x) - p(x)|| + ||p(x) - \tilde{p}(x)|| \leq \varepsilon + ||p(x)|| \left| 1 - \frac{1}{1+\varepsilon} \right| \\ &\leq \varepsilon + (1+\varepsilon)\frac{\varepsilon}{1+\varepsilon} < 2\varepsilon . \end{aligned}$$

Summarizing, we have proved that for each $\varepsilon > 0$ there exists a C^1-map $\tilde{p} : D^n \to D^n$ satisfying for every $x \in D^n$ the estimate

$$||f(x) - \tilde{p}(x)|| < 2\varepsilon .$$

Now if the continuous map $f : D^n \to D^n$ had no fixed point, the number

$$\mu = \inf_{x \in D^n} ||f(x) - x||$$

would be strictly positive. Choose for $\varepsilon = \mu/2$ a smooth approximation $\tilde{p}$ with the properties stated above; by Theorem 17 this map then has a fixed point $x_0 \in D^n$, for which in turn

$$||f(x_0) - \tilde{p}(x_0)|| \;=\; ||f(x_0) - x_0|| \;<\; \mu$$

would have to hold. But this contradicts the definition of μ. □

Brouwer's fixed point theorem can be viewed as an existence result for real, non-linear systems of equations. We state a possible application, which plays an important role in *Galerkin's method.*

Theorem 19. *Let $g_1, \ldots, g_n : D^n(R) \to \mathbb{R}$ be continuous functions defined on the ball $D^n(R)$ of radius R, and suppose that for all points $x = (x^1, \ldots, x^n) \in S^{n-1}(R)$ in the sphere $S^{n-1}(R)$ the following inequality holds:*

$$\sum_{i=1}^{n} g_i(x) \cdot x^i \;\geq\; 0\,.$$

Then the system of equations

$$g_1(x) \;=\; g_2(x) \;=\; \ldots \;=\; g_n(x) \;=\; 0$$

has at least one solution in $D^n(R)$.

Proof. We combine the functions to define a map $g : D^n(R) \to \mathbb{R}^n$, $g(x) := (g_1(x), \ldots, g_n(x))$. If $g(x) \neq 0$ holds for all points $x \in D^n(R)$, we can consider the map $f : D^n(R) \to S^{n-1}(R)$,

$$f(x) \;:=\; -R\,\frac{g(x)}{||g(x)||}\,,$$

whose image lies in the sphere $S^{n-1}(R)$. By the fixed point theorem, f has a fixed point in $S^{n-1}(R)$. Hence there exists a point $x_0 \in S^{n-1}(R)$ such that

$$x_0 \;=\; -R\,\frac{g(x_0)}{||g(x_0)||}\,.$$

This implies $R^2 \cdot ||g(x_0)|| = -R\,\langle g(x_0), x_0\rangle$, contradicting the assumption of the theorem. □

In particular, the assumption of the theorem is satisfied for $g_i(x) \;=\; x^i + h_i(x)$ if the functions $h_i : \mathbb{R}^n \to \mathbb{R}$ grow more slowly than linear forms,

$$|h_i(x)| \leq C_i \cdot ||x||^{1-\varepsilon}\,.$$

Under this condition the sum

$$\sum_{i=1}^{n} g_i(x) x^i \;=\; R^2 + \sum_{i=1}^{n} h_i(x) x^i$$

behaves like R^2 on the sphere $S^{n-1}(R)$ and becomes positive for sufficiently large radii.

Corollary 1. *The system of equations*

$$h_1(x) \ = \ x^1, \ \ldots, \ h_n(x) \ = \ x^n$$

has at least one solution for arbitrary continuous functions bounded by $||x||^{1-\varepsilon}$ *at infinity.*

Example 11. The system of equations $x = \sqrt[4]{1 + x^2 + y^2}$, $y = \cos(x + y)$ has in $\mathbb{R}^2$ at least one solution, e.g.,

$$x \ = \ 1.2758079, \quad y \ = \ 0.14722564 .$$

Example 12. Consider in $\mathbb{R}^2$ the following system of equations:

$$g_1(x, y) = x + e^{-(x+y)^2} = 0, \quad g_2(x, y) = y + e^{-(x-y)^2} = 0 .$$

The picture below diaplays the graph of the function $g_1(x, y) \cdot x + g_2(x, y) \cdot y$ over the set $[-2, 2] \times [-2, 2]$. It shows that this function is positive on the circle $S^1(2)$ of radius 2. Hence the above system of equations has at least one solution in the disc $D^2(2)$. A numerical computation of the solution leads to the values

$$x \ = \ -0.303122, \quad y \ = \ -0.789407 .$$

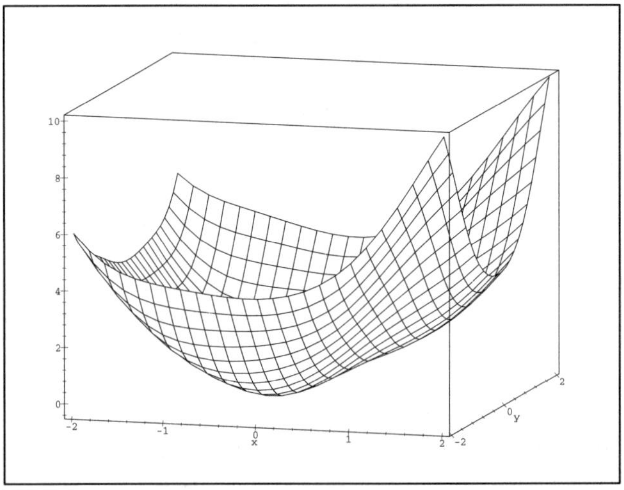

Exercises

1. Let $f : \mathbb{R}^2 - \{0\} \to \mathbb{R}^2 - \{0\}$, $f(r,\theta) = (r\cos\theta, r\sin\theta)$, be the polar coordinate map on the "punctured" plane. Prove:

a) The winding form satisfies $f^*(\frac{x\,dy - y\,dx}{x^2+y^2}) = d\theta$;

b) the radial form $x\,dx + y\,dy$ satisfies $f^*(x\,dx + y\,dy) = r\,dr$.

2. Consider on $\mathbb{R}^2 - \{0\}$ the winding form $\omega^1 = \frac{x\,dy - y\,dx}{x^2+y^2}$ as well as the following family of curves depending on the integer parameter $n \in \mathbb{Z}$:

$$c_n : \ [0,1] \longrightarrow \mathbb{R}^2 - \{0\}, \quad c_n(t) \ = \ (\cos 2\pi nt, \sin 2\pi nt)\,.$$

Compute the line integral of the winding form along the curve c_n, and conclude that the curves c_n are not homotopic in $\mathbb{R}^2 - \{0\}$ for different values of the parameter n.

3. Compute the exterior derivative of the following differential forms:

a) $xy\,dx \wedge dy + 2x\,dy \wedge dz + 2y\,dx \wedge dz$;

b) $z^2\,dx \wedge dy + (z^2 + 2y)dx \wedge dz$;

c) $13\,x\,dx + y^2 dy + xyz\,dz$;

d) $e^x \cos(y)\,dx - e^x \sin(y)\,dy$;

e) $x\,dy \wedge dz + y\,dx \wedge dz + z\,dx \wedge dy$.

4. Consider $\mathbb{R}^{2n}$ with coordinates $x^1, \ldots, x^{2n}$ and the following differential form of degree 2:

$$\omega^2 \ = \ dx^1 \wedge dx^{n+1} + dx^2 \wedge dx^{n+2} + \ldots + dx^n \wedge dx^{2n}\,.$$

Prove:

a) The form ω^2 is closed;

b) the n-th exterior power of ω^2 is related to the volume form via the formula

$$\omega^2 \wedge \ldots \wedge \omega^2 \ = \ (-1)^{n(n-1)/2} n!\, dx^1 \wedge \ldots \wedge dx^{2n}\,.$$

5. The subject of this exercise is to explain why it makes sense to denote the vector fields $x \mapsto (x, e_i)$, defined using the standard basis $e_1, \ldots, e_n$ of

$\mathbb{R}^n$, by $\partial/\partial x^i$. Writing these vector fields for the moment as E_i, every vector field can be written in the form

$$\mathcal{V} = \sum_{i=1}^{n} V^i(x) \cdot E_i\,.$$

For each function $f : U \to \mathbb{R}$ on an open set U of $\mathbb{R}^n$ we define a new function, the derivative of f in the direction of the vector field $\mathcal{V}$, by

$$(\mathcal{V}(f))(x) := (Df_x)(\mathcal{V}(x))\,.$$

Prove the formula

$$\mathcal{V}(f) = \left(\sum_{i=1}^{n} V^i \frac{\partial}{\partial x^i}\right)(f)\,,$$

which by omitting the argument f provides the explanation asked for.

6. Prove the following rules for vector fields on $\mathbb{R}^3$:

a) $\mathrm{div}(\mathcal{V}_1 \times \mathcal{V}_2) = \langle \mathrm{curl}(\mathcal{V}_1), \mathcal{V}_2\rangle - \langle \mathcal{V}_1, \mathrm{curl}(\mathcal{V}_2)\rangle$;

b) $\mathrm{curl}(\mathrm{curl}(\mathcal{V})) = \mathrm{grad}(\mathrm{div}(\mathcal{V})) - \Delta(\mathcal{V})$, where the Laplacian is to be applied componentwise to $\mathcal{V}$.

7. Compute the line integral

$$\int_C (x^2 - 2xy)dx + (y^2 - 2xy)dy$$

along the curve $C = \{(x, y) \in \mathbb{R}^2 : \ x \in [-1, 1],\, y = x^2\}$.

8. Compute the line-integral

$$\int_C \sin(y)dx + \sin(x)dy\,,$$

where C is the segment joining the points $(0, \pi)$ and $(\pi, 0)$.

9. Consider on $\mathbb{R}^3$ the differential form $\omega^2 = y\, dx\wedge dy$. Determine all 1-forms $\eta^1 = p\, dx + q\, dy$ satisfying $d\eta^1 = \omega^2$.

10. Prove the following converse to one of the statements of Example 9: If ω^1 is a 1-form defined on the open set U, and if the line integral of ω^1 is independent of the curve, then ω^1 is exact.

Hint. Prove this by explicitly constructing a "primitive function". In physics, the vector field $\mathcal{V}$ corresponding to the 1-form $\omega^1_{\mathcal{V}}$ is called *conservative*, if it does no work along any closed curve γ, $\int_\gamma \omega^1_{\mathcal{V}} = 0$. Thus, $\mathcal{V}$ is conservative if and only if $\omega^1_{\mathcal{V}}$ is exact.

11. Consider the singular 2-cube, $f : [0,2\pi] \times [0,2\pi] \to S^2 \subset \mathbb{R}^3 - \{0\}$,

$$f(u,v) = (\cos u \sin v,\ \sin u \sin v,\ \cos v),$$

as well as the 2-form $\omega^2 = (x\,dy \wedge dz + y\,dz \wedge dx + z\,dx \wedge dy)/r^3$ on $\mathbb{R}^3 - \{0\}$, where $r = (x^2+y^2+z^2)^{1/2}$ denotes the distance from the origin.

a) Prove that ω^2 is closed;

b) compute the integral of ω^2 over f;

c) conclude from the properties just proved that ω^2 is not exact on $\mathbb{R}^3 - \{0\}$, and that there is no singular 3-chain c^3 in $\mathbb{R}^3 - \{0\}$ whose boundary equals f, $\partial c^3 = f$.

12. Prove that the integral defines a unique bilinear map

$$\int : H^k_{DR}(U) \times H^{\mathrm{kub}}_k(U) \longrightarrow \mathbb{R}, \quad \Big([\omega^k],\, [s^k]\Big) \longmapsto \int_{s^k} \omega^k .$$

13. Let $\omega^1 = f(x)\,dx$ be a 1-form on the interval $[0,1]$ with $f(0) = f(1)$. Prove that there exist a real number p and a function g with $g(0) = g(1)$ by means of which ω^1 can be written as

$$\omega^1 = p\,dx + dg .$$

14 (Continuation of 13)**.** Let η^1 be a closed 1-form on $\mathbb{R}^2 - \{0\}$ and ω^1 the winding form. Prove that there exist a number p as well as a function $g : \mathbb{R}^2 - \{0\} \to \mathbb{R}$ for which

$$\eta^1 = p\,\omega^1 + dg .$$

Consequently, the winding form is the generating element of the first de Rham cohomology of $\mathbb{R}^2 - \{0\}$.

Hint. Consider the polar coordinate map f from Exercise 1 and its pullback $f^*\omega^1$. This can be written as $f^*\omega^1 = A(r,\theta)dr + B(r,\theta)d\theta$; here $B(r,\theta)d\theta$ is a 1-form on $[0,2\pi]$ (depending on the parameter r) to which the previous exercise applies.

15. Consider on $\mathbb{R}^3$ the following exact differential form known from 7:

$$\omega^2 = xy\,dx \wedge dy + 2x\,dy \wedge dz + 2y\,dx \wedge dz,$$

and the upper half-ball $A \subset S^2$:

$$A = \big\{(x,y,z) \in \mathbb{R}^3 : \ x^2+y^2+z^2 = 1,\ z \geq 0\big\}.$$

Prove that the integral of ω^2 over A vanishes.

16. Let C be the circle in $\mathbb{R}^2$ with the equation $x^2 + (y-1)^2 = 1$ in its standard parametrization. Compute the line integral

$$\int_C xy^2\, dy - yx^2\, dx$$

a) directly;

b) using Green's formula.

17. Let E be the ellipse with the equation $x^2/a^2 + y^2/b^2 = 1\, (a > 0, b > 0)$ in its standard parametrization. Compute by means of Cauchy's integral formula the integral

$$\int_E \frac{dz}{z}$$

and obtain from this the value of the integral

$$\int_0^{2\pi} \frac{dt}{a^2 \cos^2(t) + b^2 \sin^2(t)}\,.$$

18. Let $\mathcal{H}$ be an infinite-dimensional Hilbert space, and $D = \{x \in \mathcal{H} : ||x|| \leq 1\}$ its unit ball. Does D have the fixed point property?

19. A subset $A \subset X$ of a metric space X is called a *retract* of X, if there is a continuous map $r : X \to A$ such that $r(a) = a$ for all points $a \in A$. Prove that if X has the fixed point property, then so does every retract A of X.

20. The set $A = \{(x, y) \in [-1, 1]^2 :\ xy = 0\}$ has the fixed point property.

Chapter 3

Vector Analysis on Manifolds

3.1. Submanifolds of $\mathbb{R}^n$

In Chapter 2 we introduced an integration method for differential forms over sets which can be represented as images of singular chains. These sets, however, may be quite irregular, and it is rather difficult to develop a differential calculus for functions defined on them. Further notions like tangent space, vector field, etc., are not available either in their context. Hence we will now restrict the possible subsets of $\mathbb{R}^n$ to a class for which a differential as well as an integral calculus can be established in a satisfactory manner. These sets are called *manifolds*, and they are—intuitively speaking—characterized by the fact that their points can be defined locally in a continuous (differentiable) way by finitely many real parameters, that is, locally these sets look just like euclidean space. It was the fundamental idea of *B. Riemann* in his Habilitationsvortrag (1854) to introduce the notion of a manifold as the new basic concept of space into geometry. In physics, manifolds occur as configuration and phase spaces of particle systems as well as in field theory. The precise description of what a submanifold of euclidean space is supposed to be is the content of the following definition.

Definition 1. A subset M of $\mathbb{R}^n$ is called a k-dimensional *submanifold without boundary* if, for each point $x \in M$, there exist open sets $x \in U \subset \mathbb{R}^n$ and $V \subset \mathbb{R}^n$ as well as a diffeomorphism $h : U \to V$ such that the image $h(U \cap M)$ is contained in the subspace $\mathbb{R}^k \subset \mathbb{R}^n$:

$$h(U \cap M) \;=\; V \cap (\mathbb{R}^k \times \{0\}) \;=\; \{y \in V : \; y^{k+1} = \ldots = y^n = 0\}\,.$$

The set $U^* := U \cap M$ together with the map $h^* := h|_{U^*} : U^* \to V^* := V \cap \mathbb{R}^k$ is called a *chart* around the point x of the manifold. The sets U^* and V^* are open subsets of M and $\mathbb{R}^k$, respectively (see next page).

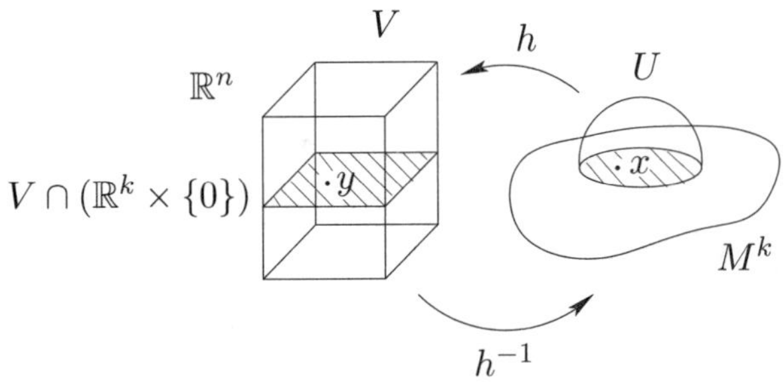

A family of charts covering the manifold M is called an *atlas*. The notion "diffeomorphism" can be understood in the sense of an arbitrary C^l-regularity ($l \geq 1$). Correspondingly, we have manifolds of regularity class C^l. For simplicity, we suppose in this chapter that all maps, manifolds, etc., are of class C^∞, and for this reason we will simply talk about smooth maps, manifolds, etc. Note that, without any change, all the statements also hold assuming only C^2-regularity.

For any two charts (h^*, U^*) and (h_1^*, U_1^*) of the manifold for which the intersection $U^* \cap U_1^*$ is not empty, one can ask how they are related. The map

$$h^* \circ (h_1^*)^{-1} : \; h_1(U^* \cap U_1^*) \longrightarrow h^*(U^* \cap U_1^*)$$

is called the *chart transition* from one chart to the other. The sets $h_1(U^* \cap U_1^*)$ and $h^*(U^* \cap U_1^*)$ are open subsets of the coordinate space $\mathbb{R}^k$, and the transition function $h^* \circ (h_1^*)^{-1}$ is obviously a diffeomorphism.

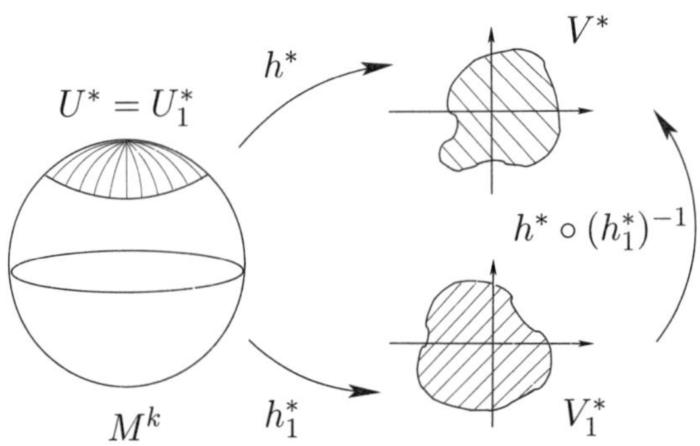

The notion of dimension also needs to be explained for manifolds. If a subset of $\mathbb{R}^k$ is mapped homeomorphically onto an open subset of the space $\mathbb{R}^l$, the dimensions of both coincide, $k = l$. Under the additional assumption of differentiability, which will always be made here, this is easy to prove: the differential of a diffeomorphism at an arbitrary point is a linear isomorphism between the tangent spaces $T_x\mathbb{R}^k$ and $T_y\mathbb{R}^l$, and this immediately implies $k = l$. The corresponding fact for homeomorphisms is a deep topological result going back to Brouwer (1910). In any case, the number k occurring in the definition of a manifold is uniquely determined and will be called the *dimension* of the manifold. Sometimes we will write the dimension of a manifold as an upper index, i.e., denote the manifold also by M^k.

In the first theorem we will prove that, under certain conditions, subsets of $\mathbb{R}^n$ defined by equations are submanifolds. This will give rise to plenty of examples.

Theorem 1. *Let $U \subset \mathbb{R}^n$ be an open subset, and let $f : U \to \mathbb{R}^{n-k}$ be a smooth map. Consider the set*

$$M = \{x \in U : f(x) = 0\}.$$

If the differential $Df(x)$ has maximal rank $(n-k)$ at each point $x \in M$ of the set M, then M is a smooth, k-dimensional submanifold of $\mathbb{R}^n$ without boundary.

Proof. The proof is based on a straightforward application of the implicit function theorem. If $x_0 \in M$ is a point from M, then there exist an open neighborhood $x_0 \in U_{x_0} \subset U$ and a diffeomorphism $h_{x_0} : U_{x_0} \to h_{x_0}(U_{x_0}) \subset \mathbb{R}^n$ such that the map $f \circ h_{x_0}^{-1} : h_{x_0}(U_{x_0}) \to \mathbb{R}^{n-k}$ is given by the formula

$$f \circ h_{x_0}^{-1}(x^1, \ldots, x^n) = (x^{k+1}, \ldots, x^n).$$

This implies $U_{x_0} \cap M = h_{x_0}^{-1}(h_{x_0}(U_{x_0}) \cap \mathbb{R}^k)$, and hence the chart around the point $x_0 \in M$ we asked for is constructed. □

Example 1. Every open subset $U \subset \mathbb{R}^n$ is an n-dimensional manifold without boundary.

Example 2. The sphere $S^n = \{x \in \mathbb{R}^{n+1} : ||x|| = 1\}$ is an n-dimensional manifold. To see this, we consider the function $f : \mathbb{R}^{n+1} \to \mathbb{R}$ defined by $f(x) = ||x||^2 - 1$. Then we have $Df(x) = (2x^1, \ldots, 2x^{n+1})$, and the rank of the $(1 \times n)$-matrix $Df(x)$ on S^n is equal to 1. Theorem 1 implies that S^n is an n-dimensional manifold.

Example 3. Consider a smooth map $f : \mathbb{R}^n \to \mathbb{R}^m$ and its *graph*

$$G(f) = \{(x, f(x)) : x \in \mathbb{R}^n\} \subset \mathbb{R}^{n+m}.$$

$G(f)$ is the zero set of the map $\Phi : \mathbb{R}^{n+m} = \mathbb{R}^n \times \mathbb{R}^m \to \mathbb{R}^m$ defined by $\Phi(x, y) = f(x) - y$. The differential $D\Phi$ has maximal rank equal to m at each point. Therefore, $G(f) \subset \mathbb{R}^{n+m}$ is an n-dimensional manifold.

Example 4. The torus of revolution is the surface in $\mathbb{R}^3$ described by the equation $(0 < r_2 < r_1)$

$$(\sqrt{x^2+y^2} - r_1)^2 \;=\; r_2^2 - z^2 \,.$$

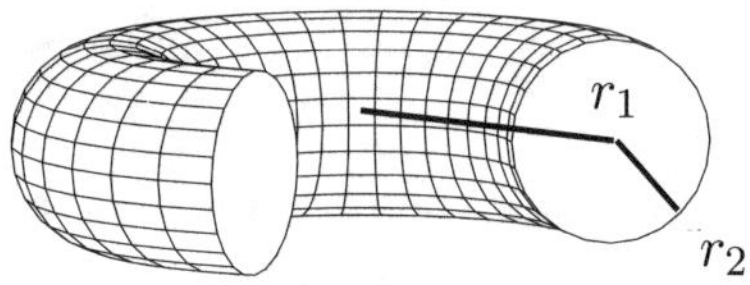

A parametrization can be obtained by the formulas

$$x \;=\; (r_1 + r_2 \cos\varphi)\cos\psi, \quad y \;=\; (r_1 + r_2\cos\varphi)\sin\psi, \quad z \;=\; r_2 \sin\varphi$$

with parameters $0 \le \varphi, \psi \le 2\pi$. The partial derivatives of the function $f(x,y,z) = (\sqrt{x^2+y^2} - r_1)^2 - r_2^2 + z^2$ are

$$\frac{\partial f}{\partial x} \;=\; 2x\Big(1 - \frac{r_1}{\sqrt{x^2+y^2}}\Big), \quad \frac{\partial f}{\partial y} \;=\; 2y\Big(1 - \frac{r_1}{\sqrt{x^2+y^2}}\Big), \quad \frac{\partial f}{\partial z} \;=\; 2z\,,$$

and it is obvious that the vector Df does not vanish at any point of the torus of revolution. Hence Theorem 1 applies and shows that the above equation defines a manifold.

Example 5. Not every set defined by an equation is a manifold. For example, consider the set in $\mathbb{R}^2$ described by the equation $x^4 = y^2$:

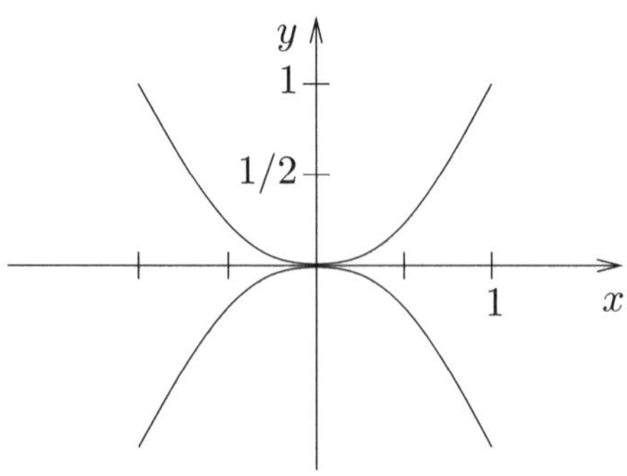

Near the point $(0,0) \in \mathbb{R}^2$ this set is not a manifold. In fact, after having deleted this point, any neighborhood of the set splits into four components and hence cannot be homeomorphic to an interval. Indeed, the assumption of Theorem 1 concerning the differential of $f(x,y) = x^4 - y^2$ is not satisfied at the point $(0,0) \in \mathbb{R}^2$.

Example 6. On the other hand, there exist manifolds in $\mathbb{R}^n$ which cannot be described by systems of equations satisfying the assumptions of Theorem 1. Later we will see that every manifold defined as in Theorem 1 has a particular property—it is orientable. An example of a non-orientable manifold is the so-called *Möbius strip.* One of its parametrizations is

$$x = \cos(u) + v\cos(u/2)\cos(u), \;\; y = \sin(u) + v\cos(u/2)\sin(u), \;\; z = v\sin(u/2)$$

with parameters $0 \leq u \leq 2\pi, \;\; -\pi \leq v \leq \pi$.

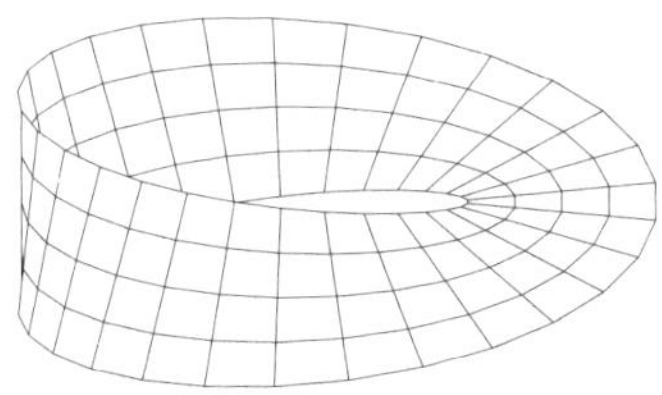

Apart from equations, manifolds can also be defined by prescribing their local parametrizations (charts). Let us explain this construction principle.

Theorem 2. *Let M be a subset of $\mathbb{R}^n$ and assume that for each point $x \in M$ there are an open set U, $x \in U \subset \mathbb{R}^n$, an open set $W \subset \mathbb{R}^k$, and a smooth map $f : W \to U$ such that the following conditions are satisfied:*

(1) $f(W) = M \cap U$;

(2) f *is bijective;*

(3) *the differential $Df(y)$ has rank k at each point $y \in W$;*

(4) $f^{-1} : M \cap U \to W$ *is continuous.*

Then M is a k-dimensional submanifold without boundary in $\mathbb{R}^n$.

Proof. For an arbitrary point $x \in M$ we choose a map $f : W \to U$ with the stated properties and denote by y the pre-image of x, $f(y) = x$. The differential has rank k, and hence we can assume without loss of generality that

$$\det \begin{bmatrix} \frac{\partial f^1}{\partial y^1} & \cdots & \frac{\partial f^k}{\partial y^1} \\ \vdots & \ddots & \vdots \\ \frac{\partial f^1}{\partial y^k} & \cdots & \frac{\partial f^k}{\partial y^k} \end{bmatrix} \neq 0 \, .$$

Consider the map defined by $g(a, b) := f(a) + (0, b)$ with $g : W \times \mathbb{R}^{n-k} \to \mathbb{R}^n$. Then the determinant of the differential of g coincides with the determinant above, and hence, in particular, it is different from zero. Applying the inverse function theorem of differential calculus, we obtain two open sets V_1 and V_2, with $(y,0) \in V_1$ and $x \in V_2$ in $\mathbb{R}^n$, for which $g : V_1 \to V_2$ is a diffeomorphism. We invert this map and denote the resulting inverse diffeomorphism by $h := g^{-1} : V_2 \to V_1$. By assumption f^{-1} is continuous, and hence there exists an open set $\mathcal{O} \subset \mathbb{R}^n$ such that

$$\{f(a) : (a,0) \in V_1\} \;=\; f(W) \cap \mathcal{O}\,.$$

Consider now the sets $\tilde{V_2} := V_2 \cap \mathcal{O}$ and $\tilde{V_1} = g^{-1}(V_2)$. Then we have

$$\tilde{V_2} \cap M \;=\; V_2 \cap \mathcal{O} \cap M \;=\; \{g(a,0) : (a,0) \in V_1\}\,,$$

and thus we obtain

$$h(\tilde{V_2} \cap M) \;=\; g^{-1}(\tilde{V_2} \cap M) \;=\; \tilde{V_1} \cap (\mathbb{R}^k \times \{0\})\,.$$

Therefore, the condition to be satisfied for each point of a manifold holds for M. □

Now we will extend the notion of manifold, taking into account also boundary points. We confine ourselves to the case that the boundary itself is a smooth manifold without boundary (no corners or edges). We define the *k-dimensional half-space* $\mathbb{H}^k$ to be the set

$$\mathbb{H}^k \;=\; \{x \in \mathbb{R}^k : x^k \geq 0\}\,.$$

Definition 2. A subset M of $\mathbb{R}^n$ is called a *k-dimensional submanifold (with boundary)* if for each point $x \in M$ one of the following conditions is satisfied:

(1) There exist open sets U and V, with $x \in U \subset \mathbb{R}^n, V \subset \mathbb{R}^n$, and a diffeomorphism $h : U \to V$ such that

$$h(U \cap M) \;=\; V \cap (\mathbb{R}^k \times \{0\})\,.$$

(2) There exist open sets U and V, with $x \in U \subset \mathbb{R}^n, V \subset \mathbb{R}^n$, and a diffeomorphism $h : U \to V$ such that

$$h(U \cap M) \;=\; V \cap (\mathbb{H}^k \times \{0\})$$

and the k-th component h^k of h vanishes at the point x, $h^k(x) = 0$.

Conditions (1) and (2) cannot be satisfied at the same time for one and the same point $x \in M$, for otherwise there would exist diffeomorphisms $h_1 : U_1 \to V_1,\; h_2 : U_2 \to V_2$ such that

$$h_1(U_1 \cap M) \;=\; V_1 \cap \mathbb{R}^k \;\text{ and }\; h_2(U_2 \cap M) \;=\; V_2 \cap \mathbb{H}^k,\; h_2^k(x) = 0\,.$$

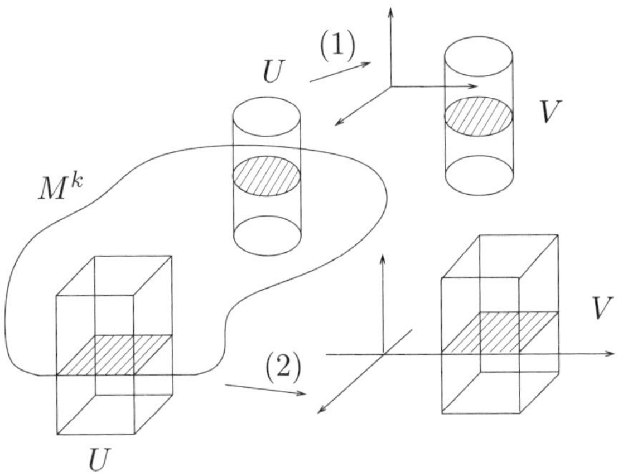

The set $h_1(U_1 \cap U_2)$ then would be an open subset in $\mathbb{R}^k$ mapped diffeomorphically onto $h_2(U_1 \cap U_2)$ by the chart transition map $h_2 \circ h_1^{-1}$. Since $h_2^k(x) = 0$, the set $h_2(U_1 \cap U_2)$ would thus contain a point from the boundary $\partial\mathbb{H}^k = \mathbb{R}^{k-1}$ of the half-space. Consequently, it could not be open in $\mathbb{R}^k$. Altogether this contradicts the inverse function theorem. This observation justifies the following

Definition 3. Let $M \subset \mathbb{R}^n$ be a manifold. A *boundary point* of M is a point $x \in M$ for which condition (2) of Definition 2 is satisfied. The set of all boundary points is denoted by ∂M and called the *boundary of* M.

Theorem 3. *Let M be a k-dimensional manifold. Then its boundary ∂M is either empty or a smooth $(k-1)$-dimensional manifold without boundary,*

$$\partial\partial M \;=\; \emptyset\,.$$

Proof. Fix a boundary point $x \in \partial M$ and choose open sets $U \subset \mathbb{R}^n$, $x \in U$, and $V \subset \mathbb{R}^n$ with

$$h(U \cap M) \;=\; V \cap (\mathbb{H}^k \times \{0\})\,.$$

For every other boundary point $x^* \in U \cap \partial M$ the k-th component of h has to vanish at x^* by the preceding observation, $h^k(x^*) = 0$. Hence we have

$$h(U \cap \partial M) \;=\; V \cap (\mathbb{R}^{k-1} \times \{0\}),$$

and thus $(h\,|_{U\cap\partial M}, U \cap \partial M)$ is a chart for the boundary ∂M. □

Example 7. The boundary of the Möbius strip is a closed space curve.

Example 8. The $(n-1)$-dimensional sphere S^{n-1} is the boundary of the n-dimensional ball D^n.

3.2. Differential Calculus on Manifolds

When a manifold is covered by charts, every chart range is an open subset of $\mathbb{R}^n$ and hence a set on which differential calculus is familiar from analysis. In this way it is possible to develop a differential calculus on manifolds.

As in the preceding section, we will now start from a k-dimensional manifold M^k and a chart $h : U \to V$ around a point x and denote by $y := h(x)$ the image of x under this chart map. Then $h^{-1} : V \to U$ is smooth and $(Dh^{-1})_y = (h^{-1})_{*,y}$ is a linear map between the tangent spaces to $\mathbb{R}^n$ (compare Definition 2, Ch. 2)

$$(h^{-1})_{*,y} : \quad T_y\mathbb{R}^n \longrightarrow T_x\mathbb{R}^n .$$

Definition 4. The *tangent space of the manifold* M^k at the point x is defined to be the image of $T_y\mathbb{R}^k$ under the map $(h^{-1})_{*,y}$:

$$T_xM^k \;:=\; (h^{-1})_{*,y}(T_y\mathbb{R}^k) \;\subset\; T_x\mathbb{R}^n .$$

The tangent space T_xM^k is a k-dimensional vector space, since the differential of the diffeomorphism h^{-1} is injective. Moreover, we have to check that the tangent space of the manifold just defined does not depend on the choice of the chart. But this is an immediate consequence of the equivalent description for the tangent space that is to follow next.

Theorem 4. *The tangent space T_xM^k consists of all vectors $(x, v) \in T_x\mathbb{R}^n$ for which there exists a smooth curve $\gamma : [0, \pm\varepsilon) \to M^k \subset \mathbb{R}^n$ such that $\gamma(0) = x$ and $\dot\gamma(0) = v$.*

Proof. A vector $\nu = (x, v) \in T_xM^k$ in the tangent space can be represented as $v = (Dh^{-1})_y(w)$ for a certain vector $w \in \mathbb{R}^k$. The image under h^{-1} of the straight line in $\mathbb{R}^k$ through y in the direction of the vector w is the curve $\gamma(t)$ we were looking for: the equality

$$\gamma(t) \;=\; h^{-1}(y + tw)$$

immediately implies $\gamma(0) = h^{-1}(y) = x$, and from the chain rule we obtain for the tangent vector

$$\frac{d\gamma(t)}{dt} \;=\; \frac{d}{dt}\left(h^{-1}(y + tw)\right)|_{t=0} \;=\; (Dh^{-1})_y(w) \;=\; v\,.$$

The converse is proved analogously. □

If the manifold M^k is defined by $(n - k)$ equations, and if, in addition, the differentials of the defining functions are linearly independent, then the tangent space has a simple description.

Theorem 5. *Let $f_1, \dots, f_{n-k} : \mathbb{R}^n \longrightarrow \mathbb{R}$ be smooth functions and suppose that*

$$df_1 \wedge \ldots \wedge df_{n-k} \neq 0 .$$

Then the tangent space $T_x M^k$ of the manifold

$$M^k = \{x \in \mathbb{R}^n : f_1(x) = \cdots = f_{n-k}(x) = 0\}$$

consists of all vectors $\nu \in T_x\mathbb{R}^n$ satisfying

$$df_1(\nu) = \ldots = df_{n-k}(\nu) = 0 .$$

In particular, the euclidean gradient fields $\mathrm{grad}(f_1), \dots, \mathrm{grad}(f_{n-k})$ *are perpendicular to the tangent space of the manifold at each point of M^k.*

Proof. Taking a curve $\gamma : (-\varepsilon, \varepsilon) \to M^k$ in M^k and differentiating the equation

$$f_1(\gamma(t)) = \ldots = f_{n-k}(\gamma(t)) = 0$$

with respect to the curve parameter t yields

$$df_1(\dot\gamma(t)) = \ldots = df_{n-k}(\dot\gamma(t)) = 0 .$$

The tangent space $T_x M^k$ is thus the subspace of all those vectors $\nu \in T_x\mathbb{R}^n$ on which all the differentials $df_1, \cdots, df_{n-k}$ vanish. Comparing the dimensions of these two vector spaces shows that they have to coincide. $\square$

The set of all tangent spaces to the manifold is called the *tangent bundle* of M^k and denoted by TM^k. It is a manifold of dimension $2k$. In fact, at least in the case that M^k is determined by equations, $f_1 = \cdots = f_{n-k} = 0$, the set TM^k is in turn defined as a subspace of $\mathbb{R}^n \times \mathbb{R}^n$ by the equations

$$f_1(x) = \ldots = f_{n-k}(x) = 0 \quad \text{and} \quad df_1(x,v) = \cdots = df_{n-k}(x,v) = 0 .$$

These are $2(n-k)$ conditions in $\mathbb{R}^{2n}$. The corresponding functional determinant (Jacobian) does not vanish, since the differentials df_i are linearly independent. The formula $\pi(x,v) := x$ defines a projection $\pi : TM^k \longrightarrow M^k$ on the tangent bundle of the manifold assigning to every vector its base point.

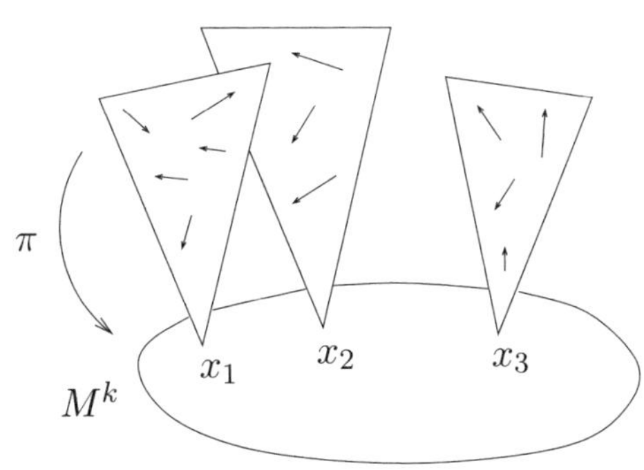

Example 9. Consider the sphere $S^n = \{x \in \mathbb{R}^{n+1} : ||x|| = 1\}$. The differential of the function $||x||^2 - 1$ is $2(x^1, \cdots, x^{n+1})$, and hence the tangent space to the sphere at any point consists of all vectors perpendicular to this point:

$$TS^n = \{(x, v) \in \mathbb{R}^{n+1} \times \mathbb{R}^{n+1} : \ ||x|| = 1, \ \langle x, v \rangle = 0\}.$$

Definition 5. Let $M^k \subset \mathbb{R}^n$ and $N^l \subset \mathbb{R}^m$ be two manifolds, and let $f : M^k \to N^l$ be a continuous map. We call f a *differentiable map* if for each chart $h^{-1} : V \to M^k$ of the manifold M^k the resulting map $f \circ h^{-1} : V \to N^l \subset \mathbb{R}^m$ defined on the open subset $V \subset \mathbb{R}^k$ is differentiable.

As in euclidean space, the differential of a smooth map can be introduced as a linear map between tangent spaces. For a tangent vector $(x, v) \in T_x M^k$ we choose a curve $\gamma : [0, \varepsilon] \to M^k$ with $\gamma(0) = x$ and $\dot{\gamma}(0) = v$. The composition $f \circ \gamma(t)$ is a curve in N^l passing through $f(x) \in N^l$, and its tangent vector describes the result of applying the differential of f to the tangent vector (x, v),

$$f_{*,x}(x, v) := \left(f(x), \ \frac{d}{dt} f \circ \gamma(0) \right).$$

The differential of a smooth map between two manifolds has all the properties which are familiar from euclidean space.

Theorem 6. *The differential $f_{*,x} : T_x M^k \to T_{f(x)} N^l$ of a smooth map is a linear map between the tangent spaces, and the differential of the superposition of two smooth maps f and g is equal to the superposition of their differentials,*

$$(g \circ f)_{*,x} = g_{*,f(x)} \circ f_{*,x}.$$

The last formula is the generalized *chain rule.*

Definition 6. A *vector field* $\mathcal{V}$ on a manifold M^k assigns to every point $x \in M^k$ a vector $\mathcal{V}(x) \in T_x M^k$ in the corresponding tangent space.

If the map $\mathcal{V} : M^k \to T\mathbb{R}^n = \mathbb{R}^n \times \mathbb{R}^n$ is smooth, then we will speak of a smooth vector field on the manifold. Vector fields can again be added and multiplied by functions, so that the vector space of smooth vector fields is a module over the ring $C^\infty(M^k)$ of all C^∞-functions on M^k.

Example 10. The formula $\mathcal{V}(x) = (x, \ (x^2, -x^1, 0))$ defines a vector field on the 2-dimensional sphere (see the figure on the next page).

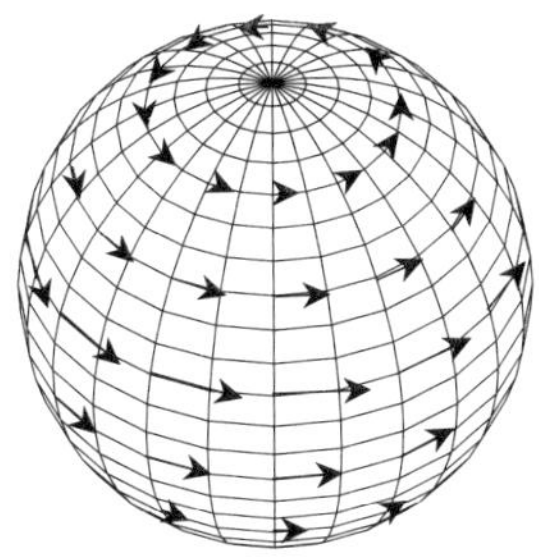

For a chart map $h : V \to M^k \subset \mathbb{R}^n$, which this time and and sometimes also later will be denoted by h instead of h^{-1}, $h_*(\partial/\partial y^i)$ are vector fields tangent to M^k defined on the subset $h(V) \subset M^k$, and they provide a basis in each tangent space. For simplicity and as long as it is clear to which chart we refer, these vector fields on the manifold will also be denoted by $\partial/\partial y^i$. On the subset $h(V) \subset M^k$, every other vector field $\mathcal{V}$ can be represented as their linear combination

$$\mathcal{V}(y) = \sum_{i=1}^{k} V^i(y) \frac{\partial}{\partial y^i} .$$

Here $V^i(y)$ are functions defined on the set $h(V)$; using the chart map, now and then they will also be considered as functions on the parameter set V. These functions are called the *components of the vector field* $\mathcal{V}$ with respect to the fixed chart.

Example 11. In euclidean coordinates on $\mathbb{R}^2$, consider the vector field

$$\mathcal{V} = x^1 \frac{\partial}{\partial x^2} - x^2 \frac{\partial}{\partial x^1} ,$$

depicted on the next page. Introducing in $\mathbb{R}^2 - \{0\}$ *polar coordinates* by the formula

$$h(r, \varphi) = (r \cos\varphi, r \sin\varphi), \quad 0 < r < \infty,\ 0 \leq \varphi \leq 2\pi ,$$

we see that $\mathcal{V}$ corresponds to the vector field $\partial/\partial\varphi$. In fact, differentiating the map h, we obtain the formulas

$$(1)\quad \frac{\partial}{\partial r} = h_* \left(\frac{\partial}{\partial r} \right) = \cos\varphi \frac{\partial}{\partial x^1} + \sin\varphi \frac{\partial}{\partial x^2} = \frac{1}{r} \left(x^1 \frac{\partial}{\partial x^1} + x^2 \frac{\partial}{\partial x^2} \right) ,$$

$$(2)\quad \frac{\partial}{\partial \varphi} = h_* \left(\frac{\partial}{\partial \varphi} \right) = -r \sin\varphi \frac{\partial}{\partial x^1} + r \cos\varphi \frac{\partial}{\partial x^2} = -x^2 \frac{\partial}{\partial x^1} + x^1 \frac{\partial}{\partial x^2} .$$

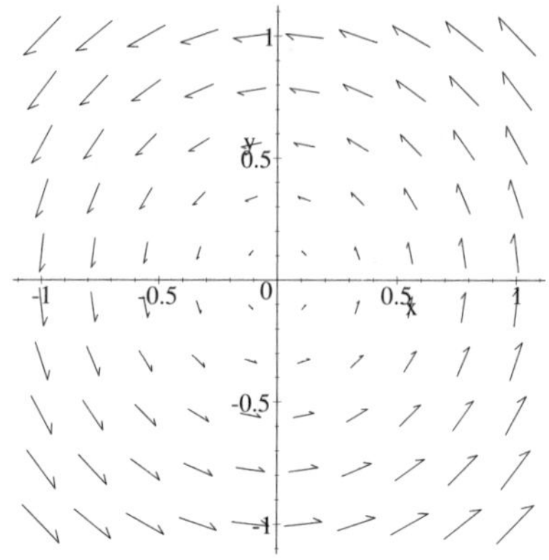

Differentiable functions $f : M^k \to \mathbb{R}$ can be differentiated with respect to a vector field $\mathcal{V}$. At a fixed point $x \in M^k$ we choose a curve $\gamma : [0, \varepsilon] \to M^k$ satisfying the initial conditions $\gamma(0) = x$ and $\dot{\gamma}(0) = \mathcal{V}(x)$. The *derivative of f at the point x in the direction of $\mathcal{V}(x)$* is now defined by the formula

$$\mathcal{V}(f)(x) \;:=\; \frac{d}{dt} f \circ \gamma(t)_{|t=0} \,.$$

The result is a C^∞-function $\mathcal{V}(f)$ defined on the manifold M^k. In the next theorem we summarize the properties of this differentiation:

Theorem 7.

(1) *$(\mathcal{V} + \mathcal{W})(f) = \mathcal{V}(f) + \mathcal{W}(f)$;*

(2) *$\mathcal{V}(f_1 + f_2) = \mathcal{V}(f_1) + \mathcal{V}(f_2)$;*

(3) *$\mathcal{V}(f_1 \cdot f_2) = \mathcal{V}(f_1) \cdot f_2 + f_1 \cdot \mathcal{V}(f_2)$;*

(4) *If the vector field $\mathcal{V}$ is represented as $\mathcal{V} = \sum V^i(y)\partial/\partial y^i$ in some local chart, then*

$$\mathcal{V}(f) \;=\; \sum_{i=1}^{k} V^i(y) \frac{\partial (f \circ h)}{\partial y^i} \,.$$

Proof. We will prove (4); all the other claims follow immediately. If the point $x \in M^k$ corresponds to the point $y \in V$ under the chain map $h : V \to M^k$, then $\gamma(t) = h(y + t(V^1(y), \ldots, V^k(y)))$ is a curve in M^k satisfying the initial conditions $\gamma(0) = x$ and $\dot{\gamma}(0) = \mathcal{V}(x)$. The formula for $\mathcal{V}(f)(x)$ to be proved then follows from the chain rule:

$$\mathcal{V}(f)(x) = \frac{d}{dt} f \circ g \left(y + t(V^1(y), \ldots, V^k(y)) \right) = \sum_{i=1}^{k} V^i(y) \frac{\partial (f \circ h)}{\partial y^i} . \qquad \square$$

Next we will discuss the notion of Riemannian metric on a submanifold M^k of euclidean space. The scalar product in $\mathbb{R}^n$ is denoted by $\langle v, w\rangle$. We restrict it to the tangent spaces of the submanifold.

Definition 7. Let $M^k \subset \mathbb{R}^n$ be a submanifold. In each tangent space $T_x M^k$ the formula

$$g_x\left((x,v),\,(x,w)\right) \;:=\; \langle v, w\rangle$$

defines a scalar product. The family $\{g_x\}$ of all these scalar products is called the *Riemannian metric* of M^k.

In a chart $h : V \to M^k$, the Riemannian metric is locally described by the functions g_{ij} defined on the set V,

$$g_{ij}(y) \;=\; g_{h(y)}\left(\frac{\partial}{\partial y^i}, \frac{\partial}{\partial y^j}\right) \;=\; \left\langle \frac{\partial h}{\partial y^i}, \frac{\partial h}{\partial y^j}\right\rangle .$$

Thus, the components $g_{ij}(y)$ of the Riemannian metric are the scalar products of the partial derivatives of the chart map h considered as being vector-valued. When it is clear to which chart we refer, h is often omitted (as in the examples to follow). The $(k \times k)$ matrix (g_{ij}) is symmetric and positive definite for each $y \in V$. By $g(y)$ we will denote its determinant, which always is a positive function. Recall that we already agreed in Chapter 1 for bilinear forms to denote by $(g^{ij}(y))$ the inverse of the matrix $(g_{ij}(y))$.

Example 12. In euclidean coordinates on $\mathbb{R}^n$, the entries $g_{ij}(y) = \delta_{ij}$ are constant and equal to one for coinciding indices $i = j$, and vanish in the other cases.

Example 13. In polar coordinates on $\mathbb{R}^2 - \{0\}$, equations (1) and (2) immediately lead to the following coefficients for the usual euclidean metric:

$$g_{rr} = \left\langle \frac{\partial}{\partial r}, \frac{\partial}{\partial r}\right\rangle = 1, \quad g_{r\varphi} = \left\langle \frac{\partial}{\partial r}, \frac{\partial}{\partial \varphi}\right\rangle = 0, \quad g_{\varphi\varphi} = \left\langle \frac{\partial}{\partial \varphi}, \frac{\partial}{\partial \varphi}\right\rangle = r^2 .$$

In particular, in polar coordinates on the plane we have $g(r, \varphi) = r^2$.

Example 14. On the sphere $S^2 - \{N, S\} \subset \mathbb{R}^3$ without north and south pole, we consider *spherical coordinates* defined by

$$h(\varphi, \psi) = (\cos\varphi\cos\psi,\ \sin\varphi\cos\psi,\ \sin\psi)\,, \quad 0 \le \varphi \le 2\pi,\ -\pi/2 < \psi < \pi/2.$$

Computing the tangent vector fields yields

$$\text{(3)} \qquad \frac{\partial}{\partial\varphi} \;=\; -\sin\varphi\cos\psi\frac{\partial}{\partial x^1} + \cos\varphi\cos\psi\frac{\partial}{\partial x^2}\,,$$

$$\text{(4)} \qquad \frac{\partial}{\partial\psi} \;=\; -\cos\varphi\sin\psi\frac{\partial}{\partial x^1} - \sin\varphi\sin\psi\frac{\partial}{\partial x^2} + \cos\psi\frac{\partial}{\partial x^3}\,.$$

For the coefficients of the Riemannian metric on the sphere we thus obtain

$$g_{\varphi\varphi} = \left\langle \frac{\partial}{\partial\varphi}, \frac{\partial}{\partial\varphi} \right\rangle = \cos^2\psi, \ g_{\varphi\psi} = \left\langle \frac{\partial}{\partial\varphi}, \frac{\partial}{\partial\psi} \right\rangle = 0, \ g_{\psi\psi} = \left\langle \frac{\partial}{\partial\psi}, \frac{\partial}{\partial\psi} \right\rangle = 1.$$

The determinant of the metric on the sphere in these coordinates is equal to $g(\varphi, \psi) = \cos^2\psi$; therefore, the spherical coordinates degenerate at the poles, which is why we deleted them.

We will use the Riemannian metric g of a manifold in order to associate with every smooth function a vector field, the so-called gradient. For a fixed point $x \in M^k$ and a tangent vector $\nu \in T_xM^k$ we first choose a vector field $\mathcal{V}$ such that $\mathcal{V}(x) = \nu$. The assignment $\nu \mapsto \mathcal{V}(f)(x)$ determines a linear functional $T_xM^k \to \mathbb{R}$ on the tangent space, and hence there exists a vector $\mathrm{grad}(f)(x) \in T_xM^k$ such that the equality

$$\mathcal{V}(f)(x) \;=\; g_x\left(\mathrm{grad}(f)(x), \mathcal{V}(x)\right)$$

holds for all vector fields. The vector field $\mathrm{grad}(f)$ is called the *gradient* of the smooth function $f : M^k \to \mathbb{R}$.

Theorem 8. *Let $f : M^k \to \mathbb{R}$ be a smooth function, and let $h : V \to M^k$ be a chart. Denote the coefficients of the Riemannian metric in this chart by g_{ij}. Then*

$$\mathrm{grad}(f) \;=\; \sum_{i,j=1}^{k} \frac{\partial(f \circ h)}{\partial y^i} g^{ij} \frac{\partial}{\partial y^j}\,.$$

Proof. Inserting the right-hand side into the definition, we compute

$$g\left(\sum_{i,j=1}^{k} \frac{\partial(f \circ h)}{\partial y^i} g^{ij} \frac{\partial}{\partial y^j}, \frac{\partial}{\partial y^m}\right)$$
$$= \sum_{i,j=1}^{k} \frac{\partial(f \circ h)}{\partial y^i} g^{ij} g_{jm} = \sum_{i=1}^{k} \frac{\partial(f \circ h)}{\partial y^i} \delta^i_m = \frac{\partial(f \circ h)}{\partial y^m}\,.$$

According to Theorem 7, the last expression is precisely $(\partial/\partial y^m)(f)$, which by the definition of the gradient implies the result. □

Example 15. The formula for the gradient of a function defined on an open subset $U \subset \mathbb{R}^n$ of euclidean space expressed in cartesian coordinates is

$$\mathrm{grad}(f) \;=\; \sum_{i=1}^{n} \frac{\partial f}{\partial x^i} \frac{\partial}{\partial x^i}\,. \tag{5}$$

Example 16. In polar coordinates on $\mathbb{R}^2 - \{0\}$ the coefficients of the metric are

$$g^{rr} \;=\; 1, \quad g^{r\varphi} \;=\; 0, \quad g^{\varphi\varphi} \;=\; \frac{1}{r^2}\,.$$

For a function f in the variables r, φ, the formula of the preceding theorem yields

$$\text{(6)} \qquad \operatorname{grad}(f) \;=\; \frac{\partial f}{\partial r} g^{rr} \frac{\partial}{\partial r} + \frac{\partial f}{\partial \varphi} g^{\varphi\varphi} \frac{\partial}{\partial \varphi} \;=\; \frac{\partial f}{\partial r}\frac{\partial}{\partial r} + \frac{1}{r^2}\frac{\partial f}{\partial \varphi}\frac{\partial}{\partial \varphi}\,.$$

Example 17. In spherical coordinates on $S^2 - \{N, S\}$ we have

$$g^{\varphi\varphi} \;=\; \frac{1}{\cos^2\psi}, \quad g^{\varphi\psi} \;=\; 0, \quad g^{\psi\psi} \;=\; 1\,,$$

and for the gradient of a function $f : S^2 - \{N, S\} \to \mathbb{R}$ this leads to

$$\text{(7)} \qquad \operatorname{grad}(f) \;=\; \frac{1}{\cos^2\psi}\frac{\partial f}{\partial \varphi}\frac{\partial}{\partial \varphi} + \frac{\partial f}{\partial \psi}\frac{\partial}{\partial \psi}\,.$$

For example, the height function $f(x^1, x^2, x^3) = x^3$ on the sphere may be written in spherical coordinates as $f(\varphi, \psi) = \sin\psi$, and thus

$$\operatorname{grad}(f) \;=\; \cos\psi\,\frac{\partial}{\partial \psi}\,.$$

Hence the vector field $\operatorname{grad}(f)$ on the sphere S^2 vanishes at precisely two points, at the north and the south pole.

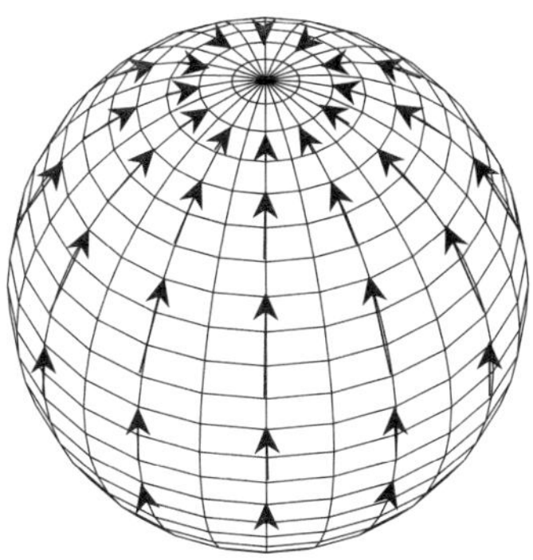

Vector fields and the Riemannian metric can be represented in various coordinates. Now we derive the transformation formulas for its components with respect to different charts. To this end, fix two charts $h : V \to M^k$ and $\tilde{h} : \tilde{V} \to M^k$ and denote the points from V by $y = (y^1, \ldots, y^k)$ and those from $\tilde{V}$ by $z = (z^1, \ldots, z^k)$. The chart transition maps will be denoted by $\phi = (\phi^1, \ldots, \phi^k) := \tilde{h}^{-1} \circ h$ and $\psi = (\psi^1, \ldots, \psi^k) := h^{-1} \circ \tilde{h}$, respectively.

Theorem 9.

(1) *Let $\mathcal{V}$ be a vector field and denote by $V^1, \dots, V^k$ its components in the chart h and by $\tilde{V}^1, \dots, \tilde{V}^k$ the components in the chart $\tilde{h}$. Then, for each index l between 1 and k,*

$$V^l(y) = \sum_{i=1}^{k} \tilde{V}^i(\phi(y)) \frac{\partial \psi^l}{\partial z^i}(\phi(y)) .$$

(2) *Denote by g_{ij} the coefficients of the Riemannian metric with respect to the chart h and by $\tilde{g}_{ij}$ those referring to the chart $\tilde{h}$. Then they are related by the formula*

$$g_{ij}(y) = \sum_{l,m=1}^{k} \frac{\partial \phi^l}{\partial y^i} \frac{\partial \phi^m}{\partial y^j} \tilde{g}_{lm}(\phi(y)) .$$

In particular, for the determinant of the Riemannian metric the following holds:

$$g(y) = \det\left[\frac{\partial \phi^m}{\partial y^r}\right]^2 \cdot \tilde{g}(\phi(y)) .$$

Proof. The first formula is a consequence of the chain rule. In fact,

$$\frac{\partial}{\partial z^i} = \tilde{h}_*\left(\frac{\partial}{\partial z^i}\right) = h_* \circ \psi_*\left(\frac{\partial}{\partial z^i}\right) = h_*\left(\sum_{l=1}^{k} \frac{\partial \psi^l}{\partial z^i} \frac{\partial}{\partial y^l}\right) = \sum_{l=1}^{k} \frac{\partial \psi^l}{\partial z^i} \frac{\partial}{\partial y^l} .$$

This leads to the following formula, describing the vector field in the chart $\tilde{h}$:

$$\mathcal{V} = \sum_{i=1}^{k} \tilde{V}^i \frac{\partial}{\partial z^i} = \sum_{l=1}^{k} \left(\sum_{i=1}^{k} \tilde{V}^i \frac{\partial \psi^l}{\partial z^i}\right) \frac{\partial}{\partial y^l} .$$

The second formula follows directly from the first:

$$\tilde{g}_{ij} = \left\langle \frac{\partial}{\partial z^i}, \frac{\partial}{\partial z^j} \right\rangle = \sum_{l,m=1}^{k} \frac{\partial \psi^l}{\partial z^i} \frac{\partial \psi^m}{\partial z^j} \left\langle \frac{\partial}{\partial y^l}, \frac{\partial}{\partial y^m} \right\rangle = \sum_{l,m=1}^{k} \frac{\partial \psi^l}{\partial z^i} \frac{\partial \psi^m}{\partial z^j} g_{ij} ,$$

if, in addition, the (formally completely equivalent) roles of h and $\tilde{h}$ are exchanged. □

We will use these transformation formulas to define the divergence of a vector field. This will be taken to be a function on the manifold determined by a local formula. Therefore, it has to be proved that the corresponding expression is independent of the choice of the chart.

Definition 8. Let $\mathcal{V} = \sum_i V^i(y)\partial/\partial y^i$ be a vector field on the manifold M^k expressed with respect to the coordinates of the chart $h : V \to M^k$, and let

$g(y) : V \to \mathbb{R}$ be the determinant of the Riemannian metric in this chart. Define a function $\operatorname{div}(\mathcal{V}) : h(V) \to \mathbb{R}$, the *divergence* of $\mathcal{V}$, by the formula

$$\operatorname{div}(\mathcal{V})(h(y)) := \frac{1}{\sqrt{g}} \sum_{i=1}^{k} \frac{\partial \left(\sqrt{g}\, V^i(y)\right)}{\partial y^i} .$$

Theorem 10. *The defining formulas for the divergence of a vector field for two arbitrary charts $h : V \to M^k$ and $\tilde{h} : \tilde{V} \to M^k$ of the manifold M^k coincide at corresponding points:*

$$\frac{1}{\sqrt{g}} \sum_{i=1}^{k} \frac{\partial\left(\sqrt{g}\, V^i(y)\right)}{\partial y^i} = \frac{1}{\sqrt{\tilde{g}}} \sum_{i=1}^{k} \frac{\partial\left(\sqrt{\tilde{g}}\, \tilde{V}^i(z)\right)}{\partial z^i} .$$

The proof uses a formula expressing the derivative of a determinant by its minors. For the sake of completeness we state it here.

Lemma 1. *Let k^2 functions $h_{ij}(x)$ be defined on an open subset V of $\mathbb{R}^k$. Denote by $H(x) := \det[h_{ij}]$ the determinant and by $\Delta_{ij}(x)$ the minors of the matrix $h_{ij}(x)$. Then*

$$\frac{\partial H(x)}{\partial x^r} = \sum_{i,j=1}^{k} \frac{\partial h_{ij}}{\partial x^r} \Delta_{ij}(x) .$$

Proof. Differentiating the determinant and applying the Leibnitz rule yields

$$\frac{\partial H(x)}{\partial x^r} = \det \begin{bmatrix} \frac{\partial h_{11}}{\partial x^r} & h_{12} & \cdots & h_{1k} \\ \vdots & \vdots & & \vdots \\ \frac{\partial h_{k1}}{\partial x^r} & h_{k2} & \cdots & h_{kk} \end{bmatrix} + \ldots + \det \begin{bmatrix} h_{11} & \cdots & h_{1,k-1} & \frac{\partial h_{1k}}{\partial x^r} \\ \vdots & & \vdots & \vdots \\ h_{k1} & \cdots & h_{k,k-1} & \frac{\partial h_{kk}}{\partial x^r} \end{bmatrix} .$$

Expanding each of the k resulting determinants with respect to the column containing the derivatives leads to the stated formula. □

Proof of Theorem 10. Applying the preceding formula, we obtain

$$\begin{aligned} \frac{1}{\sqrt{g(y)}} \sum_{i=1}^{k} \frac{\partial\left(\sqrt{g(y)}\, V^i(y)\right)}{\partial y^i} &= \sum_{i=1}^{k} \frac{\partial V^i(y)}{\partial y^i} + \sum_{i=1}^{k} V^i(y) \frac{\partial}{\partial y^i} \left(\ln \sqrt{g(y)}\right) \\ &= \sum_{i=1}^{k} \frac{\partial}{\partial y^i} \left[\sum_{j=1}^{k} \tilde{V}^j(\phi(y)) \cdot \frac{\partial \psi^i}{\partial z^j}(\phi(y)) \right] \\ &+ \sum_{i=1}^{k} V^i(y) \frac{\partial}{\partial y^i} \left[\ln \sqrt{\tilde{g}(\phi(y))} + \ln \left| \det \frac{\partial \phi^m}{\partial y^r} \right| \right] . \end{aligned}$$

Now we compute the partial derivative of the first sum and simplify the resulting expression, making use of the following formulas:

$$\sum_{i=1}^{k} \frac{\partial \phi^l}{\partial y^i} \frac{\partial \psi^i}{\partial z^j} = \delta^l_j .$$

This immediately leads to the expression

$$\begin{aligned} \frac{1}{\sqrt{g(y)}} \sum_{i=1}^{k} \frac{\partial\big(\sqrt{g(y)}\, V^i(y)\big)}{\partial y^i} &= \frac{1}{\sqrt{\tilde{g}(z)}} \sum_{i=1}^{k} \frac{\partial\big(\sqrt{\tilde{g}(z)}\, \tilde{V}^i(z)\big)}{\partial z^i} \\ &+ \sum_{i,j,l=1}^{k} \tilde{V}^j(\phi(y)) \left[\frac{\partial^2 \psi^i}{\partial z^j \partial z^l} \frac{\partial \phi^l}{\partial y^i} + \frac{\partial}{\partial z^j} \ln \left| \det \frac{\partial \phi^m}{\partial y^r} \right| \circ \psi \right] . \end{aligned}$$

Last, we have to show the following equation for each index $1 \leq j \leq k$:

$$\sum_{i,l=1}^{k} \frac{\partial^2 \psi^i}{\partial z^j \partial z^l} \frac{\partial \phi^l}{\partial y^i} + \frac{\partial}{\partial z^j} \ln \left| \det \frac{\partial \phi^m}{\partial y^r} \right| \circ \psi = 0 .$$

Using the lemma, the second summand in this equation can be written as

$$\begin{aligned} \frac{\partial}{\partial z^j} \ln \left| \det \frac{\partial \phi^m}{\partial y^r} \right| \circ \psi &= \left(\det \left[\frac{\partial \phi^m}{\partial y^r} \right] \circ \psi \right)^{-1} \sum_{i=1}^{k} \frac{\partial}{\partial y^i} \left(\det \frac{\partial \phi^m}{\partial y^r} \right) \frac{\partial \psi^i}{\partial z^j} \\ &= \left(\det \left[\frac{\partial \phi^m}{\partial y^r} \right] \circ \psi \right)^{-1} \sum_{i,\alpha,\beta=1}^{k} \frac{\partial^2 \phi^\alpha}{\partial y^i \partial y^\beta} \Delta_{\alpha\beta} \frac{\partial \psi^i}{\partial z^j} . \end{aligned}$$

The matrix $[\partial \psi^i / \partial z^j]$ is the inverse of $[\partial \phi^m / \partial y^r]$, and, by Cramer's rule,

$$\frac{\partial \psi^\beta}{\partial z^\alpha} = \Delta_{\alpha\beta} \cdot \left(\det \left[\frac{\partial \phi^m}{\partial y^r} \right] \circ \psi \right)^{-1} .$$

This yields

$$\frac{\partial}{\partial z^j} \ln \left| \det \frac{\partial \phi^m}{\partial y^r} \right| \circ \psi = \sum_{i,\alpha,\beta=1}^{k} \frac{\partial^2 \phi^\alpha}{\partial y^i \partial y^\beta} \frac{\partial \psi^i}{\partial z^j} \frac{\partial \psi^\beta}{\partial z^\alpha} ,$$

and the equation to be proved becomes

$$\sum_{i,\alpha=1}^{k} \frac{\partial^2 \psi^i}{\partial z^j \partial z^\alpha} \frac{\partial \phi^\alpha}{\partial y^i} + \sum_{i,\alpha,\beta=1}^{k} \frac{\partial^2 \phi^\alpha}{\partial y^i \partial y^\beta} \frac{\partial \psi^i}{\partial z^j} \frac{\partial \psi^\beta}{\partial z^\alpha} = 0 .$$

But this identity immediately follows from the fact that ϕ and ψ are mutually inverse maps: Differentiate the equation

$$\sum_{i=1}^{k} \frac{\partial \phi^\alpha}{\partial y^i} (\psi(z)) \cdot \frac{\partial \psi^i}{\partial z^j} = \delta^\alpha_j$$

with respect to the variables z^l, set $\alpha = l$, and finally take the sum. □

Example 18. In cartesian coordinates on $\mathbb{R}^n$, the formula for the divergence becomes

$$\text{div}(\mathcal{V}) \;=\; \sum_{i=1}^{n} \frac{\partial V^i}{\partial x^i}. \tag{8}$$

Example 19. In polar coordinates on $\mathbb{R}^2 - \{0\}$, since $g = r$, for each vector field $\mathcal{V} = V^1\partial/\partial r + V^2\partial/\partial\varphi$ the following formula holds:

$$\text{div}(\mathcal{V}) \;=\; \frac{1}{r}\left(\frac{\partial(rV^1)}{\partial r} + \frac{\partial(rV^2)}{\partial\varphi}\right) \;=\; \frac{\partial V^1}{\partial r} + \frac{1}{r}V^1 + \frac{\partial V^2}{\partial\varphi}. \tag{9}$$

Example 20. In spherical coordinates on S^2, the determinant fo the metric is $g = \cos^2\psi$ and thus for every vector field $\mathcal{V} = V^1\partial/\partial\varphi + V^2\partial/\partial\psi$ the corresponding formula becomes

(10)

$$\text{div}(\mathcal{V}) = \frac{1}{\cos\psi}\left(\frac{\partial(\cos\psi\cdot V^1)}{\partial\varphi} + \frac{\partial(\cos\psi\cdot V^2)}{\partial\psi}\right) = \frac{\partial V^1}{\partial\varphi} + \frac{\partial V^2}{\partial\psi} - \tan\psi\cdot V^2.$$

Theorem 11. *Let $\mathcal{V}$ be a vector field and f a smooth function on M^k. Then*

$$\text{div}(f\cdot\mathcal{V}) \;=\; f\cdot\text{div}(\mathcal{V}) + \mathcal{V}(f).$$

Proof. It suffices to prove the formula in an arbitrary chart. This is done by a straightforward computation:

$$\begin{aligned}\text{div}(f\cdot\mathcal{V}) \;&=\; \frac{1}{\sqrt{g}}\sum_{i=1}^{n}\frac{\partial(\sqrt{g}\cdot f\cdot V^i)}{\partial y^i} \;=\; \frac{1}{\sqrt{g}}\sum_{i=1}^{n}\left(\frac{\partial f}{\partial y^i}\sqrt{g}\,V^i + f\frac{\partial(\sqrt{g}\,V^i)}{\partial y^i}\right)\\ &=\; f\cdot\text{div}(\mathcal{V}) + \sum_{i=1}^{n} V^i\frac{\partial f}{\partial y^i} \;=\; f\cdot\text{div}(\mathcal{V}) + \mathcal{V}(f).\end{aligned}$$ □

The last operator to be introduced in this chapter is the Laplacian Δ; it acts on functions and is a second order differential operator.

Definition 9. Let $f : M^k \to \mathbb{R}$ be a smooth function. Then the *Laplacian* of f is defined as a function on M^k by the formula

$$\Delta(f) \;:=\; \text{div}(\text{grad}(f)).$$

Remark. The operator $\Delta : C^\infty(M^k) \to C^\infty(M^k)$ is a linear operator acting in the vector space of all C^∞-functions on M^k. In mathematics, there are—concerning the sign of Δ—two differing conventions. Classically and mainly in analysis, Δ is defined as above. On the other hand, in various branches of geometry and global analysis it is common to introduce the Laplacian via the formula $\Delta(f) = -\text{div}(\text{grad}(f))$ and to call this the *Laplace-Beltrami operator*. There are good reasons to do so, as we will see in connection with the classical integral formulas. Thus, studying a text, the reader has to be careful to detect which sign convention was chosen by the author. In this book we decided to use the particular choice of sign for the Laplacian stated in the definition.

Theorem 12. *The Laplacian* $\Delta : C^\infty(M^k) \to C^\infty(M^k)$ *is a linear operator. Moreover, for two functions* f_1 *and* f_2 *the following formula holds:*

$$\Delta(f_1 \cdot f_2) = f_1 \cdot \Delta(f_2) + f_2 \cdot \Delta(f_1) + 2\,\langle \mathrm{grad}(f_1), \mathrm{grad}(f_2)\rangle .$$

Proof. The linearity of Δ is obvious, and the second formula follows by combining the corresponding rules for the operators grad and div:

$$\begin{aligned} \Delta(f_1 f_2) &= \mathrm{div}(\mathrm{grad}(f_1 f_2)) \\ &= \mathrm{div}(f_1\,\mathrm{grad}(f_2) + f_2\,\mathrm{grad}(f_1)) \\ &= f_1\,\mathrm{div}(\mathrm{grad}(f_2)) + f_2\,\mathrm{div}(\mathrm{grad}(f_1)) + 2\,\langle \mathrm{grad}(f_1), \mathrm{grad}(f_2)\rangle . \end{aligned}$$

□

In local coordinates we immediately obtain from

$$\mathrm{grad}(f) = \sum_{i,j=1}^{k} \frac{\partial(f \circ h)}{\partial y^i}\, g^{ij}\, \frac{\partial}{\partial y^j}$$

the formula

$$\Delta(f) = \frac{1}{\sqrt{g}} \sum_{i,j=1}^{k} \frac{\partial}{\partial y^j}\left(\sqrt{g}\,\frac{\partial(f \circ h)}{\partial y^i}\, g^{ij}\right).$$

Again, we write out the explicit form of the Laplacian in the examples we have discussed so far.

Example 21. In cartesian coordinates on $\mathbb{R}^n$,

$$\Delta(f) = \sum_{i=1}^{n} \frac{\partial^2 f}{(\partial x^i)^2} . \tag{11}$$

Example 22. In polar coordinates on $\mathbb{R}^2$,

$$\Delta(f) = \frac{1}{r}\left(\frac{\partial}{\partial r}\left(r\frac{\partial f}{\partial r}\right) + \frac{\partial}{\partial \varphi}\left(r \cdot \frac{1}{r^2}\frac{\partial f}{\partial \varphi}\right)\right) = \frac{\partial^2 f}{\partial r^2} + \frac{1}{r}\frac{\partial f}{\partial r} + \frac{1}{r^2}\frac{\partial^2 f}{\partial \varphi^2} . \tag{12}$$

Example 23. In spherical coordinates on S^2,

$$\Delta f = \frac{1}{\cos^2\psi}\frac{\partial^2 f}{\partial \varphi^2} + \frac{\partial^2 f}{\partial \psi^2} - \tan\psi\frac{\partial f}{\partial \psi} . \tag{13}$$

In these coordinates the height function $f(x^1, x^2, x^3) = x^3$ on S^2 can be represented as $f = \sin\psi$. This leads to

$$\Delta(f) = -2\sin\psi = -2f ,$$

i. e., f is an eigenfunction of the Laplace-Beltrami operator corresponding to the eigenvalue -2. Other eigenfunctions of Δ on S^2 are constructed starting from a harmonic and homogeneous polynomial $P(x^1, x^2, x^3)$ in $\mathbb{R}^3$, $\Delta^{\mathbb{R}^3} P = 0$, and restricting this to the sphere S^2. The resulting function is an eigenfunction of the Laplacian (see Exercise 25).

3.3. Differential Forms on Manifolds

Up to now, for every manifold M^m we considered its tangent space $T_x M^m$ and vector fields. Now we go one step further and form the exterior power

$$\bigwedge\nolimits^k_x(M^m) \;:=\; \bigwedge\nolimits^k(T^*_x M^m)$$

of all k-forms $\omega^k_x : T_x M^m \times \ldots \times T_x M^m \to \mathbb{R}$ at the point $x \in M^m$.

Definition 10. A *k-form ω^k on a manifold M^m* is a family $\{\omega^k_x\}$ distinguishing a k-form $\omega^k_x \in \bigwedge^k_x(M^m)$ at each point $x \in M^m$.

The differential of a smooth map $f : N^n \to M^m$ between two manifolds allows to pull back k-forms from M^m to yield k-forms on N^n. This is accomplished via the formula

$$(f^*\omega^k)(v_1, \ldots, v_k) \;=\; \omega^k(f_*(v_1), \ldots, f_*(v_k)),$$

where $v_1, \ldots, v_k \in T_y N^n$ are tangent vectors to N^n at the point $y \in N^n$, and $f_* : T_y N^n \to T_{f(y)} M^m$ is the differential of f at this point. The k-form $f^*\omega^k$ is called the *induced form* or *pullback* of ω^k by the map f on N^n. This construction can, in particular, be applied to a chart $h : V \to M^m$ of the manifold M^m. Hence, for a fixed chart, to every k-form ω^k on M^m there corresponds a k-form $h^*(\omega^k)$ on the open set V in the space $\mathbb{R}^m$ or $\mathbb{H}^m$, respectively. If $y = (y^1, \ldots, y^m)$ are the associated coordinates, then $h^*(\omega^k)$ can be represented by means of the component functions ω_I as

$$h^*(\omega^k) \;=\; \sum_I \omega_I dy^I .$$

The summation extends over all ordered multi-indices $I = (i_1 < \ldots < i_k)$, where dy^I is a shorthand notation for the k-form $dy^I := dy^{i_1} \wedge \ldots \wedge dy^{i_k}$.

Definition 11. A k-form ω^k on the manifold M^m is called a *differential k-form* or *smooth* k-form if for each chart $h : V \to M^m$ the coefficients ω_I of the k-form $h^*(\omega^k)$ are smooth functions on the subset $V \subset \mathbb{R}^m$.

Differential forms can be added and multiplied by functions defined on M^m in an obvious way. Hence the set $\Omega^k(M^m)$ of all C^∞-forms of degree k is a module over the ring $C^\infty(M^m)$ of smooth functions on the manifold.

The exterior derivative of k-forms—familiar from the euclidean space $\mathbb{R}^n$ and discussed in Chapter 2—can now be transferred to the situation of k-forms on manifolds without difficulties, preserving all the known properties. This proceeds as follows: For a k-form ω^k on M^m and a chart $h : V \to M^m$ we first consider the induced form $h^*(\omega^k)$ and its differential $d(h^*(\omega^k))$. The latter is a $(k+1)$-form on the set V, and its pullback under the inverse chart map $h^{-1} : h(V) \to V$ is a $(k+1)$-form. This yields a $(k+1)$-form

$$d\omega^k \ := \ (h^{-1})^*(d(h^*\omega^k)),$$

defined on the open set $h(V) \subset M^m$. The construction just described is independent of the particular choice of the chart, and hence it uniquely defines a global $(k+1)$-form $d\omega^k$ on the manifold M^m. In fact, for another chart $h_1 : V_1 \to M^m$ satisfying $h(V) \cap h_1(V_1) \neq \emptyset$, we obtain on the intersection $h(V) \cap h_1(V_1)$ the equality

$$\begin{aligned} (h_1^{-1})^*(d(h_1^*\omega^k)) &= (h^{-1})^*(h_1^{-1} \circ h)^*(d((h^{-1} \circ h_1)^* h^*(\omega^k))) \\ &= (h^{-1})d((h_1^{-1} \circ h)^*(h^{-1} \circ h_1)^* h^*(\omega^k)) \\ &= (h^{-1})d(h^*\omega^k). \end{aligned}$$

This computation relies on the fact that the exterior derivative commutes with the diffeomorphism $(h_1^{-1} \circ h)$, a property discussed in Chapter 2.

Definition 12. The $(k+1)$-form $d\omega^k$ defined starting from the k-form ω^k on M^m is called the *exterior derivative* of ω^k.

All properties of the exterior derivative known from euclidean space remain valid in the situation of a manifold. The next theorem summarizes them.

Theorem 13. *For arbitrary forms $\omega^k, \omega_1^k, \eta^l$ on a manifold M^m the following properties hold:*

(1) $d(\omega^k + \omega_1^k) = d\omega^k + d\omega_1^k$;

(2) $dd\omega^k = 0$;

(3) $d(\omega^k \wedge \eta^l) = (d\omega^k) \wedge \eta^l + (-1)^k \omega^k \wedge (d\eta^l)$;

(4) *if $f : N^n \to M^m$ is a smooth map, then f^* commutes with the exterior derivative, $d(f^*\omega^k) = f^*(d\omega^k)$.*

The purely algebraic operation of forming the inner product between a vector field $\mathcal{V}$ and a k-form ω^k is also transferred into the situation that both objects are defined over a manifold.

Definition 13. We define the *inner product* of a vector field $\mathcal{V}$ and a k-form ω^k by

$$(\mathcal{V} \lrcorner\, \omega^k)(\mathcal{W}_1, \ldots, \mathcal{W}_{k-1}) \ := \ \omega^k(\mathcal{V}, \mathcal{W}_1, \ldots, \mathcal{W}_{k-1}).$$

3.4. Orientable Manifolds

An orientation of a real vector space is the choice of one of the two equivalence classes in the set of its bases discussed in the first chapter. This can, in particular, be applied to the tangent space T_xM^m of a manifold, and leads to the notion of an orientation $\mathcal{O}_x$ at the point $x \in M^m$. An orientation of a manifold M^m consists in a "continuous" choice of orientations at each of its points.

Definition 14. An *orientation* $\mathcal{O}$ *of a manifold* M^m is a family $\mathcal{O} = \{\mathcal{O}_x\}$ of orientations in all tangent spaces T_xM^m depending continuously on the point x in the following sense: At each point $x \in M^m$ there exists a chart $h : V \to M^m$ containing this point such that the basis $\{h_*(\partial/\partial y^1), \ldots, h_*(\partial/\partial y^m)\}$ is compatible with the orientation $\mathcal{O}_{h(y)}$ for every point $y \in V$.

Definition 15. A manifold M^m is called *orientable* if there exists at least one orientation on it.

First we state a necessary and at the same time also sufficient condition for the orientability of a manifold in terms of the chart transition maps.

Theorem 14. *Let* $\mathcal{O}$ *be an orientation on* M^m. *Then there exists a family* $\{(h_i, V_i)\}_{i\in I}$ *of charts with the following properties:*

(1) *The image sets* $h_i(V_i)$ *cover the manifold,*

$$M^m = \bigcup_{i\in I} h_i(V_i)\,;$$

(2) *If the intersection* $h_i(V_i) \cap h_j(V_j) \neq \emptyset$ *is non-empty, then the determinant of the differential of the chart transition map* $h_j^{-1} \circ h_i$ *is positive,*

$$\det[D(h_j^{-1} \circ h_i)] > 0\,.$$

Conversely, if there exists a family of charts with these properties, then M^m *is orientable.*

Proof. Let $\mathcal{O}$ be an orientation on M^m. Choose those of the charts $h : V \to M^m$ for which the basis $\{h_*(\partial/\partial y^1), \ldots, h_*(\partial/\partial y^m)\}$ is compatible with the orientation of the manifold. By the definition of an orientation these charts cover the manifold. For two of these charts, $h_i : V_i \to M^m$ and $h_j : V_j \to M^m$, with coordinates $y = (y^1, \ldots, y^m)$ and $z = (z^1, \ldots, z^m)$ we have at mutually corresponding points

$$\begin{aligned}\{h_{i*}(\partial/\partial y^1), \ldots, h_{i*}(\partial/\partial y^m)\} &= \mathcal{O}_{h_i(y)} = \mathcal{O}_{h_j(z)} \\ &= \{h_{j*}(\partial/\partial z^1), \ldots, h_{j*}(\partial/\partial z^m)\}.\end{aligned}$$

Therefore, these bases are compatibly oriented, the transition matrix between them is the differential $D(h_j^{-1} \circ h_i)$, and we obtain

$$\det[D(h_j^{-1} \circ h_i)] > 0 \,.$$

The converse is proved analogously. □

If the manifold $M^m \subset \mathbb{R}^n$ is described by $(n - m)$ independent equations, then it is orientable. We will prove this fact now.

Theorem 15. *Let $f_1, \ldots, f_{n-m} : U \to \mathbb{R}$ be smooth functions defined on an open subset $U \subset \mathbb{R}^n$, and assume that*

$$df_1 \wedge \ldots \wedge df_{n-m} \neq 0$$

at each point. Then the manifold

$$M^m \;=\; \{x \in U : \; f_1(x) = \ldots = f_{n-m}(x) = 0\}$$

is orientable.

Proof. Consider the euclidean gradients $\mathrm{grad}(f_1), \ldots, \mathrm{grad}(f_{n-m})$ of the functions. By assumption, these vector fields are linearly independent at each point of the set U and, moreover, perpendicular to the tangent space $T_x M^m$ at the points of the manifold (compare Theorem 5). An orientation $\mathcal{O}_x$ in $T_x M^m$ is distinguished by requiring that a basis $v_1, \ldots, v_m \in T_x M^m$ is positively oriented if and only if

$$d\mathbb{R}^n(\mathrm{grad}(f_1)(x), \ldots, \mathrm{grad}(f_{n-m})(x), v_1, \ldots, v_m) > 0.$$

Here $d\mathbb{R}^n = dx^1 \wedge \ldots \wedge dx^n$ denotes the volume form on $\mathbb{R}^n$. It is not hard to see that this condition determines an orientation on M^m. □

An oriented submanifold $M^m \subset \mathbb{R}^n$ carries a distinguished differential form of highest degree, the so-called volume form. Choose a basis $e_1, \ldots, e_m \in T_x M^m$ consisting of mutually perpendicular vectors of length one in the fixed orientation $\mathcal{O}_x$ at any tangent space $T_x M^m$. For arbitrary vectors $v_1, \ldots, v_m \in T_x M^m$ we define

$$dM^m(v_1, \ldots, v_m) \;=\; \det \begin{bmatrix} \langle v_1, e_1 \rangle, & \ldots, & \langle v_1, e_m \rangle \\ \vdots & \ddots & \vdots \\ \langle v_m, e_1 \rangle, & \ldots, & \langle v_m, e_m \rangle \end{bmatrix} .$$

This definition uniquely determines the form dM^n. Every other basis $e_1^*, \ldots, e_m^*$ in the same orientation can be represented as a linear combination $e_i^* = \sum A_{ij} e_j$ with an orthogonal matrix A of positive determinant. But then $\det(A) = 1$.

Definition 16. The m-form dM^m on the oriented manifold $M^m \subset \mathbb{R}^n$ is called the *volume form*.

Remark. The volume form dM^m is *not* the exterior derivative of an $(m-1)$-form. Nevertheless, this form is traditionally denoted by dM^m. The volume form does not vanish at any point of the manifold. Evaluating dM^m on any orthogonal basis $e_1, \ldots, e_m$ in the orientation yields

$$dM^m(e_1, \ldots, e_m) = 1.$$

On the other hand, the orientation can be reconstructed from the volume form. In fact, a basis $v_1, \ldots, v_m$ is positively oriented if

$$dM^m(v_1, \ldots, v_m) > 0.$$

Changing the orientation of the manifold results in a change of sign for the volume form.

Theorem 16. *An m-dimensional manifold M^m is orientable if it carries a nowhere vanishing differential form of degree m.*

Proof. For an orientable manifold M^m the volume form dM^m has the necessary property. Conversely, suppose that there is an m-form ω^m on M^m not vanishing at any point. Then we call a basis $v_1, \ldots, v_m$ positively oriented if

$$\omega^m(v_1, \ldots, v_m) > 0\,.$$

This determines an orientation on M^m. □

In a chart h the induced volume form $h^*(dM^m)$ is proportional to $dy^1 \wedge \ldots \wedge dy^m$, $h^*(dM^m) = f(y)dy^1 \wedge \ldots \wedge dy^m$. We compute the function $f(y)$ as follows:

$$\begin{aligned} f^2(y) &= (h^*(dM^m)(\partial/\partial y^1, \ldots, \partial/\partial y^m))^2 \;=\; \det{}^2\big[\langle h_*(\partial/\partial y^i), e_j\rangle\big] \\ &= \det\big[\langle h_*(\partial/\partial y^i), e_j\rangle\big] \cdot \det\big[\langle e_j, h_*(\partial/\partial y^l)\rangle\big] \\ &= \det\big[\langle h_*(\partial/\partial y^i), h_*(\partial/\partial y^l)\rangle\big] \;=\; \det[g_{il}]\,. \end{aligned}$$

Hence $f^2(y)$ is equal to the determinant $\det[g_{ij}(y)]$ of the Riemannian metric, and we obtain the formula

$$h^*(dM^m) \;=\; \sqrt{g(y)}\,dy^1 \wedge \ldots \wedge dy^m\,.$$

Example 24. For a surface piece $M^2 \subset \mathbb{R}^n$ with a parametrization $h : \mathbb{R}^2 \to M^2 \subset \mathbb{R}^n$, the following classical notation is frequently used:

$$\begin{aligned} E &= g_{11}(x,y) = \left\langle \frac{\partial h}{\partial x}, \frac{\partial h}{\partial x} \right\rangle, \\ G &= g_{22}(x,y) = \left\langle \frac{\partial h}{\partial y}, \frac{\partial h}{\partial y} \right\rangle, \\ F &= g_{12}(x,y) = \left\langle \frac{\partial h}{\partial x}, \frac{\partial h}{\partial y} \right\rangle. \end{aligned}$$

The formula for the volume form is then

$$dM^2 \;=\; \sqrt{EG-F^2}\,dx\wedge dy\,.$$

Example 25. The volume form of $\mathbb{R}^n$ in cartesian coordinates is

$$d\mathbb{R}^n \;=\; dx^1\wedge\ldots\wedge dx^n\,. \tag{14}$$

Example 26. In polar coordinates on $\mathbb{R}^2-\{0\}$ we have $g(r,\varphi)=r^2$, and hence the volume form is

$$d\mathbb{R}^2 \;=\; r\cdot dr\wedge d\varphi\,. \tag{15}$$

Example 27. In spherical coordinates on the 2-sphere, $g(\varphi,\psi)=\cos^2\psi$, and thus the volume form is

$$dS^2 \;=\; \cos\psi\cdot d\varphi\wedge d\psi\,. \tag{16}$$

In the case of an oriented manifold, the divergence of a vector field $\mathcal{V}$ can be expressed in terms of the exterior derivative and the volume form.

Theorem 17. *Let $M^m\subset\mathbb{R}^n$ be an oriented manifold, let dM^m be its volume form, and let $\mathcal{V}$ be a vector field with divergence* $\mathrm{div}(\mathcal{V})$. *Then*

$$d(\mathcal{V}\lrcorner\, dM^m) \;=\; \mathrm{div}(\mathcal{V})\cdot dM^m\,.$$

Proof. In local coordinates we start from

$$\mathcal{V} \;=\; \sum_{i=1}^m V^i(y)\frac{\partial}{\partial y^i} \quad\text{and}\quad dM^m \;=\; \sqrt{g}\,dy^1\wedge\ldots\wedge dy^m\,.$$

This implies

$$\mathcal{V}\lrcorner\, dM^m \;=\; \sqrt{g}\sum_{i=1}^m(-1)^{i-1}V^i dy^1\wedge\ldots\wedge\widehat{dy^i}\wedge\ldots dy^m,$$

and the formula

$$d(\mathcal{V}\lrcorner\, dM^m) \;=\; \sum_{i=1}^m\frac{\partial(\sqrt{g}\,V^i)}{\partial y^i}\,dy^1\wedge\ldots\wedge dy^m \;=\; \mathrm{div}(\mathcal{V})\cdot dM^m$$

immediately follows from the definition of the divergence of a vector field. □

Remark. As in the euclidean space $\mathbb{R}^n$, for every (oriented) manifold M^m the 1-form $\omega^1_{\mathcal{V}}$ *dual to a vector field* $\mathcal{V}$ can be described using the volume form and the Hodge operator:

$$*\omega^1_{\mathcal{V}} \;:=\; \mathcal{V}\lrcorner\, dM^m\,.$$

The divergence formula can then be written as $d(*\omega^1_{\mathcal{V}})=\mathrm{div}(\mathcal{V})\cdot dM^m$.

Non-orientable manifolds exist; proving non-orientability for a particular manifold, however, is sometimes a little more difficult than to show orientability. We state a simple criterion for the non-orientability of a manifold. This can be used, e.g., to prove that the Möbius strip is non-orientable.

Theorem 18. *Suppose that, for a manifold M^m, there exist two charts $h : V \to M^m$ and $\tilde{h} : \tilde{V} \to M^m$ with the following properties:*

(1) *V and $\tilde{V}$ are connected sets, and the intersection $h(V) \cap \tilde{h}(\tilde{V})$ is not connected;*

(2) *the determinant $\det(D(\tilde{h}^{-1} \circ h))$ has opposite sign at two points in $h^{-1}(h(V) \cap \tilde{h}(\tilde{V}))$.*

Then M^m is a non-orientable manifold.

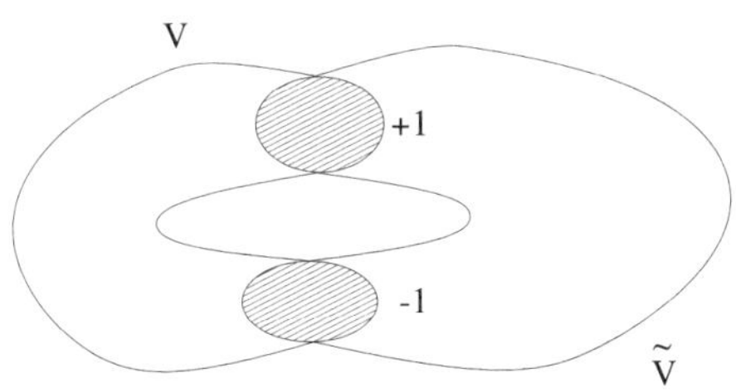

Proof. Assume that M^m is orientable. The sets V and $\tilde{V}$ are open subsets of $\mathbb{R}^m$, and hence orientable. Thus, without loss of generality, we may assume that the chart transition maps h and $\tilde{h}$ preserve the orientations. Consider the volume form dM^m, and represent it in both coordinate systems:

$$h^*(dM^m) \;=\; \sqrt{g}\, dy^1 \wedge \ldots \wedge dy^m, \quad \tilde{h}^*(dM^m) \;=\; \sqrt{\tilde{g}}\, dz^1 \wedge \ldots \wedge dz^m .$$

Note that in both these cases the sign is "+", since the chart transition maps $h, \tilde{h}$ preserve the orientation. Because

$$(\tilde{h}^{-1} \circ h)^*(dz^1 \wedge \ldots \wedge dz^m) \;=\; \det(D(\tilde{h}^{-1} \circ h))\, dz^1 \wedge \ldots \wedge dz^m,$$

we have

$$\sqrt{\tilde{g}} \cdot \det(D(\tilde{h}^{-1} \circ h)) \;=\; \sqrt{g}\,.$$

But this is a contradiction, since $\sqrt{g}$ and $\sqrt{\tilde{g}}$ are always positive. □

The orientation of a manifold induces a unique orientation on its boundary. To define it, we make use of the exterior normal vector field. Its definition relies on the following observation.

Lemma 2. *Let $U \subset \mathbb{H}^m$ be an open subset of the half-space, let $x \in U \cap \partial\mathbb{H}^m$ be a point in the boundary $\partial\mathbb{H}^m$, and let $f : U \to V$ be a diffeomorphism from U to an open subset $V \subset \mathbb{H}^m$. For a vector $\nu = (x, (v^1, \ldots, v^m)) \in T_x\mathbb{H}^m$*

at the point x with non-positive m-th component, $v^m \leq 0$, the image vector $f_*(\nu) = (f(x), (w^1, \ldots, w^m))$ also has non-positive m-th component $w^m \leq 0$.

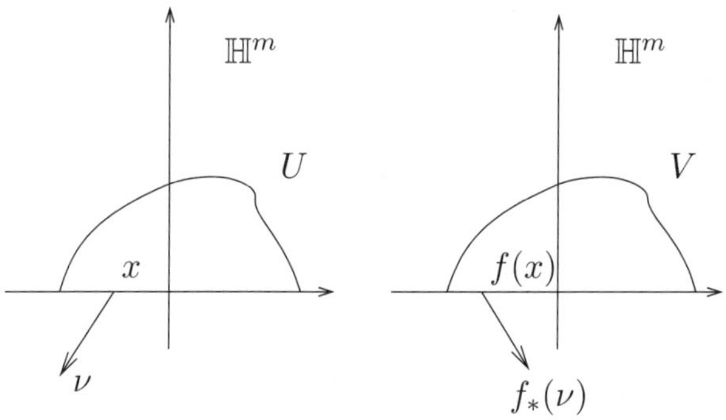

Proof. Choose a straight line $\gamma(t) = x + t \cdot \nu$ and note that, because of $v^m \leq 0$, the point $\gamma(t)$ belongs to $U \cap \mathbb{H}^m$ for sufficiently small negative values of the parameter t. Then we have $\gamma(0) = x, \dot{\gamma}(0) = \nu$, and the m-th component of $f_*(\nu)$ becomes negative:

$$\frac{d}{dt} f^m(x + t\nu) = \lim_{t \to 0^-} \frac{f^m(x + t\nu) - 0}{t} \leq 0 . \qquad \square$$

At a boundary point $x \in \partial M^m \subset M^m$ of a manifold there are two tangent spaces. On the one hand, there is the tangent space $T_x M^m$ of M^m, and, on the other hand, the boundary determines its own $(m-1)$-dimensional tangent space, $T_x(\partial M^m) \subset T_x M^m$. At each boundary point $x \in \partial M^m$, we define a unique tangent vector $\mathcal{N}(x) \in T_x M^m$ by the following conditions:

(1) $\mathcal{N}(x)$ is perpendicular to $T_x(\partial M^m)$;

(2) $\mathcal{N}(x)$ has length 1;

(3) for a chart $h : V \subset \mathbb{H}^m \to M^m$ around the point $x \in M^m$ the m-th component of the vector $(h)_*^{-1}(\mathcal{N}(x))$ is negative.

Definition 17. The vector field $\mathcal{N}$ constructed along the boundary is called the (*exterior*) *unit normal vector field* of the boundary.

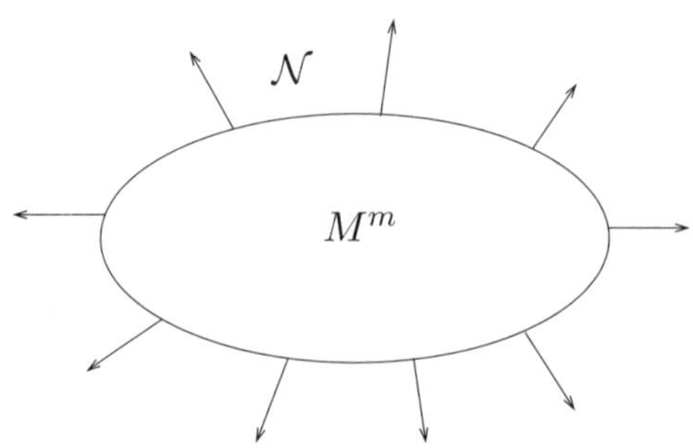

We fix an orientation of the boundary ∂M^m of an oriented manifold M^m by calling a basis $v_1, \ldots, v_{m-1} \in T_x(\partial M^m)$ positively oriented if the extended basis $\mathcal{N}(x), v_1, \ldots, v_{m-1}$ is positively oriented in $T_x M^m$. A simple argument shows that this condition actually determines an orientation of the boundary. For the corresponding volume forms we have the important formula:

Theorem 19. *The volume form of the boundary of a manifold is the inner product of the exterior normal vector field $\mathcal{N}$ with dM^m:*

$$d(\partial M^m) \;=\; \mathcal{N} \lrcorner\, dM^m .$$

Proof. Choose an orthonormal basis in the orientation of the vector space $T_x M^m$ consisting of the normal vector $e_1 := \mathcal{N}(x)$ together with additional vectors $e_2, \ldots, e_m$. For arbitrary tangent vectors $v_1, \ldots, v_{m-1}$ to the boundary we obtain

$$\begin{aligned} (\mathcal{N} \lrcorner\, dM^m)(v_1, \ldots, v_{m-1}) &= dM^m(e_1, v_1, \ldots, v_{m-1}) \\ &= \det \begin{bmatrix} \langle v_1, e_2\rangle & \ldots & \langle v_1, e_m\rangle \\ \vdots & \ddots & \vdots \\ \langle v_{m-1}, e_2\rangle & \ldots & \langle v_{m-1}, e_m\rangle \end{bmatrix} \\ &= d(\partial M^m)(v_1, \ldots, v_{m-1}), \end{aligned}$$

since the vectors $v_1, \ldots, v_{m-1}$ are perpendicular to e_1. □

In the subsequent sections of this book the boundary of a manifold will always be oriented this way.

Example 28. The exterior unit normal vector field of the sphere $S^{n-1}(R)$ of radius R, considered as the boundary of the ball $D^n(R)$, has the form

$$\mathcal{N}(x) \;=\; \frac{1}{R}\Big(x^1 \frac{\partial}{\partial x^1} + \ldots + x^n \frac{\partial}{\partial x^n}\Big).$$

The volume form of $S^{n-1}(R)$ can also be computed by means of $dS^{n-1}(R) = \mathcal{N} \lrcorner\, d\mathbb{R}^n$, yielding the formula

$$dS^{n-1}(R) \;=\; \frac{1}{R}\sum_{i=1}^{n}(-1)^{i-1} x^i dx^1 \wedge \ldots \wedge \widehat{dx^i} \wedge \ldots \wedge dx^n . \tag{17}$$

Hence the volume form of the sphere $S^{n-1}(R)$ can be represented via the embedding $i : S^{n-1}(R) \to \mathbb{R}^n$ as the form induced from the following $(n-1)$-form defined on $\mathbb{R}^n$:

$$\omega^{n-1} \;=\; \frac{1}{R}\sum_{i=1}^{n}(-1)^{i-1} x^i dx^1 \wedge \ldots \wedge \widehat{dx^i} \wedge \ldots \wedge dx^n ,$$

i.e., the following equation holds:

$$dS^{n-1}(R) \;=\; i^*(\omega^{n-1}).$$

Computing the exterior derivative of ω^{n-1} in $\mathbb{R}^n$ leads to the formula

$$d\omega^{n-1} = \frac{n}{R} d\mathbb{R}^n ,$$

and by applying Stokes' theorem from Chapter 2 we obtain the integral of the volume form:

$$\int_{S^{n-1}(R)} dS^{n-1}(R) = \int_{D^n(R)} d\omega^{n-1} = \frac{n}{R}\mathrm{vol}(D^n(R)) .$$

3.5. Integration of Differential Forms over Manifolds

The integral of an m-form over an m-dimensional manifold will be defined by dividing the manifold into small subsets which are diffeomorphic to open subsets of $\mathbb{R}^m$ or $\mathbb{H}^m$, respectively, and integrating the given form using its chart representatives one by one. The sum of the resulting values is then the total integral of the form over the manifold. From the very beginning we will confine ourselves to compact manifolds in order to have to deal with finitely many summands only and thus to avoid convergence questions for series arising otherwise. A detailed exposition of this definition of the integral requires a so-called *partition of unity* in order not to count contributions from overlapping charts twice. This is a special family of smooth functions to be discussed first. Recall that the *support* $\mathrm{supp}(\varphi)$ of a function $\varphi : M^m \to \mathbb{R}$ is the closure of the set $\{x \in M^m : \varphi(x) \neq 0\}$.

Theorem 20. *Let M^m be a compact manifold. Then there exist smooth functions and charts*

$$\varphi_i : M^m \to \mathbb{R} \quad \text{and} \quad h_i : V_i \to M^m \quad (1 \leq i \leq l)$$

with the following properties:

(1) *The support of the function φ_i is contained in $h_i(V_i)$,*

$$\mathrm{supp}(\varphi_i) \subset h_i(V_i) .$$

(2) *The functions φ_i are non-negative, and their sum is equal to one:*

$$\sum_{i=1}^{l} \varphi_i(x) \equiv 1 .$$

Proof. We choose a chart $h_x : V_x \to M^m$ around each point $x \in M^m$ satisfying $h_x(0) = x$. Here, the V_x are open subsets of $\mathbb{R}^m$ or $\mathbb{H}^m$. Choose, moreover, non-negative functions $\psi_x : \mathbb{R}^m, \mathbb{H}^m \to \mathbb{R}$ such that

(1) $\psi_x(0) = 1$;

(2) $\mathrm{supp}(\psi_x) \subset V_x$.

Setting

$$\tilde{\varphi}_x(y) = \begin{cases} 0 & \text{if } y \notin h_x(V_x)\,, \\ \psi_x(h_x^{-1}(y)) & \text{if } y \in h_x(V_x)\,, \end{cases}$$

these functions can be transferred to form a family of non-negative smooth functions $\tilde{\varphi}_x : M^m \to \mathbb{R}$ on M^m. By construction we have $\tilde{\varphi}_x(x) = 1$ and $\text{supp}(\tilde{\varphi}_x) \subset h_x(V_x)$. Taking advantage of the compactness of M^m, we finally obtain finitely many functions $\tilde{\varphi}_1, \ldots, \tilde{\varphi}_l$ and charts $(h_1, V_1), \ldots, (h_1, V_l)$ satisfying

$$M^m = \bigcup_{i=1}^{l} \{x \in M^m : \tilde{\varphi}_i(x) \neq 0\}$$

and $\text{supp}(\tilde{\varphi}_i) \subset h_i(V_i)$. The sum $\psi := \sum_j \tilde{\varphi}_j$ is positive at each point, and the functions φ_i we are looking for result from normalizing these functions $\tilde{\varphi}_i$, $\varphi_i = \tilde{\varphi}_i / \psi$. □

Now we will define the integral of an m-form ω^m over an oriented and compact manifold M^m. To this end, we choose charts $h_1, \ldots, h_l$ and a partition of unity, $\varphi_1, \ldots, \varphi_l$ subordinate to them satisfying the properties formulated in the preceding theorem. Furthermore, we suppose that the chart maps $h_i : V_i \to M^m$ preserve orientation. Then

$$h_i^*(\varphi_i \omega^m) := f_i(y) \cdot dy^1 \wedge \ldots \wedge dy^m$$

is an m-form on V_i with compact support, and we eventually define the integral:

$$\int_{M^m} \omega^m := \sum_{i=1}^{l} \int_{V_i} h_i^*(\varphi_i \omega^m) = \sum_{i=1}^{l} \int_{V_i} f_i(y) \cdot dy^1 \ldots dy^m\,.$$

We show that this definition of the integral is independent of the chosen partition of unity. Consider another atlas $(\tilde{h}_1, \tilde{V}_1), \ldots, (\tilde{h}_r, \tilde{V}_r)$ with subordinate partition of unity $\tilde{\varphi}_1, \ldots, \tilde{\varphi}_r$. If the intersection of the chart ranges is not empty, $h_i(V_i) \cap \tilde{h}_\alpha(\tilde{V}_\alpha) \neq \emptyset$, then

$$\int_{\tilde{V}_\alpha} \tilde{h}_\alpha^*(\tilde{\varphi}_\alpha \varphi_i \omega^m) = \int_{V_i} h_i^*(\tilde{\varphi}_\alpha \varphi_i \cdot \omega^m)\,.$$

This follows immediately from Theorem 7 in Chapter 2, since the determinant of the differential $D(h_i^{-1} \circ \tilde{h}_\alpha)$ is positive. Summing now over the index i, $1 \leq i \leq l$, as well as the index α, $1 \leq \alpha \leq r$, we obtain from $\sum\limits_i \varphi_i \equiv \sum\limits_\alpha \tilde{\varphi}_\alpha \equiv 1$ the relation

$$\sum_{\alpha=1}^{r} \int_{\tilde{V}_\alpha} \tilde{h}_\alpha^*(\tilde{\varphi}_\alpha \cdot \omega^m) = \sum_{i=1}^{l} \int_{V_i} h_i^*(\varphi_i \cdot \omega^m)\,.$$

Definition 18. The *m-dimensional volume* of a compact and oriented submanifold $M^m \subset \mathbb{R}^n$ is the number

$$\mathrm{vol}(M^m) \;:=\; \int_{M^m} dM^m \,.$$

Consider a curve, viewed as a 1-dimensional submanifold $M^1 \subset \mathbb{R}^n$ of $\mathbb{R}^n$. If $h : (a,b) \to M^1$ is a parametrization of the curve, then

$$\int_{M^1} dM^1 \;=\; \int_a^b ||\dot{h}(t)||\, dt \,,$$

and we recover the length of the curve. More generally, if $h : V \to M^m$ is a parametrization of M^m, then, using the coefficients of the Riemannian metric, the volume of M^m can be written as

$$\mathrm{vol}(M^m) \;=\; \int_V \sqrt{\det[g_{ij}(y)]}\, dy \,.$$

Example 29. The last formula in §3.4 shows that the $(n-1)$-dimensional volume of the sphere $S^{n-1}(R)$ and the n-dimensional volume of the ball $D^n(R)$, both of radius R, are related by

$$\mathrm{vol}(S^{n-1}(R)) \;=\; \frac{n}{R}\mathrm{vol}(D^n(R))\,.$$

Example 30. The coefficients of the Riemannian metric on the torus of revolution discussed in Example 4 are

$$g_{\psi\psi} = (r_1 + r_2\cos\varphi)^2, \quad g_{\psi\varphi} = 0, \quad g_{\varphi\varphi} = r_2^2\,.$$

Hence $\sqrt{g} = r_2(r_1 + r_2\cos\varphi)$, and we obtain

$$\mathrm{vol}(T^2) \;=\; \int_0^{2\pi}\!\!\int_0^{2\pi} r_2(r_1 + r_2\cos\varphi)d\varphi d\psi \;=\; 4\pi^2 r_1 r_2\,.$$

We will conclude this section by a remark concerning measure theory on manifolds. Let $C^0(M^m)$ be the ring of all continuous functions defined on the compact and oriented manifold M^m. Setting

$$\mu(f) \;:=\; \int_{M^m} f\cdot dM^m$$

defines a linear functional $\mu : C^0(M^m) \to \mathbb{R}$ which is positive and monotonously continuous in the sense of the following properties:

(1) If $f \in C^0(M^m)$ is a non-negative function, then $\mu(f) \geq 0$.

(2) If f_n is a monotonously increasing sequence converging to a function $f \in C^0(M^m)$, then

$$\lim_{n\to\infty} \mu(f_n) = \mu(f)\,.$$

This turns the pair $(C^0(M^m), \mu)$ into a so-called *Daniell-Stone functional* on the set M^m. Within the framework of general measure theory, one first associates with every D-S functional μ on a set X an outer measure μ^* on X. The *Carathéodory construction* then leads to a σ-algebra of subsets in X on which μ^* becomes a measure. By applying these general principles to a compact and oriented manifold, one constructs the so-called *Lebesgue measure* on M^m. It is defined on a σ-algebra containing the Borel sets, and every bounded and Borel-measurable function turns out to be integrable. For the purposes of vector analysis we do not need this extension of the notions of measure and integral to manifolds, since the functions occurring are, as a rule, at least continuous. It is, however, interesting to note that the notion of integral treated here fits into the general theory of measure and integration as sketched above. The interested reader may refer to the literature for details (see the book of K. Maurin in the bibliography).

3.6. Stokes' Theorem for Manifolds

In the preceding sections we collected all the notions necessary for the formulation of Stokes' theorem on manifolds.

Theorem 21 (Stokes' Theorem). *Let M^k be a compact, oriented manifold, and suppose that the boundary ∂M^k is endowed with the induced orientation. Then, for every $(k-1)$-form ω^{k-1} on M^k,*

$$\int_{\partial M^k} \omega^{k-1} = \int_{M^k} d\omega^{k-1} .$$

If, in particular, M^k has no boundary, then for every $(k-1)$-form ω^{k-1}

$$\int_{M^k} d\omega^{k-1} = 0 .$$

Proof. In Theorem 20, it was shown that we can choose finitely many charts $(V_1, h_1), \ldots, (V_r, h_r)$ covering the manifold M^k, together with a subordinate partition of unity $\varphi_1, \ldots, \varphi_r : M^k \to \mathbb{R}$. We label these charts in such a way that, on the one hand, for each index i less than a certain index r_0 the set V_i intersects the boundary, $V_i \cap \partial M^k \neq \emptyset$, and, on the other hand, for all indices $i > r_0$ the set V_i is disjoint from the boundary, $V_i \cap \partial M^k = \emptyset$. From $\sum_i \varphi_i = 1$ we obtain $\sum_i d\varphi_i = 0$ and use this to rewrite the exterior derivative of ω^{k-1}:

$$d\omega^{k-1} = \left(\sum_{i=1}^{r} \varphi_i\right) d\omega^{k-1} = \sum_{i=1}^{r} \varphi_i d\omega^{k-1} + \sum_{i=1}^{r} d\varphi_i \wedge \omega^{k-1} = \sum_{i=1}^{r} d(\varphi_i \omega^{k-1}).$$

This implies the equation

$$\int_{M^k} d\omega^{k-1} = \sum_{i=1}^{r} \int_{V_i} d\left(h_i^*(\varphi_i \omega^{k-1})\right).$$

V_i is an open subset of $\mathbb{R}^k$ for $r_0+1 \le i \le r$, and $h_i^*(\varphi_i\omega^{k-1})$ is a k-form on V_i whose support is completely contained in V_i. For each of these indices i we choose a k-chain c_i^k in $\mathbb{R}^k$ for which

$$\operatorname{supp} h_i^*(\varphi_i\omega^{k-1}) \subset \operatorname{Int} c_i^k \subset c_i^k \subset V_i.$$

Applying now Stokes' theorem for chains (Theorem 8, Chapter 2) to c_i^k, we obtain

$$\int_{V_i} d\left(h_i^*(\varphi_i\omega^{k-1})\right) = \int_{c_i^k} d\left(h_i^*(\varphi_i\omega^{k-1})\right) = \int_{\partial c_i^k} h_i^*(\varphi_i\omega^{k-1}) = 0,$$

since the form vanishes on the boundary of the chain. Now we consider the indices i between 1 and r_0. For any of these, V_i is an open subset of the half-space $\mathbb{H}^k$, and as before we obtain chains c_i^k in $\mathbb{H}^k$ with the same properties of the supports with respect to $h_i^*(\varphi_i\omega^{k-1})$. Applying Stokes' theorem to these as well, we obtain

$$\int_{V_i} d\left(h_i^*(\varphi_i\omega^{k-1})\right) = \int_{\partial c_i^k} h_i^*(\varphi_i\omega^{k-1}).$$

Now $h_i^*(\varphi_i\omega^{k-1})$ does not necessarily vanish any more on $\partial c_i^k \cap (\mathbb{R}^{k-1}\times\{0\})$, but only at the points of ∂c_i^k belonging to the interior of $\mathbb{H}^k$:

$$\int_{V_i} d\left(h_i^*(\varphi_i\omega^{k-1})\right) = \int_{V_i\cap\mathbb{R}^{k-1}} h_i^*(\varphi_i\omega^{k-1}).$$

The pairs $(V_i\cap\mathbb{R}^{k-1}, h_i|_{\mathbb{R}^k})$ with $i = 1, \ldots, r_0$ form a covering of the boundary ∂M^k. Hence, by the definition of the integral,

$$\int_{\partial M^k} \omega^{k-1} = \sum_{i=1}^{r_0} \int_{V_i\cap\mathbb{R}^{k-1}} h_i^*\left(\varphi_i\omega^{k-1}\right),$$

showing the equation we set out to prove. □

In the sections to follow we will be dealing with various applications of Stokes' theorem. As a generalization of the discussion in §2.5, we will first study line integrals and prove an analogue—only for 1-forms, however—of Poincaré's lemma. This holds for manifolds in which every closed path can be contracted to a point.

Definition 19. A connected manifold M^k is called *simply connected* if every two C^1-curves $c_0, c_1 : [0,1] \to M^k$ with coinciding initial and end points are homotopic.

By Theorem 9 in §2.5, whose proof immediately carries over to the case of a manifold, on a simply connected manifold the line integral of a closed 1-form ω^1 depends exclusively on the end points of the curve. Having fixed a point $x_0 \in M^k$, the line integral along a curve joining the points x_0 and x,

$$f(x) \;=\; \int_{x_0}^{x} \omega^1,$$

uniquely defines a function on the manifold. Its differential df coincides with ω^1, and we obtain

Theorem 22. *Every closed* 1*-form on a simply connected manifold is exact.*

Example 31. The winding form defined on $\mathbb{R}^2 - \{0\}$ is closed, but not exact. This shows that $\mathbb{R}^2 - \{0\}$ is not simply connected.

3.7. The Hedgehog Theorem (Hairy Sphere Theorem)

Consider two oriented compact manifolds M^k and N^k without boundary and of equal dimension. Two maps $f_0, f_1 : M^k \to N^k$ between them are called homotopic if there exists a smooth map

$$F : M^k \times [0,1] \longrightarrow N^k \text{ such that } F(x,0) \;=\; f_0(x) \text{ and } F(x,1) \;=\; f_1(x)\,.$$

We prove

Theorem 23. *Let ω^k be a k-form on N^k and let $f_0, f_1 : M^k \to N^k$ be homotopic maps. Then*

$$\int_{M^k} f_0^*(\omega^k) \;=\; \int_{M^k} f_1^*(\omega^k)\,.$$

Proof. The oriented manifold $M^k \times [0,1]$ has boundary

$$\partial(M^k \times [0,1]) \;=\; M^k \times \{1\} - M^k \times \{0\}\,,$$

where the minus sign indicates that the orientation is reversed. Therefore, Stokes' theorem implies

$$\int_{M^k} f_1^*(\omega^k) - \int_{M^k} f_0^*(\omega^k) \;=\; \int_{\partial(M^k\times[0,1])} F^*(\omega^k) \;=\; \int_{M^k\times[0,1]} dF^*(\omega^k)\,.$$

But the form $dF^*(\omega^k) = F^*(d\omega^k) = 0$ vanishes, since the k-dimensional manifold N^k carries no non-trivial $(k+1)$-forms. □

Theorem 24. *The antipodal map from the sphere to itself, $A : S^n \to S^n$, $A(x) = -x$, is homotopic to the identity Id_{S^n} only for odd dimensions n.*

Proof. Consider on $\mathbb{R}^{n+1}$ the form

$$\omega^n \;=\; \sum_{i=1}^{n+1} (-1)^{i-1} x^i \, dx^1 \wedge \ldots \wedge \widehat{dx^i} \wedge \ldots \wedge dx^{n+1}\,,$$

whose restriction to the sphere S^n is its volume form dS^n (Example 28, equation (17)). If A is homotopic to the identity Id_{S^n}, then the previous theorem implies

$$\int_{S^n} A^*(\omega^n) = \int_{S^n} \omega^n = \mathrm{vol}(S^n).$$

The induced form $A^*(\omega^n) = (-1)^{n+1}\omega^n$ is proportional to the form ω^n. Thus we obtain the condition

$$(-1)^{n+1}\mathrm{vol}(S^n) = \mathrm{vol}(S^n),$$

i. e., $(n+1)$ has to be an odd number. □

Theorem 25 (Hedgehog Theorem). *A sphere S^{2k} of even dimension has no nowhere vanishing, continuous tangent vector field.*

Proof. Suppose that there exists such a vector field on the n-dimensional sphere S^n. We first approximate this vector field by a smooth vector field $\mathcal{V}$ (Stone-Weierstrass theorem), and then normalize it so that the vector $\mathcal{V}(x)$ has length one at each point. Next we consider the resulting smooth tangent vector field as a vector-valued function $\mathcal{V} : S^n \to \mathbb{R}^{n+1}$ satisfying the following two conditions:

$$\langle x, \mathcal{V}(x)\rangle = 0, \quad ||\mathcal{V}(x)|| = 1.$$

Define the homotopy $F : S^n \times [0,1] \to S^n$ from the sphere to itself by the formula

$$F(x,t) = \cos(\pi t)\cdot x + \sin(\pi t)\cdot \mathcal{V}(x).$$

The length of $F(x,t)$ is equal to one everywhere, since x and $\mathcal{V}(x)$ are perpendicular. Moreover, $F(x,0) = x$ and $F(x,1) = -x$, i. e., F is a homotopy between the identity and the antipodal map of the sphere S^n. But then the dimension n has to be an odd number. □

In the German mathematical literature, this result is known as the "Hedgehog Theorem" („Satz vom Igel"), since its contents can be expressed figuratively by saying that a hedgehog cannot be combed in a continuous way. Because it is so vivid, we prefer this to the name "Hairy Sphere Theorem", which seems to be more common in the Anglo-Saxon world.

3.8. The Classical Integral Formulas

Now we will discuss the classical integral formulas already treated in §2.6 for chains. Compared to the preceding case, in this new formulation we benefit from having notions like divergence, gradient and Laplacian as developed in the differential calculus on manifolds at our disposal. At the same time, the orientation of the manifold and the exterior unit normal vector field on the boundary will play a special role. We will prove these integral formulas for arbitrary compact and oriented manifolds (in $\mathbb{R}^n$). This will be the final formulation of the classical integral formulas as they are needed in many branches of mathematics as well as theoretical electrodynamics and hydrodynamics. We start with the *Ostrogradski formula* relating the divergence of a vector field to its flow across the surface.

Theorem 26 (Ostrogradski Formula). *Let M^k be an oriented, compact manifold, and let $\mathcal{N}$ be the exterior unit normal vector field to its boundary. Then, for every vector field $\mathcal{V} : M^k \to TM^k$ on M^k,*

$$\int_{M^k} \mathrm{div}(\mathcal{V}) dM^k \;=\; \int_{\partial M^k} \langle \mathcal{V}, \mathcal{N} \rangle \, d(\partial M^k) \, .$$

Proof. We know from Theorem 17 that the divergence and its inner product with the volume form are related by the formula

$$\mathrm{div}(\mathcal{V}) \cdot dM^k \;=\; d(\mathcal{V} \lrcorner \, dM^k) \, .$$

A straightforward application of Stokes' theorem implies

$$\int_{M^k} \mathrm{div}(\mathcal{V}) dM^k \;=\; \int_{M^k} d(\mathcal{V} \lrcorner \, dM^k) \;=\; \int_{\partial M^k} \mathcal{V} \lrcorner \, dM^k \, .$$

Let $x \in \partial M^k$ be a point of the boundary. We decompose the vector $\mathcal{V}(x)$ into one part that is proportional to the exterior normal vector, and a vector $\mathcal{W}(x)$ that belongs to the tangent space $T_x \partial M^k$ to the boundary:

$$\mathcal{V}(x) \;=\; \langle \mathcal{V}(x), \mathcal{N}(x) \rangle \, \mathcal{N}(x) + \mathcal{W}(x) \, .$$

Moreover, note that the restriction of the inner product $\mathcal{W} \lrcorner \, dM^k$ to the boundary ∂M^k vanishes identically. This implies that for the inner product of $\mathcal{V}$ with the volume form dM^k, on the boundary ∂M^k

$$\mathcal{V} \lrcorner \, dM^k \;=\; \langle \mathcal{V}, \mathcal{N} \rangle \, \mathcal{N} \lrcorner \, dM^k + \mathcal{W} \lrcorner \, dM^k \;=\; \langle \mathcal{V}, \mathcal{N} \rangle \, \mathcal{N}(x) \lrcorner \, dM^k \, .$$

Hence, we obtain

$$\int_{M^k} \mathrm{div}(\mathcal{V}) dM^k \;=\; \int_{\partial M^k} \langle \mathcal{V}, \mathcal{N} \rangle \, \mathcal{N} \lrcorner \, dM^k \, .$$

By Theorem 19, the inner product $\mathcal{N}(x) \lrcorner \, dM^k$ coincides with the volume form of the boundary, $d(\partial M^k)$. □

As a direct application of the Ostrogradski formula we obtain *Gauss' theorem.*

Theorem 27 (Gauss' Theorem). *Let $\mathcal{V}$ be a vector field, let f be a function on the oriented, compact manifold M^k, and let $\mathcal{N}$ be the exterior unit normal vector field of the boundary. Then*

$$\int_{M^k} \langle \mathcal{V}, \mathrm{grad}(f) \rangle \, dM^k + \int_{M^k} f \cdot \mathrm{div}(\mathcal{V}) dM^k \;=\; \int_{\partial M^k} f \cdot \langle \mathcal{V}, \mathcal{N} \rangle \, d(\partial M^k) \, .$$

Proof. This equation immediately follows from the Ostrogradski formula together with the rule from Theorem 11

$$\mathrm{div}(f \cdot \mathcal{V}) \;=\; f \cdot \mathrm{div}(\mathcal{V}) + \mathcal{V}(f) \;=\; f \cdot \mathrm{div}(\mathcal{V}) + \langle \mathcal{V}, \, \mathrm{grad}(f) \rangle \, . \qquad \square$$

In a similar way we derive Green's formulas in versions that are not confined to $\mathbb{R}^2$. First we generalize *Green's first formula.*

Theorem 28 (Green's First Formula)**.** *Let $f, \, g : M^k \to \mathbb{R}$ be smooth functions on the compact, oriented manifold M^k. Then*

$$\int\limits_{M^k} f \cdot \Delta(g) \, dM^k + \int\limits_{M^k} \langle \mathrm{grad}(f), \, \mathrm{grad}(g) \rangle \, dM^k = \int\limits_{\partial M^k} f \cdot \langle \mathrm{grad}(g), \, \mathcal{N} \rangle \, d(\partial M^k).$$

Proof. By the definition of the Laplacian, we have

$$\int_{M^k} f \cdot \Delta(g) \, dM^k \;=\; \int_{M^k} f \cdot \mathrm{div}(\mathrm{grad}(g)) \, dM^k \, .$$

Now apply Gauss' theorem. $\square$

Applying Green's first formula twice leads to *Green's second formula.*

Theorem 29 (Green's Second Formula)**.** *Let $f, \, g : M^k \to \mathbb{R}$ be two smooth functions. Then*

$$\int\limits_{M^k} [g \cdot \Delta(f) - f \cdot \Delta(g)] \, dM^k = \int\limits_{\partial M^k} [g \cdot \langle \mathrm{grad}(f), \, \mathcal{N} \rangle - f \cdot \langle \mathrm{grad}(g), \, \mathcal{N} \rangle \,] \, d(\partial M^k).$$

$\square$

Remark. The scalar product $\langle \mathrm{grad}(f), \mathcal{N} \rangle$ defined only on the boundary is often denoted by the symbol $\partial f / \partial \mathcal{N}$, since it is the derivative of the function f in the direction of the exterior normal vector. This leads to a different formulation of Green's second formula:

$$\int_{M^k} [g \cdot \Delta(f) - f \cdot \Delta(g)] \, dM^k \;=\; \int_{\partial M^k} \left[g \cdot \frac{\partial f}{\partial \mathcal{N}} - f \cdot \frac{\partial g}{\partial \mathcal{N}} \right] \, d(\partial M^k) \, .$$

Corollary 1. *Let M^k be a compact, oriented manifold without boundary. Then*

(1) $\displaystyle\int\limits_{M^k} \mathrm{div}(\mathcal{V}) \, dM^k \;=\; 0$ *for every vector field $\mathcal{V}$;*

(2) $\displaystyle\int\limits_{M^k} g \Delta(f) \, dM^k = \int\limits_{M^k} f \Delta(g) \, dM^k = - \int\limits_{M^k} \langle \mathrm{grad}(f), \, \mathrm{grad}(g) \rangle \, dM^k$
for any two functions $f, \, g \in C^\infty(M^k)$.

$\square$

Hence the Laplacian is symmetric with respect to the L^2-scalar product. Moreover, the choice of sign we adopted implies that it is non-positive.

Corollary 2 (Hopf's Theorem). *Let M^k be a compact, connected, oriented manifold without boundary and assume that the function $f : M^k \to \mathbb{R}$ satisfies at each point the condition $\Delta(f)(x) \geq 0$. Then the function f is constant.*

Proof. Integrating the assumption $\Delta(f)(x) \geq 0$ over M^k and applying the symmetry of the Laplacian just proved, we first obtain

$$0 \;\leq\; \int_{M^k} 1 \cdot \Delta(f) \cdot dM^k \;=\; \int_{M^k} f \cdot \Delta(1) \cdot dM^k \;=\; 0\,,$$

i.e., the Laplacian of f vanishes identically, $\Delta(f) = 0$. Inserting $f = g$ into Green's first formula (Theorem 28) and taking into account that the boundary integral vanishes by the assumption concerning M^k, this implies

$$\int_{M^k} |\mathrm{grad}(f)|^2 \cdot dM^k \;=\; -\int_{M^k} f \cdot \Delta(f) \cdot dM^k \;=\; 0\,,$$

and hence $\mathrm{grad}(f) = 0$. Thus f is constant, since M^k is connected. □

Concluding this section, we formulate *Stokes' theorem* in its classical form on $\mathbb{R}^3$. Contrary to the preceding theorems involving the generalizations of divergence and gradient to manifolds as introduced in the second section of this chapter, this only involves the notion of curl on open subsets of $\mathbb{R}^3$ from §2.3.

Theorem 30 (Stokes' Theorem—Classical Version). *Let $M^2 \subset \mathbb{R}^3$ be a compact, oriented surface, let $\mathcal{V}$ be a vector field defined on an open subset $M^2 \subset U \subset \mathbb{R}^3$, let $\mathcal{N} : M^2 \to S^2$ be the exterior unit normal field to the surface M^2, and let $\mathcal{T} : \partial M^2 \to T(\partial M^2)$ be the unit tangent vector field on the boundary curve ∂M^2 with the induced orientation. Then*

$$\int_{M^2} \langle \mathrm{curl}(\mathcal{V}), \mathcal{N}\rangle \; dM^2 \;=\; \int_{\partial M^2} \langle \mathcal{V}, \mathcal{T}\rangle \; d(\partial M^2)\,.$$

Proof. Consider the 1-form

$$\omega^1_{\mathcal{V}} \;:=\; V^1 dx^1 + V^2 dx^2 + V^3 dx^3$$

on U associated with the vector field $\mathcal{V} = V^1\partial/\partial x^1 + V^2\partial/\partial x^2 + V^3\partial/\partial x^3$ as explained in §2.3 and its derivative

$$d\omega^1_{\mathcal{V}} =$$

$$\left[\frac{\partial V^2}{\partial x^1} - \frac{\partial V^1}{\partial x^2}\right] dx^1 \wedge dx^2 + \left[\frac{\partial V^3}{\partial x^1} - \frac{\partial V^1}{\partial x^3}\right] dx^1 \wedge dx^3 + \left[\frac{\partial V^3}{\partial x^2} - \frac{\partial V^2}{\partial x^3}\right] dx^2 \wedge dx^3.$$

Recall that the curl of $\mathcal{V}$ corresponds to the 1-form $*d\omega^1_{\mathcal{V}}$. If, on the other hand, $h : W \to M^2$ is a parametrization of the surface with components h^1, h^2, h^3 and coordinates y^1, y^2 from W, then for the exterior normal vector $\mathcal{N}$ to the surface we have the relation

$$\mathcal{N}(y^1, y^2) \;=\; \frac{\partial h}{\partial y^1} \times \frac{\partial h}{\partial y^2} \Big/ \Big\| \frac{\partial h}{\partial y^1} \times \frac{\partial h}{\partial y^2} \Big\| \,.$$

Two arbitrary vectors $v, w \in \mathbb{R}^3$ satisfy the identity

$$||v \times w||^2 \;=\; \det \begin{bmatrix} \langle v, v\rangle & \langle v, w\rangle \\ \langle v, w\rangle & \langle w, w\rangle \end{bmatrix} ,$$

and hence the preceding equation implies

$$\Big\| \frac{\partial h}{\partial y^1} \times \frac{\partial h}{\partial y^2} \Big\| \;=\; \sqrt{g}\,.$$

For the first component of the normal vector $\mathcal{N}$ written in the coordinates y^1, y^2 this reads, e.g., as

$$N^1 dM^2 \;=\; N^1 \sqrt{g}\, dy^1 \wedge dy^2 \;=\; \frac{1}{\sqrt{g}} \left[\frac{\partial h^2}{\partial y^1}\frac{\partial h^3}{\partial y^2} - \frac{\partial h^2}{\partial y^2}\frac{\partial h^3}{\partial y^1} \right] \sqrt{g}\, dy^1 \wedge dy^2 \,.$$

On the other hand, it is easy to compute the pullback by h of the forms dx^2 and dx^3:

$$h^*(dx^2) \;=\; \frac{\partial h^2}{\partial y^1} dy^1 + \frac{\partial h^2}{\partial y^2} dy^2, \quad h^*(dx^3) \;=\; \frac{\partial h^3}{\partial y^1} dy^1 + \frac{\partial h^3}{\partial y^2} dy^2 \,.$$

A direct comparison implies the following formula, which is independent of the coordinates y^1, y^2:

$$N^1 dM^2 \;=\; dx^2 \wedge dx^3 \,.$$

Similarly one proves

$$N^2 dM^2 \;=\; -\,dx^1 \wedge dx^3, \quad N^3 dM^2 \;=\; dx^1 \wedge dx^2 \,.$$

The scalar product of the curl of $\mathcal{V}$ with the unit normal field $\mathcal{N}$ multiplied by the volume form dM^2 is thus simply the differential of $\omega^1_{\mathcal{V}}$:

$$\begin{aligned} &\langle \mathrm{curl}(\mathcal{V}), \mathcal{N}\rangle\; dM^2 \\ &= \left[\frac{\partial V^3}{\partial x^2} - \frac{\partial V^2}{\partial x^3}\right] N^1 dM^2 + \left[\frac{\partial V^1}{\partial x^3} - \frac{\partial V^3}{\partial x^1}\right] N^2 dM^2 + \left[\frac{\partial V^2}{\partial x^1} - \frac{\partial V^1}{\partial x^2}\right] N^3 dM^2 \\ &= d\omega^1_{\mathcal{V}}\,. \end{aligned}$$

Therefore, Stokes' theorem can be applied in the format

$$\int\limits_{M^2} \langle \mathrm{curl}(\mathcal{V}), \mathcal{N}\rangle\, dM^2 = \int\limits_{M^2} d\omega^1_{\mathcal{V}} = \int\limits_{\partial M^2} \omega^1_{\mathcal{V}} = \int\limits_{\partial M^2} [V^1 dx^1 + V^2 dx^2 + V^3 dx^3].$$

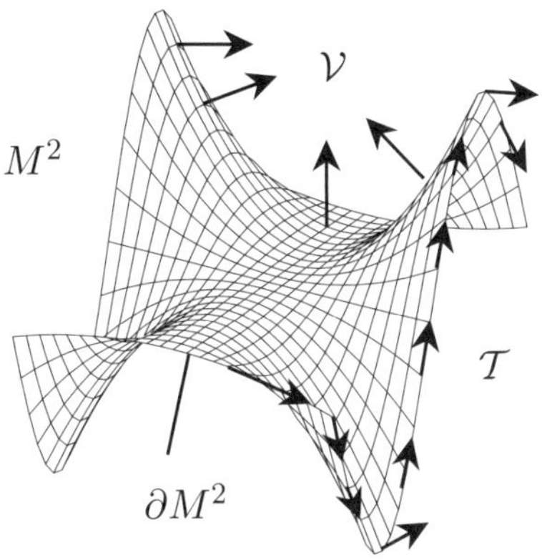

If, however, $\mathcal{T}$ is the unit tangent vector field to the curve ∂M^2, then $d(\partial M^2)(\mathcal{T}) = 1$, and hence

$$T^1 d(\partial M^2) \;=\; dx^1, \quad T^2 d(\partial M^2) \;=\; dx^2, \quad T^3 d(\partial M^2) \;=\; dx^3 \,.$$

Now we can rewrite the line integral above as

$$\int_{\partial M^2} [V^1 dx^1 + V^2 dx^2 + V^3 dx^3] \;=\; \int_{\partial M^2} \langle \mathcal{V}, \mathcal{T} \rangle \, d(\partial M^2) \,,$$

and, summarizing, we arrive at Stokes' classical integral formula. □

3.9. The Lie Derivative and the Interpretation of the Divergence

The aim of this section is to interpret the divergence of a vector field geometrically as the infinitesimal volume distortion of its flow. First we recall some results from the local theory of ordinary differential equations and introduce the flow on a manifold as well as the Lie derivative of forms. Then we compute this Lie derivative by means of the exterior derivative, which, in a special case, leads to the interpretation of the divergence mentioned in the title.

Let $\mathcal{V}$ be a vector field on the manifold M^k. An *integral curve* of $\mathcal{V}$ is a curve $\gamma : (a, b) \to M^k$ whose tangent vector $\dot{\gamma}(t) = \gamma_*(\partial/\partial t)$ at each point coincides with the value of the vector field there:

$$\dot{\gamma}(t) \;=\; \mathcal{V}(\gamma(t)) \,.$$

The well-known existence theorem for autonomous differential equations states that for every initial point $x \in M^k$ there exists a maximal integral curve $\gamma_x : (a_x, b_x) \to M^k$ defined on an interval, containing the number $0 \in \mathbb{R}$, satisfying

$$\gamma_x(0) \;=\; x\,.$$

Moreover, this maximal integral curve is uniquely determined by the initial condition. Denote by $\Sigma_{\mathcal{V}}$ the set

$$\Sigma_{\mathcal{V}} \;=\; \big\{(t,x) \in \mathbb{R} \times M^k : a_x < t < b_x\big\}$$

and call the map $\Phi : \Sigma_{\mathcal{V}} \to M^k$ defined by the formula

$$\Phi_t(x) \;=\; \Phi(t,x) \;:=\; \gamma_x(t)$$

the *flow of the vector field* $\mathcal{V}$. If the maximal integral curves are defined for all values $t \in \mathbb{R}$ of the parameter (i.e., if $\Sigma_{\mathcal{V}} = \mathbb{R} \times M^k$), then the vector field $\mathcal{V}$ is called *complete.* A complete vector field $\mathcal{V}$ determines a one-parameter group $\Phi_t : M^k \to M^k$ of diffeomorphisms from the manifold to itself:

$$\Phi_{t_0+t_1} \;=\; \Phi_{t_0} \circ \Phi_{t_1}\,.$$

This relation between diffeomorphisms is a direct consequence of the uniqueness of the integral curve for a fixed initial condition. Conversely, if a *one-parameter group of diffeomorphisms* is given, then we obtain a vector field by looking at the tangents to the trajectories $\Phi_t(x)$ of a point:

$$\mathcal{V}(x) \;:=\; \frac{d}{dt}(\Phi_t(x))_{|t=0}\,.$$

Example 32. The one-parameter transformation group in $\mathbb{R}^2$ of the vector field

$$\mathcal{V} \;=\; -x^2\frac{\partial}{\partial x^1} + x^1\frac{\partial}{\partial x^2}$$

is determined by the differential equations

$$\dot{x}^1 \;=\; -x^2, \quad \dot{x}^2 = x^1$$

and coincides with the group of all rotations in the plane.

Example 33. The one-parameter transformation group of the vector field

$$\mathcal{V} \;=\; x^1\frac{\partial}{\partial x^1} + \cdots + x^n\frac{\partial}{\partial x^n}$$

in $\mathbb{R}^n$ is determined by the differential equations

$$\dot{x}^1 \;=\; x^1, \quad \ldots, \quad \dot{x}^n \;=\; x^n$$

and coincides with the group of all dilatations in $\mathbb{R}^n$.

Suppose that the manifold M^k is a closed subset of $\mathbb{R}^n$. Then M^k is a complete metric space: Every Cauchy sequence in M^k converges. For bounded vector fields we will prove

Theorem 31. *Every bounded vector field on a complete manifold without boundary is a complete vector field. In particular, every vector field on a compact manifold without boundary is complete.*

Proof. Denote the maximal length of the vector field $\mathcal{V}$ by K,

$$K \;:=\; \sup\left\{||\mathcal{V}(x)|| :\; x \in M^k\right\}.$$

Let $\gamma : \;(a,b) \to M^k$ be a maximal integral curve of the vector field, and assume that $b < \infty$. We apply the mean value theorem of differential calculus to the vector-valued function $\gamma : (a,b) \;\to M^k \subset \mathbb{R}^n$. It states that for every two values of the parameter, $a < t_1 < t_2 < b$, there exists one more value, $t_1 \le t^* \le t_2$, such that

$$||\gamma(t_1) - \gamma(t_2)|| \;\le\; |t_2 - t_1| \cdot ||\dot{\gamma}(t^*)|| \;=\; |t_2 - t_1| \cdot ||\mathcal{V}(\gamma(t^*))|| \;\le\; |t_2 - t_1| \cdot K\,.$$

Since b is finite, any sequence $t_i \in (a,b)$ converging to b is a Cauchy sequence in $\mathbb{R}$. The last estimate shows that the image sequence $\gamma(t_i) \in M^k$ is also a Cauchy sequence. By assumption, M^k is a complete manifold, and hence the sequence $\gamma(t_i)$ converges in M^k. This observation shows that the left-hand limit,

$$\lim_{t \to b^-} \gamma(t) \;=:\; x^*,$$

exists. The point $x^* \in M^k$ is an inner point of the manifold, since there are no boundary points by assumption. We choose a chart $h : \mathbb{R}^k \to M^k$ around the point $x^* \;=\; h(0)$ and represent the vector field $\mathcal{V}$ in the corresponding coordinates,

$$\mathcal{V} \;=\; V^1 \partial/\partial y^1 + \ldots + V^k \partial/\partial y^k\,.$$

The solution $\gamma(t)$ of the differential equations

$$\dot{y}^1 \;=\; V^1(y^1, \ldots, y^k), \quad \ldots, \quad \dot{y}^k \;=\; V^k(y^1, \ldots, y^k)$$

converges for $t \to b$ to the point $0 \in \mathbb{R}^k$. Hence the integral curve $\gamma(t)$ can be extended beyond the parameter b. This contradicts the maximality of b resulting from the assumption $b < \infty$. □

Now we turn to the Lie derivative of a differential form with respect to a vector field. If ω^i is an i-form, then, pulling it back by the flow of the vector field, we obtain a 1-parameter family of forms, $\Phi_t^*(\omega^i)$. The derivative of that with respect to the parameter at $t = 0$ is the so-called Lie derivative. This concept involves the flow of the vector field only in a small neighborhood of zero and is thus well-defined for arbitrary vector fields.

Definition 20. The *Lie derivative* $\mathcal{L}_{\mathcal{V}}(\omega^i)$ of an i-form ω^i is the form

$$\mathcal{L}_{\mathcal{V}}(\omega^i) \;:=\; \frac{d}{dt}(\Phi_t^*(\omega^i))_{t=0} \;=\; \lim_{t \to 0} \frac{\Phi_t^*(\omega^i) - \omega^i}{t}\,.$$

Theorem 32. *The Lie derivative of a differential form is expressed in terms of the exterior derivative and the inner product as*

$$\mathcal{L}_{\mathcal{V}}(\omega^i) \;=\; d(\mathcal{V} \lrcorner\, \omega^i) + \mathcal{V} \lrcorner\, (d\omega^i)\,.$$

Proof. The exterior product $\omega^i \wedge \eta^j$ is bilinear, and the product rule is proved as usual,

$$\mathcal{L}_{\mathcal{V}}(\omega^i \wedge \eta^j) \;=\; \mathcal{L}_{\mathcal{V}}(\omega^i) \wedge \eta^j + \omega^i \wedge \mathcal{L}_{\mathcal{V}}(\eta^j)\,.$$

Furthermore, the exterior derivative d commutes with the induced maps Φ_t^*, and hence we have

$$\mathcal{L}_{\mathcal{V}}(d\omega^i) \;=\; d(\mathcal{L}_{\mathcal{V}}(\omega^i))\,.$$

Denote the right-hand side of the equation to be proved by $\mathcal{L}^*_{\mathcal{V}}(\omega^i)$. Then

$$\begin{aligned}\mathcal{L}^*_{\mathcal{V}}(d\omega^i) &= d(\mathcal{V} \lrcorner\, d\omega^i) + \mathcal{V} \lrcorner\, (dd\omega^1) \;=\; d(\mathcal{V} \lrcorner\, d\omega^i)\\ &= d(\mathcal{V} \lrcorner\, d\omega^i + d(\mathcal{V} \lrcorner\, \omega^i)) \;=\; d(\mathcal{L}^*_{\mathcal{V}}(\omega^i))\,.\end{aligned}$$

The first, purely algebraic equation,

$$\mathcal{V} \lrcorner\, (\omega^i \wedge \eta^j) \;=\; (\mathcal{V} \lrcorner\, \omega^i) \wedge \eta^j + (-1)^i \omega^i \wedge (\mathcal{V} \lrcorner\, \eta^j),$$

leads to the identity

$$\begin{aligned}&\mathcal{L}^*_{\mathcal{V}}(\omega^i \wedge \eta^j) \;=\; d(\mathcal{V} \lrcorner\, (\omega^i \wedge \eta^j)) + \mathcal{V} \lrcorner\, d(\omega^i \wedge \eta^j)\\ &= d((\mathcal{V} \lrcorner\, \omega^i) \wedge \eta^j + (-1)^i \omega^i \wedge (\mathcal{V} \lrcorner\, \eta^j)) + \mathcal{V} \lrcorner\, (d\omega^i \wedge \eta^j + (-1)^i \omega^i \wedge d\eta^j)\\ &= d(\mathcal{V} \lrcorner\, \omega^i) \wedge \eta^j + (-1)^{i-1}(\mathcal{V} \lrcorner\, \omega^i) \wedge d\eta^j + (-1)^i d\omega^i \wedge (\mathcal{V} \lrcorner\, \eta^j)\\ &\quad + \omega^i \wedge d(\mathcal{V} \lrcorner\, \eta^j) + (\mathcal{V} \lrcorner\, d\omega^i) \wedge \eta^j + (-1)^{i+1} d\omega^i \wedge (\mathcal{V} \lrcorner\, \eta^j)\\ &\quad + (-1)^i (\mathcal{V} \lrcorner\, \omega^i) \wedge d\eta^j + \omega^i \wedge (\mathcal{V} \lrcorner\, d\eta^j)\\ &= (d(\mathcal{V} \lrcorner\, \omega^i) + \mathcal{V} \lrcorner\, d\omega^i) \wedge \eta^j + \omega^i \wedge (d(\mathcal{V} \lrcorner\, \eta^j) + \mathcal{V} \lrcorner\, d\eta^j)\\ &= \mathcal{L}^*_{\mathcal{V}}(\omega^i) \wedge \eta^j + \omega^i \wedge \mathcal{L}^*_{\mathcal{V}}(\eta^j)\,.\end{aligned}$$

Lastly, $\mathcal{L}$ and $\mathcal{L}^*$ coincide for functions,

$$\mathcal{L}_{\mathcal{V}}(f) \;=\; \mathcal{V}(f) \;=\; \mathcal{V} \lrcorner\, df \;=\; \mathcal{L}^*_{\mathcal{V}}(f)\,,$$

and thus the formula is proved. □

We apply the formula of the preceding theorem to the volume form of an oriented manifold M^k. From the definition of the divergence of a vector field we obtain

$$\mathcal{L}_{\mathcal{V}}(dM^k) \;=\; d(\mathcal{V} \lrcorner\, dM^k) \;=\; \operatorname{div}(\mathcal{V}) \cdot dM^k\,.$$

Fixing a point $x \in M^k$ and choosing a neighborhood $U(x,\varepsilon) \subset M^k$ of radius $\varepsilon > 0$, we obtain by integration

$$\int_{U(x,\varepsilon)} \operatorname{div}(\mathcal{V}) \cdot dM^k = \frac{d}{dt}\left(\int_{U(x,\varepsilon)} \Phi_t^*(dM^k)\right)\Big|_{t=0} = \frac{d}{dt}\Big(\operatorname{vol}(\Phi_t(U(x,\varepsilon)))\Big)\Big|_{t=0}.$$

Dividing both sides by the volume of the set $U(x,\varepsilon)$ and taking the limit for $\varepsilon \to 0$, we see that the mean value theorem of integral calculus immediately implies the formula

$$\operatorname{div}(\mathcal{V})(x) \;=\; \lim_{\varepsilon\to 0}\left[\frac{1}{\operatorname{vol}(U(x,\varepsilon))}\frac{d}{dt}\Big(\operatorname{vol}\big(\Phi_t(U(x,\varepsilon))\big)\Big)_{|t=0}\right].$$

This formula allows a geometric interpretation of the divergence of a vector field: It is the volume distortion of its flow in infinitesimally small time and on infinitesimally small domains surrounding the point x.

Corollary 3. *The divergence of a vector field vanishes if and only if its flow $\Phi_t : M^k \to M^k$ consists of volume-preserving diffeomorphisms.*

Proof. For a measurable set $A \subset M^k$ we denote by $h_A(t) := \operatorname{vol}(\Phi_t(A))$ the volume of the set $\Phi_t(A)$. The derivative of the volume change is computed as follows:

$$\begin{aligned}\frac{d}{dt}(h_A(t))_{t=t_1} &= \frac{d}{dt}\left(\int_{\Phi_{t+t_1}(A)} dM^k\right)_{t=0} = \int_{\Phi_{t_1}(A)}\frac{d}{dt}(\Phi_t^*(dM^k))_{t=0}\\ &= \int_{\Phi_{t_1}(A)} \mathcal{L}_{\mathcal{V}}(dM^k) = \int_{\Phi_{t_1}(A)} \operatorname{div}(\mathcal{V})\cdot dM^k.\end{aligned}$$

From this we see that the flow Φ_t preserves the volume if and only if the divergence of the vector field $\mathcal{V}$ vanishes identically. □

The Lie derivative of a vector field $\mathcal{W}$ with respect to another vector field $\mathcal{V}$ is defined just like the Lie derivative for forms; nevertheless, it is called the commutator of both vector fields.

Definition 21. Let $\mathcal{V}$ and $\mathcal{W}$ be two vector fields, and denote by Φ_t the flow of the vector field $\mathcal{V}$. The *commutator* $[\mathcal{V},\mathcal{W}]$ is defined as the vector field

$$[\mathcal{V},\mathcal{W}] \;=\; \lim_{t\to 0}\frac{(\Phi_{-t})_*(\mathcal{W})-\mathcal{W}}{t}.$$

We prove an important formula relating the commutator of two vector fields to the exterior derivative of 1-forms.

Theorem 33. *Let ω^1 be a 1-form, and let $\mathcal{V},\mathcal{W}$ be two vector fields. Then*

$$d\omega^1(\mathcal{V},\mathcal{W}) \;=\; \mathcal{V}(\omega^1(\mathcal{W})) - \mathcal{W}(\omega^1(\mathcal{V})) - \omega^1([\mathcal{V},\mathcal{W}]).$$

Here $\mathcal{V}(\omega^1(\mathcal{W}))$ is the derivative of the function $\omega^1(\mathcal{W})$ in the direction of the vector field $\mathcal{V}$.

Proof. If $x \in M^k$ is a fixed point, then by the definition of the commutator $[\mathcal{V}, \mathcal{W}]$

$$[\mathcal{V}, \mathcal{W}](x) = \lim_{t \to 0} \frac{(\Phi_{-t})_*(\mathcal{W}(\Phi_t(x))) - \mathcal{W}(x)}{t}.$$

The formula to be proved then follows from Theorem 32 by means of the following computation:

$$\begin{aligned}
\omega^1([\mathcal{V}, \mathcal{W}])(x) &= \lim_{t \to 0} \frac{(\Phi_{-t})^*(\omega^1)(\mathcal{W}(\Phi_t(x))) - \omega^1(\mathcal{W}(x))}{t} \\
&= \lim_{t \to 0} \frac{(\Phi_{-t})^*(\omega^1) - \omega^1}{t}(\mathcal{W}(\Phi_t(x))) + \lim_{t \to 0} \omega^1 \left(\frac{\mathcal{W}(\Phi_t(x)) - \mathcal{W}(x)}{t} \right) \\
&= -(\mathcal{L}_{\mathcal{V}}\omega^1)(\mathcal{W})(x) + \mathcal{V}(\omega^1(\mathcal{W}))(x) \\
&= -\mathcal{W}(\omega^1(\mathcal{V}))(x) - d\omega^1(\mathcal{V}, \mathcal{W})(x) + \mathcal{V}(\omega^1(\mathcal{W}))(x).
\end{aligned}$$

□

If, in particular, ω^1 is a closed form, then we obtain

$$\omega^1([\mathcal{V}, \mathcal{W}]) = \mathcal{V}(\omega^1(\mathcal{W})) - \mathcal{W}(\omega^1(\mathcal{V})).$$

In local coordinates the vector fields are represented by their components,

$$\mathcal{V} = \sum_{i=1}^{k} V^i \frac{\partial}{\partial y^i}, \quad \mathcal{W} = \sum_{j=1}^{k} W^j \frac{\partial}{\partial y^j}.$$

Applying the above formula to the closed form dy^l gives us

$$dy^l([\mathcal{V}, \mathcal{W}]) = \sum_{i=1}^{k} V^i \frac{\partial W^l}{\partial y^i} - \sum_{i=1}^{k} W^i \frac{\partial V^l}{\partial y^i}.$$

This leads to the following local expression for the commutator of two vector fields:

$$[\mathcal{V}, \mathcal{W}] = \sum_{l=1}^{k} \left(\sum_{i=1}^{k} \left(V^i \frac{\partial W^l}{\partial y^i} - W^i \frac{\partial V^l}{\partial y^i} \right) \right) \frac{\partial}{\partial y^l}.$$

Example 34. The commutator of the vector fields $\mathcal{V} = y \cdot \partial/\partial x, \mathcal{W} = \partial/\partial y$ defined on $\mathbb{R}^2$ is the vector field $[\mathcal{V}, \mathcal{W}] = -\partial/\partial x$.

The algebraic properties of the commutator of vector fields are summarized in the following theorem.

Theorem 34. *Let $\mathcal{U}, \mathcal{V}, \mathcal{W}$ be vector fields, and let f be a smooth function on the manifold M^k. Then the following identities hold:*

(1) $[\mathcal{U} + \mathcal{V}, \mathcal{W}] = [\mathcal{U}, \mathcal{W}] + [\mathcal{V}, \mathcal{W}]$;

(2) $[\mathcal{V}, \mathcal{W}] = -[\mathcal{W}, \mathcal{V}]$;

(3) $[\mathcal{U}, [\mathcal{V}, \mathcal{W}]] + [\mathcal{V}, [\mathcal{W}, \mathcal{U}]] + [\mathcal{W}, [\mathcal{U}, \mathcal{V}]] = 0$ *(Jacobi identity)*;

(4) $[f \cdot \mathcal{V}, \mathcal{W}] = f \cdot [\mathcal{V}, \mathcal{W}] - \mathcal{W}(f) \cdot \mathcal{V}$.

Proof. These formulas immediately result from the local expression for the commutator. We present the calculation for the last of the formulas:

$$
\begin{aligned}
[f \cdot \mathcal{V}, \mathcal{W}] &= \sum_{i,l=1}^{k} \left(f \cdot V^i \frac{\partial W^l}{\partial y^i} - W^i \frac{\partial (f \cdot V^l)}{\partial y^i} \right) \frac{\partial}{\partial y^l} \\
&= f \cdot \sum_{i,l=1}^{k} \left(V^i \frac{\partial W^l}{\partial y^i} - W^i \frac{\partial V^l}{\partial y^i} \right) \frac{\partial}{\partial y^l} - \left(\sum_{i=1}^{k} W^i \frac{\partial f}{\partial y^i} \right) \left(\sum_{l=1}^{k} V^l \frac{\partial}{\partial y^l} \right) \\
&= f \cdot [\mathcal{V}, \mathcal{W}] - \mathcal{W}(f) \cdot \mathcal{V}. \quad \square
\end{aligned}
$$

In the following sense, the commutator is a natural operation between vector fields.

Theorem 35. *For a diffeomorphism $\Phi : M^k \to M^k$ and vector fields $\mathcal{V}, \mathcal{W}$ on the manifold M^k the differential Φ_* is compatible with the commutator:*

$$\Phi_* ([\mathcal{V}, \mathcal{W}]) = [\Phi_*(\mathcal{V}), \Phi_*(\mathcal{W})] .$$

Proof. Note first that for every 1-form ω^1 the functions $\Phi^*(\omega^1)(\mathcal{W})$ and $\omega^1(\Phi_*(\mathcal{W}))$ are related by the formula

$$\Phi^*(\omega^1)(\mathcal{W}) = \omega^1(\Phi_*(\mathcal{W})) \circ \Phi^{-1} .$$

Theorem 33 then implies

$$
\begin{aligned}
\omega^1 (\Phi_*[\mathcal{V}, \mathcal{W}]) &= (\Phi^*\omega^1) ([\mathcal{V}, \mathcal{W}]) \circ \Phi \\
&= \mathcal{V} \left((\Phi^*\omega^1)(\mathcal{W}) \right) \circ \Phi - \mathcal{W} \left((\Phi^*\omega^1)(\mathcal{V}) \right) \circ \Phi - d(\Phi^*\omega^1)(\mathcal{V}, \mathcal{W}) \circ \Phi \\
&= (\Phi_*\mathcal{V}) \left(\omega^1(\Phi_*\mathcal{W}) \right) - (\Phi_*\mathcal{W}) \left(\omega^1(\Phi_*\mathcal{V}) \right) - \left(d\omega^1 \right) (\Phi_*\mathcal{V}, \Phi_*\mathcal{W}) \\
&= \omega^1 ([\Phi_*(\mathcal{V}), \Phi_*(\mathcal{W})]) .
\end{aligned}
$$

This equation holds for every 1-form ω^1, and hence the theorem follows. $\square$

Remark. Requiring that Φ be a diffeomorphism is often too strong. The compatibility of the commutator with the differential can sometimes be proved under weaker assumptions. For example, if the map $\Phi : M \to N$ is smooth, the vector field $\mathcal{V}$ on M is said to be Φ-*related* to the vector field $\mathcal{W}$ on N if at each point $m \in M$

$$\Phi_{*,m} \big(\mathcal{V}(m) \big) = \mathcal{W} \big(\Phi(m) \big).$$

If one can now show that, together with the pairs $(\mathcal{V}_1, \mathcal{W}_1)$ and $(\mathcal{V}_2, \mathcal{W}_2)$, the commutator pair $([\mathcal{V}_1, \mathcal{V}_2], [\mathcal{W}_1, \mathcal{W}_2])$ is necessarily also Φ-related, the same compatibility property follows immediately. To see this, apply the commutator $[\mathcal{W}_1, \mathcal{W}_2]$ to a function $g : N \to \mathbb{R}$:

$$\begin{aligned}[\mathcal{W}_1, \mathcal{W}_2]_{\Phi(m)}(g) &= \mathcal{W}_1\big(\Phi(m)\big)(\mathcal{W}_2\, g) - \mathcal{W}_2\big(\Phi(m)\big)(\mathcal{W}_1\, g) \\ &= \Phi_{*,m}\big(\mathcal{V}_1(m)\big)(\mathcal{W}_2\, g) - \Phi_{*,m}\big(\mathcal{V}_2(m)\big)(\mathcal{W}_1\, g) \\ &= \mathcal{V}_1(m)\big((\mathcal{W}_2\, g) \circ \Phi\big) - \mathcal{V}_2(m)\big((\mathcal{W}_1\, g) \circ \Phi\big).\end{aligned}$$

On the other hand, $(\mathcal{W}_i\, g) \circ \Phi = \mathcal{V}_i(g \circ \Phi)$, since

$$\mathcal{V}_i(n)(g \circ \Phi) \;=\; \Phi_{*,n}\big(\mathcal{V}_i(n)\big)(g) \;=\; \mathcal{W}_i\big(\Phi(n)\big)(g) \;=\; (\mathcal{W}_i\, g)\big(\Phi(n)\big).$$

Thus we obtain

$$[\mathcal{W}_1, \mathcal{W}_2]_{\Phi(m)}(g) \;=\; [\mathcal{V}_1, \mathcal{V}_2]_m(g \circ \Phi) \;=\; \Phi_{*,m}[\mathcal{V}_1, \mathcal{V}_2]_m(g),$$

i.ė., the vector fields $[\mathcal{W}_1, \mathcal{W}_2]$ and $[\mathcal{V}_1, \mathcal{V}_2]$ are, in fact, Φ-related.

The commutator of two vector fields measures the extent to which their flows do or do not commute. This explains the name for the vector field $[\mathcal{V}, \mathcal{W}]$.

Theorem 36. *Let $\mathcal{V}$ and $\mathcal{W}$ be two complete vector fields on the manifold M^k and denote by Φ_t and Ψ_s, respectively, their flows. Then the commutator $[\mathcal{V}, \mathcal{W}]$ vanishes if and only if $\Phi_t \circ \Psi_s = \Psi_s \circ \Phi_t$ for all $-\infty < s, t < \infty$.*

Proof. Because

$$\frac{d}{dt}((\phi_{-t})_*\mathcal{W})\big|_{t=t_1} \;=\; \lim_{h\to 0} \frac{(\Phi_{-t_1-h})_*\mathcal{W} - (\Phi_{-t_1})_*\mathcal{W}}{h} \;=\; (\Phi_{-t_1})_*([\mathcal{V}, \mathcal{W}]),$$

the commutator $[\mathcal{V}, \mathcal{W}]$ vanishes if and only if the vector field $\mathcal{W}$ is invariant under the flow Φ_t, $(\Phi_t)_*\mathcal{W} = \mathcal{W}$. This condition is in turn equivalent to the commutativity $\Phi_t \circ \Psi_s = \Psi_s \circ \Phi_t$ of the diffeomorphism Φ_t with the flow Ψ_s of $\mathcal{W}$. □

3.10. Harmonic Functions

A function $f : M^k \to \mathbb{R}$ is called *harmonic* if it is a solution of the homogeneous Laplace equation $\Delta(f) = 0$. As a special case of Hopf's theorem (Corollary 2), we have the following:

Theorem 37. *Every harmonic function on a compact, connected, and oriented manifold without boundary is constant.*

If the boundary of the manifold M^k is not empty, there exist two particularly important boundary value problems for harmonic functions.

The Dirichlet Problem: Assume that a function $\varphi : \partial M^k \to \mathbb{R}$ is given. We look for a harmonic function $f : M^k \to \mathbb{R}$ whose values on the boundary ∂M^k coincide with those of φ:

$$\Delta(f) = 0 \quad \text{in } M^k \quad \text{and} \quad f\,|_{\partial M^k} = \varphi\,.$$

The Neumann Problem: Assume that a function $\varphi : \partial M^k \to \mathbb{R}$ is given. We look for a harmonic function $f : M^k \to \mathbb{R}$ whose normal derivative on the boundary coincides with φ:

$$\Delta(f) = 0 \quad \text{in } M^k \quad \text{and} \quad \frac{\partial f}{\partial \mathcal{N}} = \varphi \quad \text{on } \partial M^k\,.$$

A solution of the Neumann problem is never unique. For each solution f the sum $f + C$ is, for an arbitrary constant C, a solution of this problem, too. This is the only degree of freedom, since we have

Theorem 38. *Let M^k be a compact, connected, and oriented manifold in $\mathbb{R}^n$, and let $\varphi : \partial M^k \to \mathbb{R}$ be a smooth function.*

(1) *The Dirichlet problem has at most one solution.*

(2) *If f_1, f_2 both are solutions of the Neumann problem, then $f_1 - f_2$ is constant.*

(3) *The vanishing of the mean value of φ is a necessary condition for the solvability of the Neumann problem:*

$$\int_{\partial M^k} \varphi \cdot d(\partial M^k) = 0\,.$$

Proof. If f_1, f_2 both are solutions of the Dirichlet problem, then the difference $u := f_1 - f_2$ satisfies the equations

$$\Delta(u) = 0 \quad \text{and} \quad u\,|_{\partial M^k} = 0\,.$$

From the first Green formula we obtain

$$0 = \int_{M^k} u \cdot \Delta(u) \cdot dM^k = -\int_{M^k} ||\text{grad}(u)||^2 \cdot dM^k + \int_{\partial M^k} u \cdot \frac{\partial u}{\partial \mathcal{N}} \cdot d(\partial M^k)\,,$$

and hence the gradient of u vanishes. Since we have $u\,|_{\partial M^k} = 0$, the function u has to vanish identically. In the case of the Neumann problem the argument runs along the same lines. i.e., starting from the equations

$$\Delta(u) = 0 \quad \text{and} \quad \frac{\partial u}{\partial \mathcal{N}} = 0 \quad \text{on} \quad \partial M^k$$

and the Green formula, we again conclude that $\text{grad}(u) = 0$. Thus the difference $u = f_1 - f_2$ is constant. If the Neumann problem has at least one solution for any given function $\varphi : \partial M^k \to \mathbb{R}$, then, applying the classical integral formulas, we obtain

$$0 = \int_{M^k} \Delta(f) \cdot dM^k = \int_{\partial M^k} \frac{\partial f}{\partial \mathcal{N}} \cdot d(\partial M^k) = \int_{\partial M^k} \varphi \cdot d(\partial M^k) . \square$$

The Dirichlet and the Neumann problems have, for a given boundary condition $\varphi : \partial M^k \to \mathbb{R}$ (satisfying $\int \varphi = 0$ in the case of the Neumann problem) a solution. We will not prove this existence result here for general manifolds, but confine ourselves to the case of the ball $D^n(R) \subset \mathbb{R}^n$ of radius $R > 0$. For these spaces, there is an explicit classical solution formula which will be derived here. For the sake of simplicity we only discuss the case of dimensions $n \geq 3$ and leave it to the reader to complete the discussion in dimension $n = 2$, which differs only slightly from the one below. We start with a few preparations. By

$$r = \sqrt{(x^1)^2 + \ldots + (x^n)^2}$$

we denote the distance from the point $(x^1, \ldots, x^n) \in \mathbb{R}^n$ to the point $0 \in \mathbb{R}^n$.

Lemma 3. *Let $u(x)$ be a harmonic function defined on the set $\mathbb{R}^n - \{0\}$, and assume that u depends only on the radius r. Then there exist constants C_1 and C_2 such that*

$$u(x) = C_1 + \frac{C_2}{r^{n-2}} .$$

Proof. By assumption there exists a smooth function $h : (0, \infty) \to \mathbb{R}$ satisfying $u(x) = h(r(x))$. Differentiating this, we obtain the formulas

$$\frac{\partial u}{\partial x^i} = h^{'}(r)\frac{x^i}{r}, \quad \frac{\partial^2 u}{\partial^2 x^i} = h^{''}(r)\Big(\frac{x^i}{r}\Big)^2 + h^{'}(r)\frac{r^2 - (x^i)^2}{r^3} ,$$

and the latter implies the following differential equation for the function $h(r)$:

$$0 = \Delta(u) = \sum_{i=1}^{n} \frac{\partial^2 u}{\partial^2 x^i} = h^{''}(r) + \frac{n-1}{r} h^{'}(r) .$$

For $n \geq 3$ the function $h(r) = C_1 + C_2 \cdot r^{2-n}$ is the solution of this ordinary differential equation. $\square$

Let $y \in \mathbb{R}^n$ be a fixed point. The function $u(x) = ||x - y||^{2-n}$ defined on the set $\mathbb{R}^n - \{y\}$ is a translate of r^{2-n}, and hence a harmonic function. On the other hand, the partial derivatives of a harmonic function are harmonic functions as well. Thus

$$u(x) - 2\sum_{i=1}^{n} \frac{\partial u}{\partial x^i} \cdot y^i = \frac{||x||^2 - ||y||^2}{||x - y||^n}$$

is a harmonic function defined on $\mathbb{R}^n - \{y\}$. We will use this family of harmonic functions in the observations to follow.

Lemma 4. *Let $||x|| < 1$ be any point in the interior of the unit ball D^n. Then*

$$\int_{S^{n-1}} \frac{1 - ||x||^2}{||x-y||^n} \cdot dS^{n-1}(y) = \mathrm{vol}(S^{n-1}).$$

Proof. First we prove that the function

$$h(x) := \int_{S^{n-1}} \frac{1 - ||x||^2}{||x-y||^n} \cdot dS^{n-1}(y)$$

depends only on the radius $r = r(x)$. In fact, for a linear orthogonal map $T : \mathbb{R}^n \to \mathbb{R}^n$ we obtain, from $|\det(DT)| = |\det(T)| = 1$ and the corresponding transformation formula for the volume form of the sphere,

$$T^*(dS^{n-1}) = T^*(\mathcal{N} \lrcorner\, d\mathbb{R}^n) = \mathcal{N} \lrcorner\, (T^*(d\mathbb{R}^n)) = \mathcal{N} \lrcorner\, d\mathbb{R}^n = dS^{n-1},$$

the property to be proved:

$$\begin{aligned} h(Tx) &= \int_{S^{n-1}} \frac{1 - ||Tx||^2}{||Tx-y||^n} \cdot dS^{n-1}(y) = \int_{S^{n-1}} \frac{1 - ||x||^2}{||x - T^{-1}(y)||^n} \cdot dS^{n-1}(y) \\ &= \int_{S^{n-1}} \frac{1 - ||x||^2}{||x-z||^n} \cdot dS^{n-1}(z) = h(x). \end{aligned}$$

If, in addition, $h(x)$ is a harmonic function, then

$$\Delta(h)(x) = \int_{S^{n-1}} \Delta_x \left(\frac{1 - ||x||^2}{||x-y||^n} \cdot dS^{n-1}(y) \right) = 0.$$

The first lemma implies that there are constants C_1, C_2 with $h(x) = C_1 + C_2 \cdot r^{2-n}$. At the point $x = 0$ the function $h(x)$ is a regular function and satisfies $h(0) = \mathrm{vol}(S^{n-1})$. Thus the constant C_2 vanishes, and C_1 is equal to $\mathrm{vol}(S^{n-1})$. $\square$

The solution of the Dirichlet problem for the unit ball $D^n \subset \mathbb{R}^n$ together with an explicit formula are the subjects of the next theorem.

Theorem 39. *For every continuous function $\varphi : S^{n-1} \to \mathbb{R}$, the formula*

$$f(x) = \frac{1}{\mathrm{vol}(S^{n-1})} \int_{S^{n-1}} \frac{1 - ||x||^2}{||x-y||^n} \cdot \varphi(y) \cdot dS^{n-1}(y)$$

defines a harmonic function in the interior of the unit ball, and at a boundary point $z \in S^{n-1}$

$$\lim_{t \to 1-} f(t \cdot z) = \varphi(z).$$

Proof. The function $(1 - ||x||^2)||x - y||^{-n}$ is harmonic with respect to the variable x, and hence the function $f(x)$ is harmonic, too. We prove that $f(z)$ coincides with $\varphi(z)$ on the boundary S^{n-1} in the way stated above. Denote by $m := \sup\{|\varphi(y)| : y \in S^{n-1}\}$ the maximum of the modulus of φ and fix a positive number $\varepsilon > 0$. The continuous function φ is uniformly continuous on the compact set S^{n-1}. Hence there exists a number $\delta > 0$ such that for any two points $y, z \in S^{n-1}$ in the sphere, $||y - z|| < \delta$ implies the estimate $|\varphi(y) - \varphi(z)| < \varepsilon$. We decompose the sphere $S^{n-1} = D_1 \cup D_2$ into the parts

$$D_1 = \{y \in S^{n-1} : ||y - z|| < \delta\}, \quad D_2 = \{y \in S^{n-1} : ||y - z|| \geq \delta\}.$$

For $y \in D_2$ and $0 \leq t \leq 1$ we estimate the distance from y to the line segment between 0 and $z \in S^{n-1}$:

$$\begin{aligned} ||y - tz||^2 &= 1 - 2t\langle y, z\rangle + t^2 = (t - \langle y, z\rangle)^2 + 1 - \langle y, z\rangle^2 \geq 1 - \langle y, z\rangle^2 \\ &\geq 1 - \langle y, z\rangle = \frac{1}{2}(2 - 2\langle y, z\rangle) = \frac{1}{2}||y - z||^2 \geq \frac{\delta^2}{2}. \end{aligned}$$

We then split the difference

$$f(t \cdot z) - \varphi(z) = \frac{1}{\mathrm{vol}(S^{n-1})} \int_{S^{n-1}} \frac{1 - ||tz||^2}{||tz - y||^n}(\varphi(y) - \varphi(z)) \cdot dS^{n-1}(y)$$

into the integrals over D_1 and D_2. The modulus of the first integral can be increased to

$$\Big| \int_{D_1} \Big| \leq \varepsilon \cdot \int_{S^{n-1}} \frac{1 - ||tz||^2}{||tz - y||^n} \cdot dS^{n-1}(y) = \varepsilon \cdot \mathrm{vol}(S^{n-1}).$$

We treat the second integral using the inequality obtained before:

$$\Big| \int_{D_2} \Big| \leq 2m \cdot (1 - t^2)\Big(\frac{\sqrt{2}}{\delta}\Big)^n \cdot \mathrm{vol}(S^{n-1}).$$

In summary, for every positive number $\varepsilon > 0$ there exists a number $\delta > 0$ such that for all $0 \leq t < 1$ the following inequality holds:

$$|f(t \cdot z) - \varphi(z)| \leq \varepsilon + 2m \left(\frac{\sqrt{2}}{\delta}\right)^n (1 - t^2).$$

Hence the upper limit is bounded by ε,

$$\lim_{t \to 1-} \sup |f(t \cdot z) - \varphi(z)| \leq \varepsilon.$$

The last inequality holds for all positive numbers $\varepsilon > 0$, and this in turn implies

$$\lim_{t \to 1-} f(t \cdot z) = \varphi(z). \qquad \square$$

Next we will discuss several consequences of the solution formula for the Dirichlet problem. For a harmonic function $f : \overline{D^n}(R) \to \mathbb{R}$ defined on the closed ball of radius $R > 0$, the function $\tilde{f}(z) := f(R \cdot z)$ is also harmonic on the unit ball. Applying the previous theorem and returning to the variable $x = R \cdot z \in D^n(R)$ finally leads to the *Poisson formula for harmonic functions*:

$$f(x) = \frac{R^{n-2}}{\mathrm{vol}(S^{n-1})} \int_{S^{n-1}} \frac{R^2 - ||x||^2}{||x - R \cdot y||^n} f(R \cdot y) \cdot dS^{n-1}(y) \, .$$

Evaluating the Poisson formula at the point $x = 0$ leads to *Gauss' mean value theorem for harmonic functions.*

Theorem 40. *The value of a harmonic function f at the center of the ball coincides with the mean value of the harmonic function on the boundary of the ball:*

$$f(0) = \frac{1}{\mathrm{vol}(S^{n-1})} \int_{S^{n-1}} f(R \cdot y) \cdot dS^{n-1}(y) \, .$$

We will use Gauss' mean value theorem in the proof of the *maximum principle for harmonic functions.*

Theorem 41. *Every harmonic function f on the connected manifold $M^n \subset \mathbb{R}^n$ attaining its maximal value in the interior of M^n is constant.*

Proof. Denote by m the maximal value of f and by $\Omega = \{x \in M^n \backslash \partial M^n : f(x) = m\}$ the set of all inner points of M^n at which f attains this maximal value. By assumption Ω is a non-empty closed set in $M^n \backslash \partial M^n$. Hence it suffices to prove that Ω is an open subset of $M^n \backslash \partial M^n$. Choose a point $x_0 \in \Omega$ and a radius R_0 such that the ball $D^n(x_0, R_0)$ with center x_0 and radius R_0 is completely contained in $M^n \backslash \partial M^n$. By Gauss' mean value theorem

$$f(x_0) = \frac{1}{\mathrm{vol}(S^{n-1})} \int_{S^{n-1}} f(x_0 + R_0 \cdot y) \cdot dS^{n-1}(y) \leq m = f(x_0) \, .$$

This implies that f is constant and equal to m on the sphere with center x_0 and radius R_0. This observation can also be applied to each radius $R_0^* \leq R_0$ below R_0. Together this implies that $f \equiv m$ is constant on the ball $D^n(x_0, R_0)$, i.e., $D^n(x_0, R_0)$ is contained in Ω. Thus Ω is an open subset. $\square$

Eventually we will prove one more application of the Poisson formula, *Liouville's theorem for harmonic functions.*

Theorem 42. *Every harmonic function $f : \mathbb{R}^n \to \mathbb{R}$ bounded from below (above) is constant.*

Proof. Changing f, if necessary, by adding a constant, we can suppose without loss of generality that the function f is non-negative. Fix a point $x_0 \in \mathbb{R}^n$ and choose the radius R such that x_0 lies in the ball $D^n(0,R)$. By the Poisson formula, Gauss' mean value theorem, and the assumption $f \geq 0$ we have

$$\begin{aligned} f(x_0) &= \frac{R^{n-2}}{\mathrm{vol}(S^{n-1})}\int_{S^{n-1}} \frac{R^2-||x_0||^2}{||x_0-R\cdot y||^n}\cdot f(R\cdot y)\cdot dS^{n-1}(y) \\ &\leq \frac{R^{n-2}}{\mathrm{vol}(S^{n-1})}\int_{S^{n-1}} \frac{R^2-||x_0||^2}{\big|||x_0||-||R\cdot y||\big|^n}\cdot f(R\cdot y)\cdot dS^{n-1}(y) \\ &= \frac{R^{n-2}(R^2-||x_0||^2)}{\mathrm{vol}(S^{n-1})\big|||x_0||-R\big|^n}\int_{S^{n-1}} f(R\cdot y)\cdot dS^{n-1}(y) \\ &= \frac{R^{n-2}(R^2-||x_0||^2)}{\mathrm{vol}(S^{n-1})\big|||x_0||-R\big|^n}\cdot f(0)\,. \end{aligned}$$

Taking the limit for $R \to \infty$ yields for all points $x_0 \in \mathbb{R}^n$ the estimate

$$f(x_0) \leq f(0)\,,$$

and the maximum principle for harmonic functions implies that f is constant. □

3.11. The Laplacian on Differential Forms

This section is a supplement to the vector analysis on manifolds as explained before. We will discuss the Laplacian acting on forms and its properties. To this end, we will start from an oriented manifold M^m of dimension m together with a family of non-degenerate scalar products

$$g_x : T_xM^m \times T_xM^m \to \mathbb{R}$$

and assume that these scalar products depend smoothly on the point. This is understood to mean that for any two smooth vector fields $\mathcal{V}$ and $\mathcal{W}$ on M^m the function $g(\mathcal{V},\mathcal{W})$ is smooth. Such a family $\{g_x\}_{x\in M^m}$ is called a *pseudo-Riemannian metric* on M^m. In general, it has to be neither positive definite nor the restriction of the scalar product on the ambient euclidean space.

In this situation the tangent space is a real oriented vector space, and we denote by k the index of the scalar product. According to Chapter 1, at each point there exist a volume form $dM^m(x) \in \bigwedge_x^m(M^m)$ and a Hodge operator

$$* : \textstyle\bigwedge_x^i(M^m) \to \bigwedge_x^{m-i}(M^m)\,.$$

Together we obtain an m-form dM^m on the manifold, as well as a $*$-operator associating with every i-form a smooth $(m-i)$-form. The algebraic rules from Chapter 1 can still be applied. We will use these and Stokes' formula to determine the operator $\delta : \Omega^{i+1}(M^m) \to \Omega^i(M^m)$ adjoint to the exterior derivative $d : \Omega^i(M^m) \to \Omega^{i+1}(M^m)$ acting on i-forms. As is well-known, the *adjoint operator* is characterized by the requirement

$$\int_{M^m} (d\omega^i, \eta^{i+1}) \cdot dM^m \;=\; \int_{M^m} (\omega^i, \delta\eta^{i+1}) \cdot dM^m \,.$$

Here ω^i and η^{i+1} are forms with compact support disjoint from the boundary of M^m, and $(d\omega^i, \eta^{i+1})$ denotes the scalar product induced on forms. By Theorem 5 in Chapter 1 we have

$$\begin{aligned}(d\omega^i, \eta^{i+1}) \cdot dM^m &= (-1)^k d\omega^i \wedge (*\eta^{i+1}) \\ &= (-1)^k d(\omega^i \wedge (*\eta^{i+1})) + (-1)^{k+i+1}\omega^i \wedge (d * \eta^{i+1}) \,.\end{aligned}$$

Integrating this equation and once again applying the algebraic rules, we obtain

$$\begin{aligned}\int_{M^m} (d\omega^i, \eta^{i+1}) \cdot dM^m &= (-1)^{k+i+1}(-1)^{i(m-i)+k} \int_{M^m} \omega^i \wedge ** (d * \eta^{i+1}) \\ &= (-1)^{i(m-i)+i+1+k} \int_{M^m} (\omega^i, *d * \eta^{i+1}) \cdot dM^m \,.\end{aligned}$$

Thus we have proved

Theorem 43. *The operator δ adjoint to the exterior derivative $d : \Omega^i(M^m) \to \Omega^{i+1}(M^m)$ on a pseudo-Riemannian manifold of index k is given by*

$$\delta(\eta^{i+1}) = (-1)^{k+mi+1} * d * \eta^{i+1} \,.$$

Definition 22. Let M^m be a pseudo-Riemannian manifold. The *Hodge-Laplace operator* acting on i-forms, $\Delta : \Omega^i(M^m) \to \Omega^i(M^m)$, is defined by

$$\Delta \;:=\; d\delta + \delta d \,.$$

Example 35. For a function $\omega^0 = f$, in the sense of this definition,

$$\Delta f \;=\; \delta df \;=\; (-1)^{k+1} * d * df \,,$$

and, in the case of a positive definite scalar product, Δ coincides with the Laplace-Beltrami operator (see the remark following Theorem 11, §3.2). Considering, on the other hand, the pseudo-euclidean scalar product of signature $(n, 1)$ in $\mathbb{R}^{n+1}$, we obtain for the Hodge-Laplace operator the expression

$$\Delta f \;=\; \frac{\partial^2 f}{\partial^2 x^1} + \ldots + \frac{\partial^2 f}{\partial^2 x^n} - \frac{\partial^2 f}{\partial^2 x^{n+1}} \,.$$

$\mathbb{R}^{n+1}$ with this pseudo-Riemannian metric is called *Minkowski space.* Hence, its Laplacian on functions is the wave operator (also known as the *d'Alembertian*).

Definition 23. An i-form ω^i defined on a pseudo-Riemannian manifold is called *harmonic* if $\Delta(\omega^i) = 0$; it is called *co-closed* if $\delta\omega^i = 0$.

In the following theorem we collect the properties of the Hodge-Laplace operator immediately resulting from the definition.

Theorem 44.

(1) *Let ω^i, η^i be forms with compact support disjoint from the boundary of the manifold. Then*

$$\int_{M^m} (\Delta(\omega^i), \eta^i) \cdot dM^m = \int_{M^m} (\omega^i, \Delta(\eta^i)) \cdot dM^m .$$

(2) $\Delta \circ d = d \circ \Delta$ *and* $\Delta \circ \delta = \delta \circ \Delta$.

Finally, consider a compact manifold without boundary endowed with a Riemannian metric ($k = 0$). Then the second order differential equation $\Delta(\omega^i) = 0$ is equivalent to two first order differential equations.

Theorem 45. *Let M^m be a compact, oriented manifold without boundary, endowed with a positive definite metric. Then the following conditions are equivalent:*

(1) *ω^i is a harmonic form, $\Delta(\omega^i) = 0$;*

(2) *ω^i is a closed and co-closed form, $d\omega^i = 0 = \delta\omega^i$.*

Proof. If ω^i is harmonic, then

$$0 = \int_{M^m} (\omega^i, \Delta(\omega^i)) \cdot dM^m = \int_{M^m} (d\omega^i, d\omega^i) \cdot dM^m + \int_{M^m} (\delta\omega^i, \delta\omega^i) \cdot dM^m .$$

This implies $d\omega^i = 0 = \delta\omega^i$, since the metric is positive definite. □

Corollary 4. *Let M^m be a compact, oriented manifold without boundary, endowed with a positive definite Riemannian metric. If every closed 1-form on M^m is exact, then there exist no non-trivial harmonic forms.*

Proof. In particular, a harmonic 1-form ω^1 is closed, and hence ω^1 can be represented as the differential of a function, $\omega^1 = df$. Since

$$0 = \delta\omega^1 = \delta df = \Delta(f) ,$$

f is a harmonic function, and by Hopf's theorem (Corollary 2) it is constant on each connected component of M^m. Therefore, its differential vanishes, $\omega^1 = df = 0$. □

Example 36. The assumption of this corollary is, for example, satisfied for manifolds M^m which can be represented as the union $M^m = U \cup V$ of two open sets $U, V \subset M^m$, where

(1) the intersection $U \cap V$ is connected,

(2) U and V are diffeomorphic to star-shaped domains in $\mathbb{R}^m$.

In fact, Poincaré's lemma implies the existence of two functions $f_U : U \to \mathbb{R}$ and $f_V : V \to \mathbb{R}$ such that $\omega^1 = df_U$ over U and $\omega^1 = df_V$ over V. On the intersection we then have

$$d(f_U - f_V) \;=\; 0\,,$$

and hence $f_U - f_V = C$ is constant. Now

$$f(x) \;=\; \begin{cases} f_V(x) + C & \text{if } x \in V\,, \\ f_U(x) & \text{if } x \in U\,, \end{cases}$$

is a uniquely defined function on M^m satisfying $\omega^1 = df$. For $n \geq 2$ the sphere S^n allows a decomposition of the required shape, and hence S^n has no harmonic 1-form with respect to any Riemannian metric.

Example 37. Consider the 2-dimensional torus in $\mathbb{R}^4$,

$$T^2 \;=\; \{(x^1, x^2, x^3, x^4) \in \mathbb{R}^4 : (x^1)^2 + (x^2)^2 = 1, \quad (x^3)^2 + (x^4)^2 = 1\}\,,$$

and denote by $i : T^2 \to \mathbb{R}^4$ its embedding into $\mathbb{R}^4$. The 1-forms induced on T^2

$$\omega^1 \;:=\; i^*(-x^2 dx^1 + x^1 dx^2) \quad \text{and} \quad \eta^1 \;:=\; i^*(-x^4 dx^3 + x^3 dx^4)$$

are harmonic. To prove this, we use the parametrization $h\,:\,[0, 2\pi] \times [0, 2\pi] \to T^2$ defined by the formula

$$h(\varphi, \psi) \;=\; (\cos\varphi, \sin\varphi, \cos\psi, \sin\psi)\,.$$

In this chart,

$$h^*(\omega^1) \;=\; d\varphi \quad \text{and} \quad h^*(\eta^1) \;=\; d\psi\,,$$

and hence ω^1 and η^1 are closed forms on T^2. Computing the coefficients of the Riemannian metric leads to the matrix

$$\begin{bmatrix} g_{\varphi\varphi} & g_{\varphi\psi} \\ g_{\psi\varphi} & g_{\psi\psi} \end{bmatrix} = \begin{bmatrix} 1 & 0 \\ 0 & 1 \end{bmatrix}.$$

From this, it is easy to see that ω^1 and η^1 correspond to each other via the Hodge operator of the torus T^2:

$$*\omega^1 \;=\; \eta^1, \quad *\eta^1 \;=\; -\omega^1\,.$$

But then, ω^1 and η^1 are also co-closed, $\delta\omega^1 = *d*\omega^1 = *d\eta^1 = 0$.

Harmonic forms play an essential role in topology and global differential geometry. Here we formulate the main result of Hodge theory; a proof would go beyond the scope of this elementary text. For a compact, oriented Riemannian manifold without boundary, (M^m, g), we introduce the following vector spaces:

(1) the space $\mathcal{H}^i(M^m, g)$ of all harmonic i-forms,

$$\mathcal{H}^i(M^m, g) \ = \ \{\omega^i : \Delta(\omega^i) = 0\} \, ;$$

(2) the space $\mathcal{E}^i(M^m)$ of exact differential forms,

$$\mathcal{E}^i(M^m) \ = \ \{\omega^i : \text{ there exists an } (i-1)\text{-form } \eta^{i-1} \text{ such that } d\eta^{i-1} = \omega^i\} \, ;$$

(3) the space $\mathcal{CE}^i(M^m, g)$ of co-exact differential forms

$$\mathcal{CE}^i(M^m, g) = \{\omega^i : \text{there exists an } (i+1)\text{-form } \eta^{i+1} \text{ such that } \delta\eta^{i+1} = \omega^i\}.$$

Theorem 46 (Hodge's Theorem).

(1) *The space $\mathcal{H}^i(M^m, g)$ is a finite-dimensional vector space. It is isomorphic to the de Rham cohomology $H^i_{DR}(M^m)$. For a fixed metric, each cohomology class of closed forms contains precisely one harmonic form.*

(2) *With respect to the L^2-scalar product, the spaces $\mathcal{H}^i, \mathcal{E}^i$ and $\mathcal{CE}^i$ are orthogonal subspaces of the space $\Omega^i(M^m)$ of all square-integrable i-forms and decompose it into the sum*

$$\Omega^i \ = \ \mathcal{H}^i \oplus \mathcal{E}^i \oplus \mathcal{CE}^i \, .$$

From the Hodge decomposition we immediately obtain *Helmholtz's theorem* for 3-dimensional compact manifolds without boundary. Geometrically speaking, it states that each vector field can be represented as the sum of a gradient field, a curl field, and a harmonic field. In particular, the vector field is the sum of a curl-free and a divergence-free vector field. In the final chapter we will discuss and prove an analogous result for vector fields on $\mathbb{R}^3$ which decrease sufficiently rapidly at infinity within the framework of electrodynamics in detail.

Corollary 5. *For each vector field $\mathcal{V}$ on an oriented, compact, 3-dimensional manifold M^3 without boundary, there exist a function f, a vector field $\mathcal{W}$ and a harmonic 1-form η^1 such that*

$$\omega^1_{\mathcal{V}} \ = \ \omega^1_{\mathrm{grad}(f)} + \omega^1_{\mathrm{curl}(\mathcal{W})} + \eta^1 \, .$$

In particular, $\omega^1_{\mathcal{V}}$ is the sum of a closed and a co-closed 1-form.

Proof. Applying Hodge's theorem to the 1-form associated with the vector field $\mathcal{V}$, we conclude that there exist a function f, a 2-form β^2 and a harmonic 1-form η^1, together satisfying the equation

$$\omega^1_{\mathcal{V}} \ = \ df + \delta\beta^2 + \eta^1 \, .$$

From §2.3, we know that the gradient of a function corresponds to the differential df,

$$df \ = \ \omega^1_{\mathrm{grad}(f)} \, .$$

Let $\mathcal{W}$ be the vector field corresponding to the 1-form $*\beta^2$,

$$*\beta^2 \;=:\; \omega^1_{\mathcal{W}} .$$

The curl of the vector field $\mathcal{W}$ is the vector field determined by the equation

$$d\omega^1_{\mathcal{W}} \;=\; *\omega^1_{\mathrm{curl}(\mathcal{W})} .$$

But this implies

$$\delta\beta^2 \;=\; *d*\beta^2 \;=\; *d\omega^1_{\mathcal{W}} \;=\; **\,\omega^1_{\mathrm{curl}(\mathcal{W})} \;=\; \omega^1_{\mathrm{curl}(\mathcal{W})} ,$$

since in dimension $m = 3$ for a positive definite Riemannian metric the relation $** = \mathrm{Id}$ holds on the space of 1-forms. This completes the proof. □

Exercises

1. a) The equation

$$\frac{(x^1)^2}{a_1^2} + \ldots + \frac{(x^n)^2}{a_n^2} = 1, \qquad \text{where } a_1, \ldots, a_n > 0 \text{ are constants},$$

defines an $(n-1)$-dimensional submanifold of $\mathbb{R}^n$, the ellipsoid.

b) The equation $x^2 + y^2 = r^2$ ($r > 0$ fixed) defines a two-dimensional submanifold of $\mathbb{R}^3$, the cylinder.

c) Consider for every $c \in \mathbb{R}$ the subset of $\mathbb{R}^3$ defined by

$$M_c \;=\; \left\{ (x, y, z) \in \mathbb{R}^3 :\; x^2 + y^2 - z^2 = c \right\} .$$

For which values of the parameter c is M_c a two-dimensional submanifold of $\mathbb{R}^3$?

2. Prove or disprove the converse of Theorem 1: Let $U \subset \mathbb{R}^n$ be open, let $f : U \to \mathbb{R}^{n-k}$ be smooth, and denote by M the zero-set of f, $M := \{x \in U : f(x) = 0\}$. If the estimate $\mathrm{rk}\, Df(x) < n-k$ holds for at least one point $x \in M$, then M is not a submanifold.

3. Prove that the formula

$$x^2 + y^2 \;=\; a^2 \cosh^2(z/a), \quad a > 0 \text{ constant} ,$$

defines a two-dimensional submanifold of $\mathbb{R}^3$, which is, moreover, parametrized by

$$h(v^1, v^2) \;=\; \left(a\cosh(v^1/a)\cos v^2,\, a\cosh(v^1/a)\sin v^2,\, v^1\right) .$$

This surface is called the *catenoid* in differential geometry.

4. Let $f : \mathbb{C} \to \mathbb{C}$ be a complex polynomial $f = \sum_{i=0}^{k} a_i z^i$ without double zeroes. Consider for every natural number $l \geq 2$ the set

$$M = \{ (z, w) \in \mathbb{C}^2 : w^l - f(z) = 0 \}.$$

Prove that M is a two-dimensional submanifold of $\mathbb{C}^2 \cong \mathbb{R}^4$. The assignment $(z, w) \mapsto w$ uniquely defines a map $G : M \to \mathbb{C}$ satisfying $G^l(z, w) = f(z)$. The set M is a so-called *Riemann surface.* The l-th root of $f(z)$ is uniquely defined on it.

5. Consider the *special linear group* $\mathrm{SL}(2, \mathbb{R})$ of all real-valued (2×2) matrices of determinant 1:

$$\mathrm{SL}(2, \mathbb{R}) = \{ A \in \mathcal{M}_2(\mathbb{R}) : \det A = 1 \}.$$

Prove that $\mathrm{SL}(2, \mathbb{R})$ is a three-dimensional manifold, and that its tangent space at the neutral element E is determined by

$$T_E \, \mathrm{SL}(2, \mathbb{R}) = \{ A \in \mathcal{M}_2(\mathbb{R}) : \operatorname{tr} A = 0 \}.$$

Hint: Apply Theorem 4 and the (well-known) formula $\det(\exp A) = e^{\operatorname{tr} A}$.

6. Prove that the *orthogonal group* $\mathrm{O}(n, \mathbb{R})$ consisiting of all orthogonal $(n \times n)$ matrices

$$\mathrm{O}(n, \mathbb{R}) = \{ A \in \mathcal{M}_n(\mathbb{R}) : A \cdot A^t = E \}$$

is a submanifold of $\mathbb{R}^{n^2}$, and compute its dimension. Is the subgroup $\mathrm{SO}(n, \mathbb{R})$ of all orthogonal matrices of determinant 1 also a submanifold? If it is, compute its dimension, and describe its relation to $\mathrm{O}(n, \mathbb{R})$.

7. The set of all unit tangent vectors $T_1 S^2$ to the sphere S^2,

$$T_1 S^2 = \{ (x, v) \in \mathbb{R}^3 \times \mathbb{R}^3 : ||x|| = ||v|| = 1, \ \langle x, v \rangle = 0 \},$$

is a submanifold of $\mathbb{R}^3 \times \mathbb{R}^3$ which is diffeomorphic to the manifold

$$\{ (z^1, z^2, z^3) \in \mathbb{C}^3 : |z^1|^2 + |z^2|^2 + |z^3|^2 = 1, \ (z^1)^2 + (z^2)^2 + (z^3)^2 = 0 \}.$$

8. The product of a manifold M without boundary with an arbitrary manifold N again is a manifold, and it has the boundary $\partial (M \times N) = M \times \partial N$.

9. The tangent bundle TM of every manifold M is always an orientable manifold.

10. The topological boundary $\mathrm{Fr}(A)$ of a subset $A \subset \mathbb{R}^n$ is the intersection of the closure of A with the closure of its complement,

$$\mathrm{Fr}(A) \; := \; \overline{A} \cap \overline{\mathbb{R}^n \backslash A}\,.$$

Prove:

a) If $M^m \subset \mathbb{R}^m$ is a closed m-dimensional submanifold of $\mathbb{R}^m$, then $\partial M^m = \mathrm{Fr}(M^m)$.

b) If $M^m \subset \mathbb{R}^n$ $(m < n)$ is a submanifold with or without boundary of strictly smaller dimension, then always $\mathrm{Fr}(M^m) = M^m$.

11. Find three vector fields of length one on the sphere S^3 which are mutually othogonal at each point.

12. Construct a vector field of length one on each sphere of odd dimension.

13. Compute the representation of the following vector field in polar coordinates:

$$\mathcal{V}(x, y) \;=\; \sqrt{x^2 + y^2}\;(\partial/\partial x + \partial/\partial y)\;.$$

14. Consider on the sphere S^2 the vector field

$$\mathcal{V}(x, y, z) \;=\; ((x, y, z), (-y, x, 0))\,.$$

Prove that $\mathcal{V}$ cannot be the gradient field of any function $f : S^2 \to \mathbb{R}$. On which more general fact is this example based?

15. Compute the local coefficients (g_{ij}) of the Riemannian metric as well as the volume form of the following submanifolds in the respective charts:

a) for the sphere S^n in the chart determined by the stereographic projection $h : \mathbb{R}^n \to S^n$:

$$h(y^1, \ldots, y^n) \;=\; \left(\frac{2\,y^1}{|y|^2 + 1}, \ldots, \frac{2\,y^n}{|y|^2 + 1}, \frac{|y|^2 - 1}{|y|^2 + 1}\right)\;;$$

b) for the pseudosphere, i.e., the surface of revolution defined by the tractrix in the parametrization

$$h(u^1, u^2) \;=\; \left(a \sin u^1 \cos u^2,\; a \sin u^1 \sin u^2,\; a(\cos u^1 + \ln \tan(u^1/2))\right)\;.$$

16. Compute the formulas for the gradient of a function, for the divergence of a vector field, and for the Laplacian on the sphere S^2 in the coordinates defined by stereographic projection.

17. On an open subset of $\mathbb{R}^3$ (to be determined!) we introduce coordinates by the chart

$$h(r,\varphi,\psi) \;=\; (r\cos\varphi\cos\psi, r\sin\varphi\cos\psi, r\sin\varphi)$$

on the parameter domain $r \in (0,\infty)$, $\varphi \in [0,2\pi]$, $\psi \in (-\psi/2,\psi/2)$. Compute the coefficients of the Riemannian metric, and the formulas for the gradient, the divergence, and the Laplacian in these coordinates.

18. Let M^k and N^k be k-dimensional manifolds. The map $f : M^k \to N^k$ is called *angle-preserving* or *conformal* if, for two arbitrary vectors $w_1, w_2 \in T_xM^k$ and their image vectors $\nu_i = f_{*,x}(w_i) \in T_{f(y)}N^k$, the respective angles coincide:

$$\frac{\langle \nu_1,\nu_2\rangle}{||\nu_1||\cdot||\nu_2||} \;=\; \frac{\langle w_1,w_2\rangle}{||w_1||\cdot||w_2||}\,.$$

If, in particular, (h,U) is a chart of the manifold M^k, the chart h is angle-preserving if and only if the representation of the Riemannian metric in the corresponding coordinates is a multiple of the k-dimensional unit matrix E:

$$\big(g_{ij}(y)\big) \;=\; \lambda(y)\cdot E, \quad \lambda(y) \;>\; 0\,.$$

19. Prove the orientability of the sphere S^n, using stereographic projection and Theorem 14.

20. Let M^k be a manifold, $h : U \to M^k$ a chart, and $\gamma : [a,b] \to h(U) \subset M^k$ a curve in M^k completely contained in the image set of h. Represent the curve γ in the coordinates (h,U) as $h^{-1}\circ\gamma(t) = (y^1(t),\ldots,y^k(t))$. Prove that the length of the curve can be computed by means of the coefficients of the Riemannian metric g_{ij} in the chart (h,U) via the formula

$$l(\gamma) \;=\; \int\limits_a^b \left[\sum_{i,j=1}^{k} g_{ij}(\gamma(t))\frac{dy^i(t)}{dt}\,\frac{dy^j(t)}{dt}\right]^{1/2} dt\,.$$

21. Let $f : [a,b] \to \mathbb{R}_+$ be a positive function and let $M^2 = \{(x,y,z) \in \mathbb{R}^3 : y^2+z^2 = f^2(x)\}$ be the corresponding surface of revolution in $\mathbb{R}^3$. Prove the volume formula

$$\mathrm{vol}(M^2) \;=\; 2\pi\int_a^b f(x)\sqrt{1+(f'(x))^2}\,dx\,.$$

22. Compute the following surface integrals:

a) $\displaystyle\int_{M^2}(x+y+z)\cdot dM^2$ for $M^2 = \big\{(x,y,z \in \mathbb{R}^3) : x^2+y^2+z^2 = a^2, z \geq 0\big\}$;

b) $\displaystyle\int_{M^2} (x^2+y^2)\cdot dM^2$, where M^2 is the boundary of the subset of $\mathbb{R}^3$ described by the inequality $\sqrt{x^2+y^2} \leq z \leq 1$.

23. Compute the following surface integrals:

a) $\displaystyle\int_{S^2(R)} (x\,dy\wedge dz + y\,dz\wedge dx + z\,dx\wedge dy)$;

b) $\displaystyle\int_{M^2} [(y-z)dy\wedge dz + (z-x)dz\wedge dx + (x-y)dx\wedge dy]$, where M^2 is the boundary of the subset of $\mathbb{R}^3$ described by the inequalities $x^2+y^2 \leq z^2$, $0 \leq z \leq 1$.

24. Let the 1-form ω^1 on the Riemannian manifold M^k be given in the chart $h: V \to M^k$ by $\omega^1 = \sum_i f_i\,dy^i$. Prove the following formula for the adjoint operator:

$$\delta(\omega^1) \;=\; \frac{-1}{\sqrt{g}} \sum_{i,j=1}^{k} \frac{\partial}{\partial y^i}\left(\sqrt{g}\,g^{ij} f_j\right).$$

25. Let M^k be a compact, oriented manifold without boundary. A function $f: M^k \to \mathbb{R}$ is called an *eigenfunction* of the Laplacian with respect to the eigenvalue $\lambda \in \mathbb{R}$, if $\Delta f = \lambda f$. Prove:

a) If f_1 and f_2 are eigenfunctions with respect to different eigenvalues, $\lambda_1 \neq \lambda_2$, then

$$\langle f_1, f_2\rangle_{L^2} \;=\; \int_{M^k} f_1(x) f_2(x)\,dM^k \;=\; 0.$$

b) Let $S^2 \subset \mathbb{R}^3$ be the sphere of radius 1, $\Delta^{\mathbb{R}^3}$ the Laplacian on $\mathbb{R}^3$, Δ^{S^2} the Laplacian on S^2, $f: \mathbb{R}^3 \to \mathbb{R}$ a smooth function, and set $r = \sqrt{x^2+y^2+z^2}$. Then

$$\left(\Delta^{\mathbb{R}^3} f\right)\big|_{S^2} \;=\; \frac{\partial^2 f}{(\partial r)^2} + 2\frac{\partial f}{\partial r} + \Delta^{S^2}(f|_{S^2}).$$

c) Let $P: \mathbb{R}^3 \to \mathbb{R}$ be a homogeneous polynomial of degree m which is harmonic, i.e. P satisfies $\Delta^{\mathbb{R}^3} P(x) = 0$. Prove that $P|_{S^2}$ is an eigenfunction of the Laplacian on S^2, and determine the corresponding eigenvalue.

26. Let $M^k \subset \mathbb{R}^n$ be a compact, oriented manifold, and assume that $f: M^k \times [0,\infty) \to \mathbb{R}$ is smooth. The heat equation is

$$\Delta_x f(x,t) \;=\; \frac{\partial f(x,t)}{\partial t}.$$

Prove that if f is a solution of the heat equation satisfying $f(x,0)=0$ for all $x \in M^k$ and $f(y,t)=0$ for all boundary points $y \in \partial M^k$ and all times $t \in [0,t_0]$, then f vanishes identically on the set $M^k \times [0,t_0]$.

27. Let M^{n-1} be the boundary of a compact n-dimensional submanifold of $\mathbb{R}^n$, and $\mathcal{N}(x)$ the exterior unit normal vector field. If $\mathcal{V} \in \mathbb{R}^n$ is a constant vector, then

$$\int_{M^{n-1}} \langle \mathcal{N}(x), \mathcal{V} \rangle \cdot dM^{n-1} = 0 .$$

28. Prove the following commutator relations for the Lie derivative:

$$[\mathcal{L}_{\mathcal{V}}, i_{\mathcal{W}}] = i_{[\mathcal{V},\mathcal{W}]} \quad \text{and} \quad [\mathcal{L}_{\mathcal{V}}, \mathcal{L}_{\mathcal{W}}] = \mathcal{L}_{[\mathcal{V},\mathcal{W}]} \, .$$

Chapter 4

Pfaffian Systems

4.1. Geometric Distributions

For smooth functions $f_1, \dots, f_{m-k} : \mathbb{R}^m \to \mathbb{R}$ with linearly independent differentials, the equations

$$M^k_{c_1, \dots, c_{m-k}} = \{x \in \mathbb{R}^m : f_1(x) = c_1, \dots, f_{m-k}(x) = c_{m-k}\}$$

define a smooth k-dimensional manifold. Linearizing this, in general nonlinear, system of equations by passing to the tangent bundle, we see that this manifold is described by the system

$$TM^k_{c_1, \dots, c_{m-k}} = \{\nu \in T\mathbb{R}^m : df_1(\nu) = 0, \dots, df_{m-k}(\nu) = 0\}.$$

These equations determine at each point of $\mathbb{R}^m$ a k-dimensional subspace of the tangent space to $\mathbb{R}^m$. The resulting family of subspaces can also be described by different systems of 1-forms $\omega_1, \dots, \omega_{m-k}$. For example, if (h_{ij}) is a matrix of functions with nowhere vanishing determinant, the 1-forms $\omega_i = \sum_{j=1}^{m-k} h_{ij} \cdot df_j$ satisfy

$$TM^k_{c_1, \dots, c_{m-k}} = \{\nu \in T\mathbb{R}^m : \omega_1(\nu) = 0, \dots, \omega_{m-k}(\nu) = 0\}.$$

The level surfaces, however, cannot be recovered from the knowledge of the forms ω_i alone. In general, the problem arises under which conditions linearly independent 1-forms $\omega_1, \dots, \omega_{m-k}$ describe a family of k-dimensional manifolds via the system of equations (Pfaffian system)

$$\omega_1 = \dots = \omega_{m-k} = 0.$$

Frobenius' theorem provides a complete answer to this question. This chapter is devoted to its proof and the discussion of various applications. In

Chapter 5, we will apply it to prove the fundamental theorems of curve and surface theory.

Definition 1. A k-dimensional *geometric distribution* (or *Pfaffian system*) on M^m is a family $\mathcal{E}^k = \{E^k(x)\}$ consisting of k-dimensional subspaces $E^k(x) \subset T_xM^m$ in the tangent spaces to M^m depending smoothly on the points in the following sense: For each point $x_0 \in M^m$, there exist a neighborhood $x_0 \in U \subset M^m$ and vector fields $\mathcal{V}_1, \ldots, \mathcal{V}_k$ defined on U such that $E^k(x)$ coincides with the linear hull of the vectors $\mathcal{V}_1(x), \ldots, \mathcal{V}_k(x)$ at every point $x \in U$.

Example 1. Every nowhere vanishing vector field $\mathcal{V}$ on M^m induces a one-dimensional distribution $E^1(x)$ formed by all multiples of the vector $\mathcal{V}(x)$. Conversely, every one-dimensional distribution $\mathcal{E}^1$ is locally determined by a nowhere vanishing vector field.

Example 2. The linearly independent 1-forms $\omega_1, \ldots, \omega_{m-k}$ on M^m determine a k-dimensional distribution by

$$E^k(x) \;=\; \{\nu \in T_xM^m : \; \omega_1(\nu) = \ldots = \omega_{m-k}(\nu) = 0\}\,.$$

In analogy to the integral curve for a vector field, we now introduce the notion of an integral manifold for a distribution.

Definition 2. Let $\mathcal{E}^k$ be a k-dimensional distribution on M^m. A k-dimensional submanifold $N^k \subset M^m$ is called an *integral manifold* of $\mathcal{E}^k$ if the tangent spaces of N^k coincide with the spaces of the distribution:

$$T_xN^k \;=\; E^k(x) \quad \text{for all } x \in N^k\,.$$

Definition 3. The k-dimensional distribution $\mathcal{E}^k$ is called *integrable* if it admits at least one integral manifold through each point $x \in M^m$.

Example 3. For a distribution $\mathcal{E}^k$ defined by linearly independent 1-forms $\omega_1, \ldots, \omega_{m-k}$ as in Example 2, a submanifold $i : N^k \to M^m$ is an integral manifold of $\mathcal{E}^k$ if and only if the restrictions of the forms to N^k vanish,

$$i^*(\omega_1) \;=\; \ldots \;=\; i^*(\omega_{m-k}) \;=\; 0\,.$$

The local existence theorem for solutions of ordinary differential equations can be formulated as follows:

Proposition 1. *Every one-dimensional geometric distribution is integrable.*

Example 4. Consider the nowhere vanishing 1-form $\omega = x \cdot dy + dz$ on $\mathbb{R}^3$ together with the 2-dimensional distribution determined by ω,

$$E^2 \;=\; \big\{\nu \in T\,\mathbb{R}^3 : \; \omega(\nu) = 0\big\}\,.$$

We show that this distribution is not integrable.

Proof. Suppose that $\mathcal{E}^2$ is integrable. Then there have to exist an open set $W \subset \mathbb{R}^2$ and a smooth map $h : W \to \mathbb{R}^3$ such that $h^*(\omega) = 0$ and $\operatorname{rank}(D(h)) \equiv 2$. For example, one can choose for h a chart of the integral manifold. In the coordinates of $\mathbb{R}^3$, the map $h = (h^1, h^2, h^3)$ consists of three functions, and the condition $h^*(\omega) = 0$ is expressed on $W \subset \mathbb{R}^2$ by the equation

$$0 \;=\; h^1 dh^2 + dh^3 .$$

Differentiating this expression and forming the exterior product with the 1-form dh^2, we obtain

$$0 \;=\; dh^1 \wedge dh^2, \quad 0 \;=\; dh^2 \wedge dh^3 .$$

We then multiply $0 = h^1 dh^2 + dh^3$ once again by dh^1 and take into account the equation we already derived, $dh^1 \wedge dh^2 = 0$, to arrive at

$$dh^1 \wedge dh^3 \;=\; 0 .$$

In summary, all twofold products vanish, $dh^i \wedge dh^j = 0$, and this contradicts the assumption that the differential $D(h)$ of the map $h : W \to \mathbb{R}^3$ has the maximal rank two. Hence the 2-dimensional distribution in $\mathbb{R}^3$ defined by $\omega = x \cdot dy + dz$ cannot have an integral manifold. □

For higher-dimensional distributions on a manifold the problem thus arises: Under which conditions do they turn out to be integrable? The answer to this question forms the content of Frobenius' theorem. In order to formulate it, we also need the notion of an involutive distribution.

Definition 4. A distribution $\mathcal{E}^k$ on the manifold M^m is called *involutive* if, for every pair of vector fields $\mathcal{V}, \mathcal{W}$ on M^m whose values $\mathcal{V}(x), \mathcal{W}(x) \in E^k(x)$ at each point belong to the distribution, the commutator $[\mathcal{V}, \mathcal{W}](x) \in E^k(x)$ again has values in $\mathcal{E}^k$.

Theorem 1 (Frobenius's Theorem). *Let $\mathcal{E}^k$ be a k-dimensional distribution on the manifold M^m defined by the $(m-k)$ linearly independent 1-forms $\omega_1, \ldots, \omega_{m-k}$:*

$$\mathcal{E}^k \;=\; \big\{\nu \in TM^m : \; \omega_1(\nu) = \ldots = \omega_{m-k}(\nu) = 0\big\} .$$

Then the following conditions are equivalent:

(1) *$\mathcal{E}^k$ is integrable;*

(2) *$\mathcal{E}^k$ is involutive;*

(3) *for every point $x_0 \in M^m$, there exist a neighborhood $x_0 \in U \subset M^m$ and 1-forms θ_{ij} defined on U such that*

$$d\omega_i \;=\; \sum_{j=1}^{m-k} \theta_{ij} \wedge \omega_j$$

for $1 \le i \le m-k$*;*

(4) *for all indices* $1 \le i \le m-k$*, the following exterior products vanish:*

$$d\omega_i \wedge (\omega_1 \wedge \ldots \wedge \omega_{m-k}) \;=\; 0\,.$$

Remark. The fourth condition occurring in Frobenius's theorem is called the *integrability condition* for the geometric distribution or the corresponding Pfaffian system.

Remark. A one-dimensional distribution is determined by $m-1$ 1-forms $\omega_1, \ldots, \omega_{m-1}$, and then $d\omega_i \wedge (\omega_1 \wedge \ldots \wedge \omega_{m-1})$ is an $(m+1)$-form on the m-dimensional manifold. This has to be zero for trivial reasons, and the integrability condition of Frobenius' theorem is automatically satisfied; see Proposition 1.

We first prove some of the simpler equivalences in Frobenius' theorem and proceed as follows: (2) $\Longleftrightarrow$ (3) $\Longleftrightarrow$ (4) and the implication (1) $\Longrightarrow$ (4). The next section will be concerned with the proof of the central assertion of Frobenius' Theorem, i. e. the implication (3) $\Longrightarrow$ (1).

Proof of the equivalence (3) $\Longleftrightarrow$ (4)**.** Suppose that there exist local 1-forms θ_{ij} such that $d\omega_i = \sum\limits_{j=1}^{m-k} \theta_{ij} \wedge \omega_j$. Then

$$d\omega_i \wedge (\omega_1 \wedge \ldots \wedge \omega_{m-k}) \;=\; \sum_{j=1}^{m-k} \theta_{ij} \wedge \omega_j \wedge (\omega_1 \wedge \ldots \wedge \omega_{m-k}) \;=\; 0\,,$$

since the exterior square of any 1-form vanishes. Conversely, assume that condition (4) is satisfied. In a neighborhood $U \subset M^m$ of the point x_0 we extend the family of linearly independent 1-forms $\omega_1, \ldots, \omega_{m-k}$ by adding 1-forms $\eta_1, \ldots, \eta_k$ so that the combined family, $\{\omega_1, \ldots, \omega_{m-k}, \eta_1, \ldots, \eta_k\}$, forms a basis for $\bigwedge_x^1(M^m)$ at each point x of U. The 2-form $d\omega_i$ $(1 \le i \le m-k)$ can thus be represented as

$$d\omega_i \;=\; \sum_{\alpha,\beta=1}^{m-k} C_{\alpha\beta} \cdot \omega_\alpha \wedge \omega_\beta \;+\; \sum_{\alpha=1}^{m-k} \sum_{j=1}^{k} D_{\alpha j} \cdot \omega_\alpha \wedge \eta_j \;+\; \sum_{j,l=1}^{k} E_{jl} \cdot \eta_j \wedge \eta_l$$

with functions $C_{\alpha\beta}$, $D_{\alpha j}$ and E_{jl}. The condition $d\omega_i \wedge (\omega_1 \wedge \ldots \wedge \omega_{m-k}) = 0$ implies

$$\sum_{j,l=1}^{k} E_{jl} \cdot \eta_j \wedge \eta_l \wedge (\omega_1 \wedge \ldots \wedge \omega_{m-k}) \;=\; 0\,,$$

and hence the coefficients vanish, $E_{jl} = 0$. If we introduce the 1-forms

$$\theta_{i\alpha} \;:=\; -\sum_{\beta=1}^{m-k} C_{\alpha\beta} \cdot \omega_\beta - \sum_{j=1}^{k} D_{\alpha j} \cdot \eta_j\,,$$

the exterior derivative $d\omega_i$ takes the desired shape,

$$d\omega_i \;=\; \sum_{\alpha=1}^{m-k} \theta_{i\alpha} \wedge \omega_\alpha \,. \qquad \square$$

Proof of the implication (3) $\Longrightarrow$ (2). For any two vector fields $\mathcal{V}$ and $\mathcal{W}$ with values in the distribution $\mathcal{E}^k$, we have $\omega_i(\mathcal{V}) = \omega_i(\mathcal{W}) = 0$. From $d\omega_i(\mathcal{V},\mathcal{W}) = \mathcal{V}(\omega_i(\mathcal{W})) - \mathcal{W}(\omega_i(\mathcal{V})) - \omega_i([\mathcal{V},\mathcal{W}])$ we obtain

$$d\omega_i(\mathcal{V},\mathcal{W}) \;=\; -\omega_i([\mathcal{V},\mathcal{W}])\,.$$

If now condition (3) holds for the distribution $\mathcal{E}^k$, then

$$d\omega_i(\mathcal{V},\mathcal{W}) \;=\; \sum_{j=1}^{m-k} \theta_{ij} \wedge \omega_j(\mathcal{V},\mathcal{W}) \;=\; 0\,,$$

and hence all 1-forms $\omega_1, \ldots, \omega_{m-k}$ vanish on the commutator $[\mathcal{V},\mathcal{W}]$. Therefore, this vector field takes values in $\mathcal{E}^k$, i. e., the distribution $\mathcal{E}^k$ is involutive. $\square$

Proof of the implication (2) $\Longrightarrow$ (4). Let $\mathcal{E}^k$ be a k-dimensional involutive distribution. The form $d\omega_i \wedge (\omega_1 \wedge \ldots \wedge \omega_{m-k})$ has degree $(m-k+2)$. Inserting $(m-k+2)$ vector fields into this form, we can assume, without loss of generality, that two of these vector fields have values in $\mathcal{E}^k$. Denote them by $\mathcal{V}$ and $\mathcal{W}$. Because of the involutivity of $\mathcal{E}^k$, we have

$$\omega_j(\mathcal{V}) \;=\; \omega_j(\mathcal{W}) \;=\; 0 \quad \text{and} \quad d\omega_i(\mathcal{V},\mathcal{W}) \;=\; -\omega_i([\mathcal{V},\mathcal{W}]) \;=\; 0\,.$$

Therefore, the exterior product $d\omega_i \wedge (\omega_1 \wedge \ldots \wedge \omega_{m-k})$ vanishes on every $(m-k+2)$-tuple of vector fields. $\square$

Proof of the implication (1) $\Longrightarrow$ (4). The proof of this implication proceeds like the one before. Let $\mathcal{E}^k$ be an integrable distribution, $x_0 \in M^m$ a fixed point, and $h : W \to M^m$ a parametrization of an integral manifold through this point. Then we have $h^*(\omega_i) = 0$ $(1 \le i \le m-k)$, and this implies $h^*(d\omega_i) = 0$. Thus the 2-form $d\omega_i$ vanishes on the k-dimensional subspace $\mathcal{E}^k(x_0)$,

$$d\omega_i \,|_{E^k(x_0)\times E^k(x_0)} \;=\; 0\,.$$

Inserting again $(m-k+2)$ vectors into the form $d\omega_i \wedge (\omega_1 \wedge \ldots \wedge \omega_{m-k})$, we conclude that at least two of these vectors—call them $\mathcal{V}$ and $\mathcal{W}$—lie in the subspace $E^k(x_0)$. Hence $\omega_j(\mathcal{V}) = \omega_j(\mathcal{W}) = 0$ and $d\omega_i(\mathcal{V},\mathcal{W}) = 0$. As above, we conclude that the form $d\omega_i \wedge (\omega_1 \wedge \ldots \wedge \omega_{m-k}) = 0$ vanishes. $\square$

4.2. The Proof of Frobenius' Theorem

The implication (3) $\Rightarrow$ (1), which is the core of Frobenius' theorem, is a local statement. Hence, without loss of generality, we can suppose that the manifold M^m is an open subset of $\mathbb{R}^m$. We will prove a slightly more general result from which the proof of this implication immediately follows.

Theorem 2. *Let $\omega_1, \ldots, \omega_{m-k}$ be linearly independent 1-forms on an open subset $M^m \subset \mathbb{R}^m$ such that*

$$d\omega_i = \sum_{j=1}^{m-k} \theta_{ij} \wedge \omega_j$$

for certain 1-forms θ_{ij}. Then there exist at each point $x \in M^m$ a neighborhood $U \subset M^m$ of x and functions h_{ij} and f_j defined on the set U satisfying

$$\omega_i = \sum_{j=1}^{m-k} h_{ij} \cdot df_j \, .$$

Proof of the implication (3) $\Rightarrow$ (1)**.** Let $\mathcal{E}^k$ be a distribution with the property stated in condition (3). By Theorem 2, we can represent the forms ω_i in a neighborhood U of an arbitrary point $x_0 \in M^m$ as

$$\omega_i = \sum_{j=1}^{m-k} h_{ij} \cdot df_j$$

for certain functions. By assumption, the 1-forms $\omega_1, \ldots, \omega_{m-k}$ are linearly independent. Thus the differentials $df_1, \ldots, df_{m-k}$ are linearly independent as well, and the set

$$N^k = \{x \in U : \ f_1(x) = f_1(x_0), \ldots, \ f_{m-k}(x) = f_{m-k}(x_0)\}$$

is a submanifold containing the point $x_0 \in M^m$. At an arbitrary point $x \in N^k$, we determine the tangent space:

$$\begin{aligned} T_x N^k &= \{\nu \in TM^m : \ df_1(\nu) = \ldots = df_{m-k}(\nu) = 0\} \\ &\subset \{\nu \in TM^m : \ \omega_1(\nu) = \ldots = \omega_{m-k}(\nu) = 0\} = E^k(x) \, . \end{aligned}$$

For dimensional reasons, the vector spaces coincide, i.e., N^k is an integral manifold of the distribution $\mathcal{E}^k$ through the point $x_0 \in M^m$, and thus the integrability of the distribution $\mathcal{E}^k$ is proved. □

Proof of Theorem 2. We proceed in two steps. First we reduce the proof to the case of a system of 1-forms $\omega_1, \ldots, \omega_{m-k}$ in a special normal form. To this end, we represent the euclidean space $\mathbb{R}^m$ as a product $\mathbb{R}^m = \mathbb{R}^k \times \mathbb{R}^{m-k}$ and denote its points accordingly by $y = (y^1, \ldots, y^k) \in \mathbb{R}^k$ and $z =$

$(z^1, \ldots, z^{m-k}) \in \mathbb{R}^{m-k}$. It is sufficient to prove the claim in Theorem 2 for 1-forms of the following special type:

$$\omega_i \;=\; dz^i - \sum_{j=1}^{k} A_{ij}(y,z)\cdot dy^j \qquad (1 \le i \le m-k)\,.$$

In fact, representing the linearly independent 1-forms ω_i in coordinates,

$$\omega_i \;=\; \sum_{\alpha=1}^{m} p_{i\alpha}\cdot dx^\alpha,$$

we may first assume that the determinant of the part of the matrix (p_{ij}) containing the first $(m-k)$ columns and rows does not vanish. Let (q_{ij}) be the inverse of this matrix, and consider the 1-forms

$$\omega_i^* = \sum_{\alpha=1}^{m-k} q_{i\alpha}\cdot\omega_\alpha = \sum_{\alpha=1}^{m-k}\sum_{\beta=1}^{m} q_{i\alpha}p_{\alpha\beta}\cdot dx^\beta = dx^i - \sum_{\beta=m-k+1}^{m} dx^\beta\cdot\left(\sum_{\alpha=1}^{m-k} q_{i\alpha}p_{\alpha\beta}\right).$$

Thus the forms ω_i^* are of the special type above. Moreover,

$$\begin{aligned} d\omega_i^* &= \sum_{\alpha=1}^{m-k} dq_{i\alpha}\wedge\omega_\alpha + \sum_{\alpha=1}^{m-k} q_{i\alpha}\cdot d\omega_\alpha = \sum_{\alpha=1}^{m-k} dq_{i\alpha}\wedge\omega_\alpha + \sum_{\alpha,\beta=1}^{m-k} q_{i\alpha}\cdot\theta_{\alpha\beta}\wedge\omega_\beta \\ &= \sum_{\alpha=1}^{m-k}\left(\sum_{\beta=1}^{m-k} p_{\beta\alpha}\cdot dq_{i\beta} + \sum_{\beta,\gamma=1}^{m-k} q_{i\beta}p_{\beta\alpha}\theta_{\beta\gamma}\right)\wedge\omega_\alpha^*, \end{aligned}$$

and the system of forms $\omega_1^*, \ldots, \omega_{m-k}^*$ also satisfies the condition

$$d\omega_i^* \;=\; \sum_{\alpha=1}^{m-k} \theta_{i\alpha}^*\wedge\omega_\alpha^*$$

for certain 1-forms $\theta_{i\alpha}^*$. If Theorem 2 now holds for this system of forms, we obtain functions h_{ij}^* and f_j^* such that

$$\omega_i^* \;=\; \sum_{j=1}^{m-k} h_{ij}^*\cdot df_j^*\,.$$

Rewriting this leads to

$$\omega_i \;=\; \sum_{\alpha=1}^{m-k} p_{i\alpha}\cdot\omega_\alpha^* \;=\; \sum_{\alpha=1}^{m-k}\sum_{j=1}^{m-k} h_{\alpha j}^*\cdot p_{i\alpha}\cdot df_j^*\,.$$

Summarizing, it is sufficient to prove Theorem 2 for forms of the type

$$\omega_i \;=\; dz^i - \sum_{j=1}^{k} A_{ij}(y,z)\cdot dy^j\,.$$

Without loss of generality, we may choose the origin, $x = 0$ in $\mathbb{R}^m$, as the point $x = (y, z)$ in whose neighborhood we want to represent the forms ω_i in this special way. For fixed parameters $(y^1, \dots, y^k) \in \mathbb{R}^k$ near $0 \in \mathbb{R}^k$, we consider the ordinary differential equation with initial conditions

$$\dot{z}^i(t) = \sum_{j=1}^{k} A_{ij}(t \cdot y, z(t)) \cdot y^j, \qquad z^i(0) = z^i, \qquad 1 \le i \le m-k.$$

Denote by $F^i(t, y, z)$ the solution of this differential equation that is uniquely determined near zero. These are $m-k$ smooth functions in the variables $t \in \mathbb{R}$, $y \in \mathbb{R}^k$ and $z \in \mathbb{R}^{m-k}$ defined on a neighborhood of zero in $\mathbb{R} \times \mathbb{R}^m$. Moreover, for small values of the parameters we have $F^i(\mu \cdot t, y, z) = F^i(t, \mu \cdot y, z)$, since the functions

$$G^i(t, y, z) := F^i(\mu \cdot t, y, z)$$

satisfy the initial conditions $G^i(0, y, z) = z^i$, and they are also solutions of the differential equation corresponding to the vector $\mu \cdot y$,

$$\dot{G}^i(t, y, z) = \mu \cdot \dot{F}^i(\mu \cdot t, y, z) = \sum_{j=1}^{k} A_{ij}(\mu \cdot t \cdot y, G^1, \dots, G^{m-k}) \cdot (\mu y^i).$$

We introduce a coordinate transformation near zero in $\mathbb{R}^m = \mathbb{R}^k \times \mathbb{R}^{m-k}$ by means of the equations

$$u := y, \quad F(1, u, v) := z.$$

The Jacobian of this transformation is

$$\left.\frac{\partial(y, z)}{\partial(u, v)}\right|_0 = \begin{bmatrix} E & 0 \\ 0 & E \end{bmatrix},$$

since, by the differential equation and the initial condition,

$$\left.\frac{\partial z^i}{\partial v^j}\right|_0 = \left.\frac{\partial F^i(1, 0, v)}{\partial v^j}\right|_0 = \left.\frac{\partial v^i}{\partial v^j}\right|_0 = \delta^i_j.$$

This observation shows that both these equations determine a local diffeomorphism from the space $\mathbb{R}^m$ to itself near zero. We represent the 1-forms $\omega_1, \dots, \omega_{m-k}$ in the $\{u, v\}$-coordinates:

$$\omega_i = \sum_{j=1}^{m-k} B_{ij}(u, v) \cdot dv^j + \sum_{j=1}^{k} P_{ij}(u, v) \cdot du^j.$$

Obviously it suffices to prove that the functions $P_{ij} \equiv 0$ vanish identically. First we derive the identity

$$\sum_{j=1}^{k} P_{ij}(t \cdot y, v) \cdot y^j \equiv 0.$$

To this end, consider the map $\Phi : \mathbb{R}^1 \times \mathbb{R}^k(t,y) \longrightarrow \mathbb{R}^k \times \mathbb{R}^{m-k}(u,v)$ from the space $\mathbb{R}^1 \times \mathbb{R}^k$ with coordinates (t,y) to the space $\mathbb{R}^k \times \mathbb{R}^{m-k}$ with coordinates (u,v) defined for a fixed point $v \in \mathbb{R}^{m-k}$ by the formula $\Phi(t,y) := (ty, v)$. Then

$$\Phi^*(\omega_i) \;=\; \sum_{j=1}^{k} P_{ij}(t\cdot y, v)(t\cdot dy^j + y^j \cdot dt)\,.$$

On the other hand, with respect to the (y,z)-coordinates on $\mathbb{R}^m$, the map $\Phi : \mathbb{R}^1 \times \mathbb{R}^k(t,y) \to \mathbb{R}^k \times \mathbb{R}^{m-k}(y,z)$ is determined by

$$\Phi(t,y) \;=\; \big(t\cdot y,\, F(1,\, ty,\, v)\big) \;=\; \big(t\cdot y,\, F(t,y,v)\big)\,.$$

From the normal form,

$$\omega_i \;=\; dz^i - \sum_{j=1}^{k} A_{ij}\cdot dy^j,$$

we obtain for the induced form $\Phi^*(\omega_i)$ the new expression

$$\Phi^*(\omega_i) \;=\; \frac{\partial F^i}{\partial t}\cdot dt + \sum_{j=1}^{k} \frac{\partial F^i}{\partial y^j}\cdot dy^j - \sum_{j=1}^{k} A_{ij}(y^j\cdot dt + t\cdot dy^j)\,.$$

Comparing in $\Phi^*(\omega_i)$ the coefficient at dt and taking the differential equation into account leads to the identity

$$0 \;=\; \frac{\partial F^i}{\partial t}(t,y,v) - \sum_{j=1}^{k} A_{ij}\cdot y^j \;=\; \sum_{j=1}^{k} P_{ij}(t\cdot y,\, v)\cdot y^j\,.$$

In the final step of the proof of Frobenius' theorem, we now show that the functions $P_{ij} = 0$ vanish. To do so, we again resort to the map $\Phi(t,y,v) := (t\cdot y, v)$, but consider it this time as a map from $\mathbb{R}\times\mathbb{R}^k\times\mathbb{R}^{m-k} \ni (t,y,v)$ to $\mathbb{R}^k \times \mathbb{R}^{m-k} \ni (u,v)$ (v is no longer assumed to be constant). The identity already proved implies the formula

$$\Phi^*(\omega_i) \;=\; \sum_{j=1}^{m-k} B_{ij}(t\cdot y,\, v)\cdot dv^i + \sum_{j=1}^{k} P_{ij}(t\cdot y,\, v)\cdot t\cdot dy^j\,.$$

We denote by P^*_{ij} the function

$$P^*_{ij}(t,\, y,\, v) \;:=\; t\cdot P_{ij}(t\cdot y,\, u)\,.$$

Then

$$\Phi^*(\omega_i) \;=\; \sum_{j=1}^{k} P^*_{ij}(t,\, y,\, v)\cdot dy^j + \{\text{terms in } dv^i\}\,,$$

and for the exterior derivative we obtain the expression

$$d\Phi^*(\omega_i) \;=\; \sum_{j=1}^{k} \frac{\partial P^*_{ij}}{\partial t} \cdot dt \wedge dy^j + \{\text{terms without } dt\}\,.$$

By assumption, there exist 1-forms satisfying $d\omega_i \;=\; \sum_{j=1}^{m-k} \theta_{ij} \wedge \omega^j$, and we represent these as $\Phi^*(\theta_{ij}) \;:=\; H_{ij}dt + \{\text{terms containing } dy^i, dv^j\}$. This leads to a homogeneous differential equation for the functions we are interested in:

$$\frac{\partial P^*_{ij}}{\partial t} \;=\; \sum_{\alpha=1}^{m-k} H_{i\alpha} P^*_{\alpha j}\,,$$

with the initial condition $P^*_{ij}(0, y, v) = 0$. The only solution of this homogeneous differential equation with the given initial condition is $P^*_{ij}(t, y, v) \equiv 0$, and this immediately implies $P_{ij}(t \cdot y, v) = 0$. Thus the proof of Frobenius' theorem is completed. □

4.3. Some Applications of Frobenius' Theorem

The simplest case is that of an $(m-1)$-dimensional distribution $\mathcal{E}^{m-1}$ on an m-dimensional manifold M^m. If $\mathcal{E}^{m-1}$ is defined by one nowhere vanishing 1-form ω, the integrability of the distribution reduces to the condition that the 3-form $d\omega \wedge \omega$ vanishes,

$$d\omega \wedge \omega \;=\; 0.$$

The method to explicitly integrate this $(m-1)$-dimensional Pfaffian system is based on looking for a so-called integrating factor and an application of Poincaré's lemma.

Definition 5. An *integrating factor* for the 1-form ω is a nowhere vanishing function $f : M^m \to \mathbb{R}$ such that the 1-form $f \cdot \omega$ is closed,

$$d(f \cdot \omega) \;=\; 0\,.$$

Theorem 3. *Let ω be a nowhere vanishing 1-form on the manifold M^m.*

(1) *If there exists an integrating factor for ω, then $d\omega \wedge \omega = 0$. In this case the distribution $\mathcal{E}^{m-1}$ is integrable.*

(2) *If $d\omega \wedge \omega = 0$, then there exists an integrating factor for the 1-form ω in a neighborhood of each point in M^m.*

(3) *Locally, the integral manifolds of the distribution $\mathcal{E}^{m-1}$ are the level surfaces of the function g determined from the integrating factor f via the equations*

$$d(f \cdot \omega) \;=\; 0, \quad f \cdot \omega \;=\; dg\,.$$

Proof. The equation $d(f \cdot \omega) = 0$ implies $df \wedge \omega + f \cdot d\omega = 0$. Multiplying this equation once again by the 1-form ω leads to $f \cdot d\omega \wedge \omega = 0$. Since $f \neq 0$, we obtain $d\omega \wedge \omega = 0$ as a necessary condition for the existence of an integrating factor. If, on the other hand, $d\omega \wedge \omega = 0$, then the existence of an integrating factor follows immediately from Theorem 2, §4.2. □

In dimension $m = 2$ the 3-form vanishes, $d\omega \wedge \omega = 0$, for purely algebraic reasons. Thus

Corollary 1. *Every nowhere vanishing* 1*-form on a* 2*-dimensional manifold locally has an integrating factor.*

Example 5. Consider in $\mathbb{R}^2$ the differential equation

$$P(t,x) + Q(t,x)\,\dot{x} \;=\; 0\,.$$

Near a point $(t_0, x_0) \in \mathbb{R}^2$ at which P and Q do not vanish simultaneously, we have the 1-form

$$\omega \;=\; P \cdot dt + Q \cdot dx$$

and its integrating factor $f(t,x)$. The equivalent differential equation

$$(f \cdot P)(t,x) + (f \cdot Q)(t,x)\,\dot{x} \;=\; 0$$

is called the *total differential equation*, and the solution curves are implicitly determined by the equation

$$g(t,x) \;=\; \text{const}$$

with $dg = f \cdot \omega$. Frobenius' theorem now claims that it is always possible to solve the original differential equation by using the outlined method. It does not, however, provide an algorithm for finding the integrating factor. In simple cases this may be computed directly. If we can find, e. g., functions $F(t)$ and $G(x)$ depending only on the variables t, x and satisfying

$$\frac{\partial P(t,x)}{\partial x} - \frac{\partial Q(t,x)}{\partial t} \;=\; Q(t,x)F(t) - P(t,x)G(x)\,,$$

then $f(t,x) = e^{\int F(t)dt} e^{\int G(x)dx}$ is an integrating factor.

Example 6. Consider the differential equation

$$(2t^2 + 3tx - 4t)\dot{x} + (3x - 2tx - 3x^2) \;=\; 0\,.$$

We can choose $F(t) = 2/t$ and $G(x) = -5/x$ and obtain the integrating factor $f = t^2x^{-5}$. The solutions of this differential equation computed using this integrating factor are the curves described by the equation

$$t^3x^{-4} - \frac{1}{2}t^4x^{-4} - t^3x^{-3} \;\equiv\; \text{const}$$

(compare the figure on the next page).

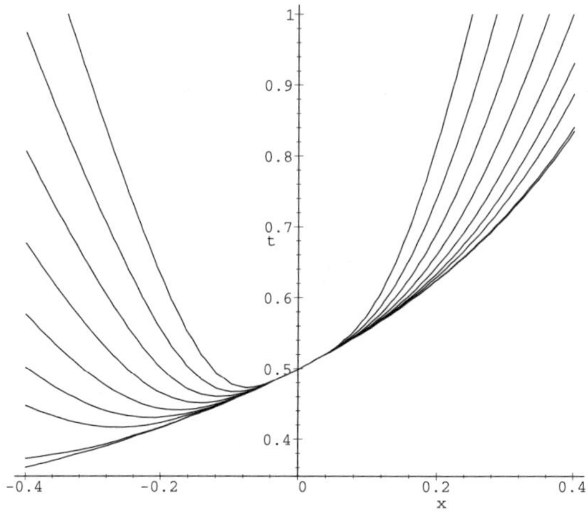

The curves $t^3 - \frac{1}{2}t^2 - t^3x = cx^4$ for different values of c

Assume now that on the manifold M^m we are given a Riemannian metric $\langle\ ,\ \rangle$ as well as an $(m-1)$-dimensional integrable distribution $\mathcal{E}^{m-1}$ described by the 1-form ω. Denote the vector field associated with the 1-form via the Riemannian metric by $\mathcal{W}$. This is uniquely determined by either of the two equivalent equations

$$*\omega \;=\; \mathcal{W} \lrcorner\, dM^m \quad \text{or} \quad \omega(\mathcal{V}) \;=\; \langle \mathcal{V}, \mathcal{W}\rangle\,.$$

Using the vector field $\mathcal{W}$, the distribution $\mathcal{E}^{m-1}$ can be described as

$$\mathcal{E}^{m-1} \;=\; \{\nu \in TM^m : \quad \langle \nu, \mathcal{W}\rangle \;=\; 0\}\,,$$

and hence $\mathcal{W}$ is orthogonal to each integral manifold N^{m-1} of the distribution. Normalizing the length of the vector field $\mathcal{W}$ to one, the volume form of each integral manifold is given by the formula

$$dN^{m-1} \;=\; \frac{1}{||\mathcal{W}||} \cdot \mathcal{W} \lrcorner\, dM^m\,.$$

We study the behavior of the integral manifold $N^{m-1} \subset M^m$ of the distribution under the flow $\Phi_t : M^m \to M^m$ of the vector field $\mathcal{W}$. Compute first the Lie derivative of the 1-form ω with respect to the associated vector field $\mathcal{W}$. Since $\omega(\mathcal{W}) = ||\mathcal{W}||^2$, the formula $\mathcal{L}_{\mathcal{W}}(\omega) = d(\mathcal{W} \lrcorner\, \omega) + \mathcal{W} \lrcorner\, (d\omega)$ implies

$$(\mathcal{L}_{\mathcal{W}}(\omega))(\mathcal{V}) \;=\; \mathcal{W}(\omega(\mathcal{V})) \;-\; \omega([\mathcal{W}, \mathcal{V}])\,.$$

In particular, for a vector field $\mathcal{V}$ tangent to the distribution this formula simplifies to

$$(\mathcal{L}_{\mathcal{W}}(\omega))(\mathcal{V}) \;=\; -\,\omega([\mathcal{W},\mathcal{V}]) \;=\; -\,\langle \mathcal{W},\, [\mathcal{W},\mathcal{V}]\rangle\,,$$

and we arrive at

Theorem 4. *The flow Φ_t of the vector field $\mathcal{W}$ maps an integral manifold of the distribution $\mathcal{E}^{m-1}$ to another integral manifold if and only if*

$$\mathcal{V}(||\mathcal{W}||^2) + d\omega(\mathcal{W},\mathcal{V}) \;=\; -\,\langle \mathcal{W},\, [\mathcal{W},\mathcal{V}]\rangle \;=\; 0$$

for every vector field $\mathcal{V}$ on M^m with values in $\mathcal{E}^{m-1}$.

Corollary 2. *Let the distribution $\mathcal{E}^{m-1}$ be defined by the closed 1-form ω. Then the flow of the dual vector field $\mathcal{W}$ transforms integral manifolds into integral manifolds if the length $||\mathcal{W}||$ is constant on every connected integral manifold.*

Corollary 3. *If the distribution $\mathcal{E}^{m-1}$ is defined by a 1-form ω of constant length, and if, moreover, the flow of the dual vector field $\mathcal{W}$ transforms integral manifolds into integral manifolds, then $d\omega = 0$. In this case $\mathcal{E}^{m-1}$ locally consists of level surfaces of a function whose gradient has constant length.*

Proof. First, Theorem 4 implies $\mathcal{W} \lrcorner\, d\omega = 0$. At a point $x \in M^m$, we choose an orthogonal basis $e_1, \ldots, e_m$ in the tangent space so that $\mathcal{W}$ is proportional to e_1, $\mathcal{W} = a \cdot e_1$. Denoting by $\sigma_1, \ldots, \sigma_m$ the dual basis, the form $\omega = a \cdot \sigma_1$ is proportional to σ_1. Represent the 2-form

$$d\omega \;=\; \sum_{i<j} b_{ij}\, \sigma_i \wedge \sigma_j$$

in this basis. Since $0 = d\omega \wedge \omega$, the 2-form $d\omega$ only contains the $m-1$ summands $d\omega = b_{12}\,\sigma_1 \wedge \sigma_2 + \ldots + b_{1m}\sigma_1 \wedge \sigma_m$. But then the condition

$$0 \;=\; \mathcal{W} \lrcorner\, d\omega \;=\; a\,(b_{12}\,\sigma_2 + \ldots + b_{1m}\,\sigma_m)$$

means that ω is a closed form. □

We compute the infinitesimal volume change of a compact integral manifold N^{m-1} under the flow Φ_t of the vector field $\mathcal{W}$. The Lie derivative of the volume form dN^{m-1} is

$$\begin{aligned}\mathcal{L}_{\mathcal{W}}(dN^{m-1}) \;&=\; \mathcal{L}_{\mathcal{W}}\Big(\frac{1}{||\mathcal{W}||} \cdot \mathcal{W} \lrcorner\, dM^m\Big) \;=\; \mathcal{W} \lrcorner\, d\Big(\frac{1}{||\mathcal{W}||} \cdot \mathcal{W} \lrcorner\, dM^m\Big)\\ &=\; \Big(\mathrm{div}(\mathcal{W}) \;-\; \frac{1}{2} \cdot \mathcal{W}(\ln ||\mathcal{W}||^2)\Big) \cdot dN^{m-1}\,,\end{aligned}$$

and hence we obtain

Theorem 5. *The derivative of the volume change of a compact integral manifold N^{m-1} of the distribution $\mathcal{E}^{m-1}$ under the flow of the vector field $\mathcal{W}$ is given by the formula*

$$\frac{d}{dt}\big(\mathrm{vol}(\Phi_t(N^{m-1}))\big)\big|_{t=0} = \int_{N^{m-1}} \Big(\mathrm{div}(\mathcal{W}) - \frac{1}{2}\cdot \mathcal{W}(\ln ||\mathcal{W}||^2)\Big)\cdot dN^{m-1}.$$

If, in particular, the distribution $\mathcal{E}^{m-1}$ consists of the level surfaces of a function $f : M^m \to \mathbb{R}$, and we choose the gradient of this function as the vector field, $\mathcal{W} = \mathrm{grad}(f)$, then $\mathrm{div}(\mathcal{W}) = \Delta(f)$ is the Laplacian of f.

Corollary 4. *The volume change of a compact level surface N^{m-1} of the function $f : M^m \to \mathbb{R}$ under the flow of the gradient vector field is given by the formula*

$$\frac{d}{dt}\big(\mathrm{vol}(\Phi_t(N^{m-1}))\big)\big|_{t=0} = \int_{N^{m-1}} \Big(\Delta(f) - \frac{1}{2}\cdot \mathrm{grad}(f) \ln ||\mathrm{grad}(f)||^2\Big)\cdot dN^{m-1}.$$

Remark. In all these formulas the Laplacian, the divergence, and the gradient are taken with respect to the manifold.

Example 7. Consider a function $f : M^m \to \mathbb{R}$ and assume that there is another function $\mu : M^m \to \mathbb{R}$ such that

$$d\big(\|\mathrm{grad}(f)\|^2\big) \;=\; 2\mu \cdot df\,.$$

By Theorems 4 and 5, the flow of the gradient vector field $\mathrm{grad}(f)$ maps level surfaces of f to level surfaces, and the volume change is described by the formula

$$\frac{d}{dt}\left(\mathrm{vol}\left(\Phi_t(N^{m-1})\right)\right)\big|_{t=0} \;=\; \int_{N^{m-1}} (\Delta(f) - \mu)dN^{m-1}\,.$$

For example, the spheres $S^{m-1}(R) \subset \mathbb{R}^m$ are the level surfaces of the function $f(x) = \|x\|^2$, and we obtain $\|\mathrm{grad}(f)\|^2 = 4\|x\|^2$ as well as $\Delta(f) = 2m$. Hence $\Delta(f) - \mu \equiv 2(m-1)$ is constant, and for the flow $\Phi_t(x) = e^{2t}\cdot x$ we obtain the following differential equation describing the evolution of the volume:

$$\frac{d}{dt}\left(\mathrm{vol}\left(\Phi_t(S^{m-1})\right)\right) \;=\; 2(m-1)\mathrm{vol}(\Phi_t(S^{m-1}))\,.$$

The second application of Frobenius' theorem will play an important role in the chapter devoted to surface theory.

Theorem 6. *Let $\Omega = (\omega_{ij})$ be a $(k\times k)$ matrix of 1-forms defined on a neighborhood of $0\in\mathbb{R}^m$, and let A_0 be an invertible $(k\times k)$ matrix. In a connected neighborhood $0\in V$ there exists a $(k\times k)$ matrix $A = (f_{ij})$ of functions satisfying*

$$\Omega \;=\; dA\cdot A^{-1} \quad \textit{and} \quad A(0) \;=\; A_0$$

if and only if

$$d\Omega \;=\; \Omega \wedge \Omega.$$

In this case, the function matrix A is uniquely determined. If, in addition, Ω is an anti-symmetric matrix ($\Omega+\Omega^t = 0$), and A_0 is an orthogonal matrix ($A_0 \cdot A_0^t = E$), then the solution $A(x)$ is also orthogonal at each point of the set V.

Proof. First, the condition $d\Omega = \Omega \wedge \Omega$ is necessary for the solvability of the equation $\Omega = dA \cdot A^{-1}$. In fact, $dA = \Omega \cdot A$ implies

$$0 \;=\; ddA \;=\; d(\Omega \cdot A) \;=\; (d\Omega)\cdot A - \Omega \wedge dA \;=\; (d\Omega - \Omega \wedge \Omega)\cdot A\,.$$

The matrix A is invertible, and hence $d\Omega = \Omega \wedge \Omega$. Uniqueness of the solution is also easy to see. For two solutions $A(x)$ and $B(x)$ we have, e. g.,

$$d(B^{-1}) \;=\; -B^{-1}\cdot(dB)\cdot B^{-1}\,,$$

and hence the differential $d(B^{-1}\cdot A)$ vanishes:

$$\begin{aligned} d(B^{-1}\cdot A) &= d(B^{-1})\cdot A + B^{-1}dA \;=\; -B^{-1}\cdot(dB)\cdot B^{-1}\cdot A + B^{-1}\cdot\Omega\cdot A \\ &= -B^{-1}\Omega\cdot B\cdot B^{-1}\cdot A + B^{-1}\cdot\Omega\cdot A \;=\; 0\,. \end{aligned}$$

Thus $B^{-1}\cdot A$ is constant, and at the point $x = 0$ it is equal to the unit matrix. This implies $A(x) = B(x)$ for all points $x \in V$. Now we prove the existence of a solution under the condition $d\Omega = \Omega \wedge \Omega$. To this end, consider the following $(k \times k)$ matrix of 1-forms on the space $\mathbb{R}^m \times \mathbb{R}^{k^2}$ with coordinates $(x^1, \ldots, x^m, z^{ij})$:

$$\Lambda \;=\; dZ - \Omega \cdot Z \;=\; \Big(dz^{ij} - \sum_{r=1}^{k} \omega_{ir}\cdot z^{rj}\Big)\,.$$

From $d\Omega = \Omega \wedge \Omega$ we obtain

$$\begin{aligned} d\Lambda &= ddZ - d\Omega \wedge Z + \Omega \wedge dZ \;=\; -\Omega\wedge\Omega\wedge Z + \Omega \wedge dZ \\ &= -\Omega\wedge\Omega\wedge Z + \Omega\wedge(\Lambda + \Omega\wedge Z) \;=\; \Omega\wedge\Lambda\,. \end{aligned}$$

The system of forms $dz^{ij} - \sum\limits_{r=1}^{k} \omega_{ir} z^{rj}$ is linearly independent in $\mathbb{R}^m \times \mathbb{R}^{k^2}$, and by Frobenius' theorem, there exists an m-dimensional integral manifold $M^m \subset \mathbb{R}^m \times \mathbb{R}^{k^2}$ through the point $(0, A_0) \in \mathbb{R}^m \times \mathbb{R}^{k^2}$. The tangent space $T_{(0,A_0)}M^m$ to this integral manifold has only the null vector in common with $\mathbb{R}^{k^2}$,

$$T_{(0,A_0)}M^m \cap \mathbb{R}^{k^2} \;=\; \{0\}\,.$$

This follows directly from the shape of the form Λ, since the tangent space to M^m is determined by the equation $\Lambda = 0$. Then the integral manifold M^m

is the graph of a map $A : W \longrightarrow R^{k^2}$ defined on an open set $0 \in W \subset \mathbb{R}^m$ satisfying the initial condition $A(0) = A_0$. From

$$A^*(\Lambda) = A^*(dZ - \Omega \cdot Z) = dA - \Omega \cdot A,$$

we see that the $(k \times k)$ matrix A is the solution of the differential equation we looked for. The remaining statements of the theorem follow from the formula

$$d(A^t \cdot A) = (dA)^t \cdot A + A^t \cdot dA = A^t \cdot \Omega^t \cdot A + A^t \cdot \Omega \cdot A\,.$$

In fact, if Ω is an anti-symmetric matrix, and $dA = \Omega \cdot A$ is a solution of the differential equation, then we immediately conclude that $d(A^t \cdot A) = 0$. □

Exercises

1. Consider on $\mathbb{R}^3$ the 1-form

$$\omega = \frac{1}{yz}\,dx + \frac{1}{xz}\,dy + \frac{1}{xy}\,dz\,.$$

Prove that the distribution defined by $\omega = 0$ is integrable. Find, moreover, an integrating factor for ω, i. e. , a function f such that $d(f \cdot \omega) = 0$.

2. Let P and Q be functions in the variables x, y, u, v. Consider on $\mathbb{R}^4$ the 1-forms

$$\omega_1 = dx - P \cdot du + Q \cdot dv, \quad \omega_2 = dy - Q \cdot du - P \cdot dv\,.$$

a) What are the conditions for the distribution $\omega_1 = \omega_2 = 0$ to be integrable?

b) Let the function $f : \mathbb{C}^2 \to \mathbb{C}$ be defined in the complex variables $z := x + iy$, $w := u + iv$ by $f(z, w) = P + i \cdot Q$. Prove that the integrability conditions for the system $\omega_1 = \omega_2 = 0$ are, in particular, satisfied when f is a holomorphic function.

3. Prove that in $\mathbb{R}^3$, a distribution defined by the form

$$\omega = P \cdot dx + Q \cdot dy + R \cdot dz$$

is integrable if the determinant

$$\det \begin{bmatrix} P & Q & R \\ \frac{\partial}{\partial x} & \frac{\partial}{\partial y} & \frac{\partial}{\partial z} \\ P & Q & R \end{bmatrix} = 0$$

vanishes. This formula is to be interpreted in the following way: Write the determinant out and apply each of the resulting differential operators to the respective function.

4. Solve the initial value problem $\dot{x} = e^x \cos t$, $x(0) = x_0$

a) by separation of variables;

b) by finding an integrating factor.

5. The method of solving a differential equation by finding an integrating factor is of interest, in particular, if separation of variables is impossible. As an example, solve the differential equation $\dot{x}(x^3 - tx) = 1$, $x(0) = x_0$. How could this equation, nevertheless, be solved by a suitable separation of variables?

6. Solve the differential equations

a) $(e^{2t} - x^2) + t \cdot \dot{x} = 0$;

b) $\sin t + e^x + \cos t \cdot \dot{x} = 0$;

c) $4\,t^3 + 6\,tx^3 + (3\,t^3 x^2 + 3)\dot{x} = 0$.

7. Let $\Omega = (\omega_{ij})$ be a $(k \times k)$ matrix of 1-forms defined on an open neighborhood of $0 \in \mathbb{R}^m$. Prove that locally there exists a $(k \times k)$ matrix A consisting of functions satisfying

$$dA \;=\; \Omega \cdot A - A \cdot \Omega$$

if and only if there exists a 1-form σ such that the matrix $\Sigma := \Omega - \sigma \cdot E$ solves the equation $d\Sigma = \Sigma \wedge \Sigma$.

Chapter 5

Curves and Surfaces in Euclidean 3-Space

5.1. Curves in Euclidean 3-Space

The notion of a curve in 3-space $\mathbb{R}^3$ is not as simple as it may seem at first glance. For a long time in the 19-th century, the common understanding was that a curve is a subset $A \subset \mathbb{R}^3$ described in a continuous way by a real parameter. In this sense, a curve was the surjective image of a continuous map $f : [0,1] \to A \subset \mathbb{R}^3$. To require the injectivity of f is too strong, in so far as it excluded curves with self-intersections. Apart from the fact that the assumption of continuity is too weak to render such a "curve" accessible to differential calculus, a much more essential reason shattered this notion of a curve.

Theorem 1 (Peano 1890). *There exists a continuous, surjective map from the interval onto the square.*

Proof. We construct the map $f : [0,1] \to [0,1] \times [0,1]$ as the limit of a uniformly converging sequence of continuous maps f_i. In the first step, we divide the interval $[0,1]$ into nine subintervals of equal length and decompose the square accordingly. The map f_1 is then to map each of these subintervals, $[(i-1)/9, i/9]$, to the diagonal of the i-th square continuously and bijectively (see the figure). The map f_2 arises in the same way when we further divide each subinterval $[(i-1)/9, i/9]$ as well as each subsquare. This results in 81 intervals and squares, and f_2 maps the small intervals continuously and bijectively onto the diagonals of the small squares

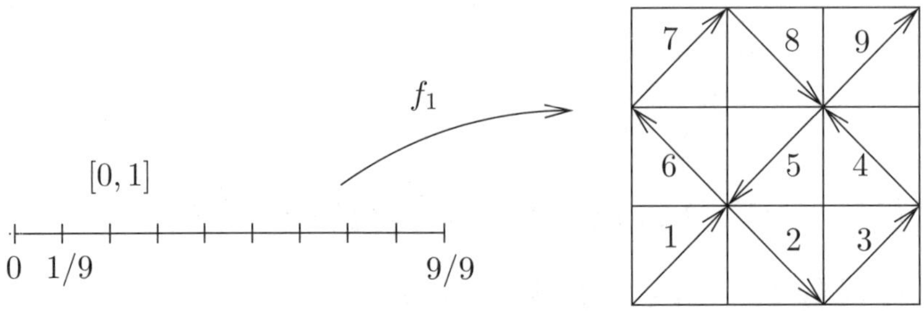

so that $f_2([(i-1)/9, i/9])$ lies in the i-th square of the first step of the decomposition. By construction, $f_1(i/9) = f_2(i/9)$. This way we obtain a sequence of continuous maps satisfying

(1) $\|f_n(t) - f_{n+1}(t)\| \le \sqrt{2}/3^n$ for all $t \in [0,1]$;

(2) $f_n(i/3^{n+1}) = f_{n+1}(i/3^{n+1}) = f_{n+2}(i/3^{n+1})$ for all n and $i \le 3^{n+1}$.

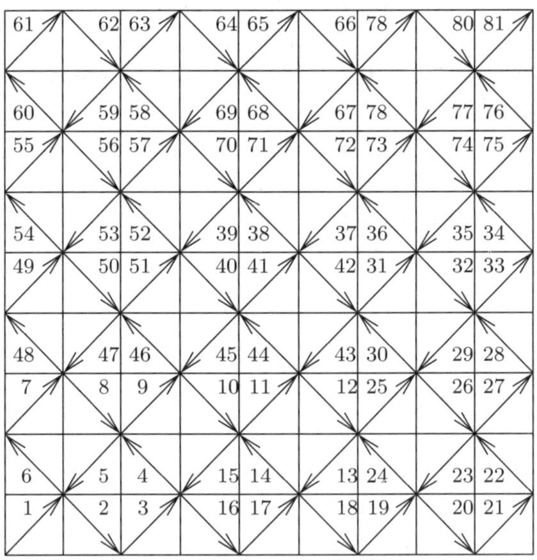

The estimate in (1) implies the Cauchy condition,

$$\|f_n(t) - f_m(t)\| \le \frac{1}{\sqrt{2}\cdot 3^{n-1}},$$

for all $t \in [0,1]$ and $m \ge n$. Thus, the sequence f_n converges uniformly to a continuous map $f : [0,1] \to [0,1] \times [0,1]$. By construction, all the points $(\frac{k}{3^n}, \frac{l}{3^m})$ belong to the image of the map f. The interval $[0,1]$ is compact, and hence the image set $f([0,1])$ is again a compact subset of the square. But the subset of the image of f mentioned before is dense in the square

$[0,1]\times[0,1]$, and hence $f([0,1]) = [0,1]\times[0,1]$, i. e. , f is continuous and surjective. $\square$

This example had an essential impact on general topology. In fact, it shows that the "dimension" of a topological space can increase under a continuous map, and that this notion has to be made more precise (topological dimension theory). We will not deal with this problem here, and instead confine ourselves to the case of smooth curves in $\mathbb{R}^3$. We could try to consider these as smooth, one-dimensional submanifolds of $\mathbb{R}^3$, but that would exclude self-intersections of curves again. For this reason, we formulate the notion of curve slightly more generally.

Definition 1. A *(parametrized) curve* is a differentiable map $\gamma : [a,b] \to \mathbb{R}^3$ from an interval to $\mathbb{R}^3$ whose derivative vanishes nowhere.

It is easy to see that sufficiently small image sets of a parametrized curve are submanifolds. Since $d\gamma(t)/dt \neq 0$, for a fixed initial point, the length function

$$L(t) \;:=\; \int_a^t \|\frac{d\gamma}{dt}(\mu)\| \cdot d\mu$$

is monotonously increasing. A point on the curve is uniquely determined by the parameter describing the length of the curve segment from the initial point to the point considered. In other words, inverting the length function, we obtain a parametrization $\gamma \circ L^{-1} : [0,\ L(\gamma)] \to \mathbb{R}^3$ of the curve γ, and the tangent vector with respect to the length parameter s has length one, $||d\gamma/ds|| = 1$. This kind of parametrization is called the *natural parametrization* of the curve. The preceding observation shows that the natural parametrization can be constructed starting from an arbitrary one. We agree on the following: If the curve is given in any parametrization $\gamma(t)$, then we denote by $\dot\gamma(t)$, etc., the derivative with respect to the parameter t; if the curve is given in its natural parametrization, then $\gamma'(s)$ is the derivative with respect to the length parameter s of the curve.

Example 1. Consider the *helix* in the parametrization

$$\gamma(t) \;=\; \big(a\cdot\cos t,\ a\cdot\sin t,\ b\cdot t\big)\,.$$

Since $||\dot\gamma(t)|| = \sqrt{a^2+b^2}$, the length function is $L(t) = \sqrt{a^2+b^2}\cdot t$. From this, we obtain the natural parametrization by passing from the parameter t to the parameter $s = L(t)$:

$$\gamma(s) \;=\; \left(a\cdot\cos\frac{s}{\sqrt{a^2+b^2}},\ a\cdot\sin\frac{s}{\sqrt{a^2+b^2}},\ \frac{b\cdot s}{\sqrt{a^2+b^2}}\right).$$

To every curve, we shall assign three linearly independent vectors at each point of general type. The first of these vectors is the normalized tangent vector.

Definition 2. If $\gamma(s)$ is a curve in its natural parametrization, the vector $\vec{t}(s) := \gamma'(s)$ is called the *unit tangent vector* to the curve.

The curvature of a curve at a point measures the angle variation of the tangent vector per length unit. More precisely:

Definition 3. Let $\mathcal{C} \subset \mathbb{R}^3$ be a curve, and let $p \in \mathcal{C}$ be a point on the curve. The *curvature* $\kappa(p)$ of the curve $\mathcal{C}$ at the point p is the limit

$$\kappa(p) \;=\; \lim_{q \to p} \frac{\phi}{\widehat{pq}}\,,$$

where ϕ is the angle between the tangent vectors at the points p and q, and $\widehat{pq}$ denotes the length of the curve segment between these points.

Theorem 2. *In the natural parametrization of the curve,* $\kappa(s) = ||\gamma''(s)||$.

Proof. Setting $p = \gamma(s)$ and $q = \gamma(s+h)$, we obtain $\widehat{pq} = h$ in the natural parametrization of the curve, and the angle ϕ is computed using the formula for the angle in an isosceles triangle,

$$\sin(\phi/2) = ||\vec{t}(s+h) - \vec{t}(s)||/2\,.$$

Thus

$$\lim_{q \to p} \frac{\phi}{\widehat{pq}} = \lim_{h \to 0} \frac{\phi/2}{\sin(\phi/2)} \cdot \lim_{h \to 0} \frac{2 \cdot \sin(\phi/2)}{h} = \lim_{h \to 0} \frac{||\vec{t}(s+h) - \vec{t}(s)||}{h} = ||\gamma''(s)||.$$

□

In an arbitrary parametrization of the curve, a straightforward parameter transformation yields the following formula (see Exercise 1):

$$\kappa(t) \;=\; \frac{||\dot{\gamma}(t) \times \ddot{\gamma}(t)||}{||\dot{\gamma}(t)||^3}\,.$$

Definition 4. The *principal normal vector* $\vec{h}$ to a curve at a point of non-vanishing curvature is the normalized derivative of the unit tangent vector,

$$\vec{h}(s) := \frac{1}{\kappa(s)} \vec{t}\,'(s) = \frac{\gamma''(s)}{||\gamma''(s)||}.$$

The vector product

$$\vec{b}(s) := \vec{t}(s) \times \vec{h}(s)$$

is called the *binormal vector* to the curve.

Thus, at each point of the curve with non-vanishing curvature, there exist three mutually orthogonal vectors $\vec{t}(s)$, $\vec{h}(s)$, and $\vec{b}(s)$ all of length one, the so-called *Frenet frame* of the curve. Finally, we introduce a last geometric characteristic of a space curve, the torsion.

Definition 5. The *torsion* of a curve $\mathcal{C} \subset \mathbb{R}^3$ of class C^3 at a point of non-vanishing curvature is the scalar product

$$\tau(s) \;=\; -\langle \frac{d\vec{b}(s)}{ds},\ \vec{h}(s)\rangle.$$

The formula for the torsion in an arbitrary parametrization is the subject of Exercise 1.

The structural equations of a curve in euclidean space are the *Frenet formulas* expressing the derivatives of the Frenet frame through this same frame.

Theorem 3 (Fundamental Theorem of Curve Theory). *Consider a curve of class C^3 with nowhere vanishing curvature in the natural parametrization. Then, in the above notation,*

$$\frac{d}{ds}\begin{bmatrix} \vec{t}(s) \\ \vec{h}(s) \\ \vec{b}(s) \end{bmatrix} \;=\; \begin{bmatrix} 0 & \kappa(s) & 0 \\ -\kappa(s) & 0 & \tau(s) \\ 0 & -\tau(s) & 0 \end{bmatrix} \cdot \begin{bmatrix} \vec{t}(s) \\ \vec{h}(s) \\ \vec{b}(s) \end{bmatrix}.$$

Let $\gamma, \gamma^* : [0, L] \to \mathbb{R}^3$ be two curves with coinciding curvature and torsion in their natural parametrizations. Then there exists a euclidean motion $A : \mathbb{R}^3 \to \mathbb{R}^3$ such that $\gamma^*(s) = A \circ \gamma(s)$. Let $\kappa(s)$ be any positive function, and let $\tau(s)$ be an arbitrary function, both defined on the interval $[0, L]$. Then there exists a curve whose curvature and torsion are the given functions.

Proof. By the definition of the principal normal vector, we have $\vec{t}\,'(s) = \kappa(s) \cdot \vec{h}(s)$. Since $\|\vec{h}(s)\| \equiv 1$, the derivative $\vec{h}'(s)$ is orthogonal to $\vec{h}(s)$. Hence it can be represented as a linear combination of $\vec{t}(s)$ and $\vec{b}(s)$. The computation

$$\langle \frac{d\vec{h}(s)}{ds},\ \vec{t}(s)\rangle \;=\; \frac{d}{ds}\langle \vec{h}(s),\ \vec{t}(s)\rangle - \langle \vec{h}(s), \frac{d\vec{t}(s)}{ds}\rangle \;=\; -\kappa(s)$$

and the definition of torsion imply the second of the Frenet formulas. But then

$$\frac{d\vec{b}(s)}{ds} \;=\; \frac{d\vec{t}(s)}{ds} \times \vec{h}(s) + \vec{t} \times \frac{d\vec{h}(s)}{ds} \;=\; -\tau(s) \cdot \vec{h}(s).$$

Viewing the vectors of the Frenet frame as row vectors,

$$A := \begin{bmatrix} \vec{t}(s) \\ \vec{h}(s) \\ \vec{b}(s) \end{bmatrix}$$

defines a (3×3) matrix A of functions on $[0, L]$ satisfying $dA \cdot A^{-1} = \Omega$, where the entries of the skew-symmetric matrix Ω of 1-forms are the curvature and the torsion of the curve:

$$\Omega = \begin{bmatrix} 0 & \kappa(s) & 0 \\ -\kappa(s) & 0 & \tau(s) \\ 0 & -\tau(s) & 0 \end{bmatrix} \cdot ds.$$

Existence and uniqueness of a curve for given curvature and torsion functions thus follow immediately from Chapter 4, Theorem 6, since the integrability condition, $d\Omega - \Omega \wedge \Omega = 0$, is satisfied for trivial reasons. Note that the solution of the system of linear differential equations, $dA = \Omega \cdot A$, is defined on the whole interval on which the curvature $\kappa(s)$ and the torsion $\tau(s)$ are given. Having first determined the matrix $A(s)$ for the prescribed functions $\kappa(s), \tau(s)$ from the equation $dA = \Omega \cdot A$, we obtain the curve $\gamma(s)$ by one more integration,

$$\gamma(s) \;=\; \int \vec{t}(s) \cdot ds\,.$$ □

Remark. The plane spanned by the tangent vector $\vec{t}$ and the principal normal vector $\vec{h}$ at a given point is called the *osculating plane* of the curve at that point. The binormal vector $\vec{b}$ is perpendicular to this plane. The third Frenet formula implies $|\tau(s)| = \|\vec{b}'(s)\|$, and hence the absolute value $|\tau(p)|$ of the torsion of a curve at a point p can be described as the limit

$$|\tau(p)| \;=\; \lim_{q \to p} \frac{\psi}{\widehat{pq}},$$

where ψ denotes the angle between the osculating planes to the curve at the points p and q (compare the proof of Theorem 2). Thus the absolute value of the torsion measures how much the curve "winds out" of the osculating plane.

Next we discuss some curves with special properties, and explain how the Frenet formulas can be applied to study them.

Definition 6. A curve whose tangents form a constant angle with a fixed direction in $\mathbb{R}^3$ is called a *slope line*[1].

Straight lines and helices are slope lines. These can be characterized by the fact that the quotient $\tau(s)/\kappa(s)$ is constant.

Theorem 4 (Lancret, 1802). *A curve of class C^3 with nowhere vanishing curvature is a slope line if and only if $\tau(s)/\kappa(s)$ is constant.*

Proof. If there exists a vector $\vec{a} \in \mathbb{R}^3$ such that $\langle \vec{t}(s), \vec{a} \rangle$ is constant, then, differentiating this equation, we obtain that the scalar product $\langle \vec{h}(s), \vec{a} \rangle$

[1]In German, such a curve is called *Böschungslinie*.

vanishes, since $\kappa(s) \neq 0$. Differentiating once again leads, using the Frenet formulas, to

$$\tau(s) \cdot \langle \vec{b}(s), \vec{a} \rangle = \kappa(s) \cdot \langle \vec{t}, \vec{a} \rangle.$$

The vector $\vec{a}$ lies in the $\{\vec{t}, \vec{b}\}$-plane, and $\langle \vec{t}, \vec{a} \rangle$ is constant. Hence $\langle \vec{b}, \vec{a} \rangle$ is also constant, and thus the quotient τ/κ is constant. Conversely, if τ/κ is constant, then consider the vector

$$\vec{a} = \frac{\tau}{\kappa} \cdot \vec{t}(s) + \vec{b}(s).$$

The Frenet formulas imply

$$\frac{d\vec{a}}{ds} = \frac{\tau}{\kappa} \cdot \frac{d\vec{t}}{ds} + \frac{d\vec{b}}{ds} = \tau \cdot \vec{h} - \tau \cdot \vec{h} = 0,$$

and hence $\vec{a}$ is a constant vector. The scalar product $\langle \vec{t}(s), \vec{a} \rangle = \tau/\kappa$ is also constant, and the curve is thus a slope line. □

Curves lying on a sphere in 3-space can be described by a similar relation between curvature and torsion.

Theorem 5. *A curve of class C^4 with nowhere vanishing curvature and torsion, $\kappa, \tau \neq 0$, lies on a sphere of radius $R > 0$ if and only if it satisfies*

$$\frac{1}{\kappa^2} + \left(\frac{\kappa'}{\kappa^2 \tau}\right)^2 = R^2$$

in the natural parametrization.

Proof. Differentiating the equation $\|\gamma(s)\|^2 \equiv R^2$, we obtain $\langle \vec{t}(s), \gamma(s) \rangle = 0$, and hence $\gamma(s)$ is a linear combination of the vectors $\vec{h}(s)$ and $\vec{b}(s)$,

$$\gamma(s) = \alpha(s) \cdot \vec{h}(s) + \beta(s) \cdot \vec{b}(s).$$

Furthermore, $\|\gamma(s)\|^2 \equiv R^2$ implies $\alpha^2(s) + \beta^2(s) \equiv R^2$. Differentiating $\langle \vec{t}(s), \gamma(s) \rangle \equiv 0$ again leads to $\kappa(s) \cdot \langle \vec{h}(s), \gamma(s) \rangle + 1 \equiv 0$, and thus

$$\alpha(s) = -\frac{1}{\kappa(s)}.$$

We differentiate the equation $\kappa(s) \cdot \langle \vec{h}(s), \gamma(s) \rangle + 1 \equiv 0$ and obtain the following relation by a simple transformation:

$$\beta(s) = -\frac{\kappa'(s)}{\kappa^2(s)\tau(s)}.$$

The asserted necessary condition for a spherical curve then immediately follows:

$$R^2 = \alpha^2(s) + \beta^2(s) = \frac{1}{\kappa^2(s)} + \left(\frac{\kappa'(s)}{\kappa^2(s)\tau(s)}\right)^2.$$

If, conversely, this relation between curvature and torsion holds for a C^4-curve, then we first differentiate it and obtain the equation

$$\frac{\tau(s)}{\kappa(s)} - \frac{d}{ds}\left(\frac{\kappa'(s)}{\kappa^2(s)\tau(s)}\right) = 0 .$$

Now consider the vector

$$\vec{a}(s) := \gamma(s) + \frac{1}{\kappa(s)} \cdot \vec{h}(s) - \frac{\kappa'(s)}{\kappa^2(s)\tau(s)} \cdot \vec{b}(s) .$$

Using the Frenet formulas and the preceding relation between curvature and torsion, we compute the derivative of the latter, and find that

$$\frac{d}{ds}\vec{a}(s) = 0 .$$

Hence $\vec{a} := \vec{a}(s)$ is constant, and

$$\|\gamma(s) - \vec{a}\|^2 = \frac{1}{\kappa^2(s)} + \left(\frac{\kappa'(s)}{\kappa^2(s)\tau(s)}\right)^2 = R^2 ,$$

i. e., the curve $\gamma(s)$ lies on the sphere of radius R with center $\vec{a}$. □

Next we turn to plane curves. Note first that these can be described as the curves with vanishing torsion.

Theorem 6. *A curve of class C^3 with nowhere vanishing curvature, $\kappa(s) \neq 0$, lies in a plane in $\mathbb{R}^3$ if and only if its torsion $\tau(s) \equiv 0$ vanishes identically.*

Proof. Let $\vec{a}$ be a vector perpendicular to the plane in $\mathbb{R}^3$ containing the curve. All the tangent vectors $\vec{t}(s)$ lie in this plane; hence $\langle\vec{t}(s), \vec{a}\rangle \equiv 0$. Since $\kappa(s) \neq 0$, we immediately obtain $\langle\vec{h}(s), \vec{a}\rangle \equiv 0$ by differentiating this equation. Thus $\vec{a}$ coincides with the binormal vector $\vec{b}(s)$. In other words, the binormal vector $\vec{b}(s)$ is constant. Then $0 = \vec{b}'(s) = -\tau(s) \cdot \vec{h}(s)$ implies that the torsion vanishes, $\tau(s) \equiv 0$. The converse is proved analogously. □

The curvature of a plane curve can be ascribed a sign. In fact, the principal normal vector is proportional to the vector obtained by rotating the tangent vector through the angle $\pi/2$ in the positive sense. The curvature of a plane curve is ascribed the positive sign if the corresponding factor is positive. Identifying $\mathbb{R}^2$ with the complex numbers, the rotation through $\pi/2$ in the positive sense corresponds to multiplication by the number $i \in \mathbb{C}$. Using the multiplication of complex numbers, this leads to

Definition 7. Let $\gamma : [0, L] \to \mathbb{C} = \mathbb{R}^2$ be a plane curve in its natural parametrization. The plane curvature $k(s)$ is the function $k : [0, L] \to \mathbb{R}$ defined by the equation

$$\frac{d}{ds}\vec{t}(s) = k(s) \cdot i \cdot \vec{t}(s) .$$

The absolute value $|k(s)|$ of the plane curvature coincides with the curvature $\kappa(s)$ of the curve viewed as a space curve.

Example 2. $\gamma(t) = (t,\ \pm t^2)$.

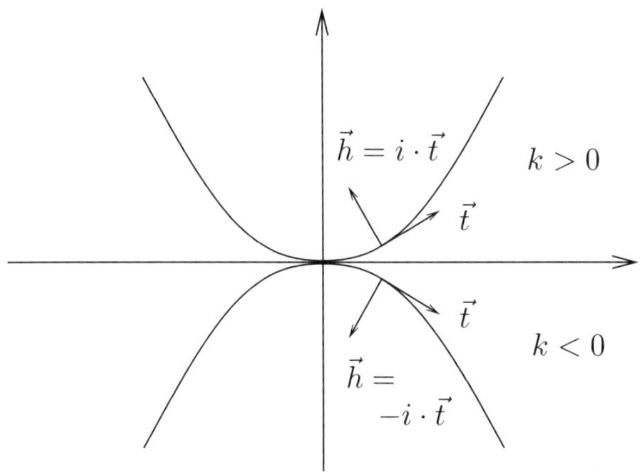

A closed curve $\gamma : [0, L] \to \mathbb{C} = \mathbb{R}^2$ is one that starts and ends at the same point and whose tangent vectors at this point coincide as well, $\gamma(0) = \gamma(L)$ and $\vec{t}(0) = \vec{t}(L)$.

Theorem 7. *Let $\gamma : [0, L] \to \mathbb{C}$ be a closed curve. Then the integral*

$$\frac{1}{2\pi} \int_\gamma k(s) ds \;=\; \frac{1}{2\pi} \int_0^L k(s) ds$$

is an integer, called the winding number of the closed curve.

Proof. Consider the map $t^* : [0, L] \to \mathbb{C}$ defined by

$$t^*(s) \;=\; \exp\left[i \int_0^s k(u) du \right] .$$

Then $dt^*(s)/ds = i \cdot k(s) \cdot t^*(s)$, and hence $\left(\vec{t}(s)/t^*(s)\right)' = 0$. The tangent vectors $\vec{t}(s)$ are thus described by the formula

$$\vec{t}(s) = C \cdot \exp\left[i \int_0^s k(u) du \right]$$

for a certain constant C. Since $\vec{t}(0) = \vec{t}(L)$, the number $\displaystyle\int_0^L k(s) ds$ is an integral multiple of 2π. □

We conclude the section on curves with the discussion of the *Fenchel inequality*, which claims that the total curvature of a closed space curve is bounded from below by 2π. We start by considering plane curves, and then generalize the result to space curves. First we need an auxiliary observation.

Lemma 1. *Let $\varphi : [a, b] \to \mathbb{R}$ be a real function of class C^1, and suppose that the function $f(t) := e^{\varphi(t)\cdot i}$ joins the numbers ± 1, i. e., $f(a) = 1,\ f(b) = -1$. Then*

(1) $\displaystyle\int_a^b |\dot{f}(\mu)| d\mu \ \geq\ \pi$;

(2) *in this estimate, equality holds if and only if the derivative $\dot{\varphi}$ does not change sign, and, moreover, $|\varphi(b) - \varphi(a)| = \pi$.*

Proof. From $\dot{f}(\mu) = i\dot{\varphi}(\mu)e^{i\varphi(\mu)}$, we obtain

$$\int_a^b |\dot{f}(\mu)| d\mu \ \geq\ \left| \int_a^b \dot{\varphi}(\mu) d\mu \right| \ =\ |\varphi(b) - \varphi(a)| \,.$$

Since $f(a) = 1$ and $f(b) = -1$, there exist integers k, l such that $\varphi(a) = 2k\pi$ and $\varphi(b) = 2l\pi + \pi$. This implies

$$\int_a^b |\dot{f}(\mu)| d\mu \ \geq\ \pi |1 + 2(k - l)| \ \geq\ \pi \,.$$

The case of equality follows immediately. □

Theorem 8. *Let $\gamma : [0, L] \to \mathbb{R}^2 = \mathbb{C}$ be a closed plane curve. The total curvature of the curve is at least 2π:*

$$\int_\gamma |k(s)| ds \ =\ \int_0^L |k(s)| ds \ \geq\ 2\pi \,.$$

Equality in this estimate occurs if and only if the closed curve γ bounds a convex region.

Proof. To every point of the curve we assign the tangent vector to the curve. This determines a map from the interval to the unit circle S^1 satisfying $\vec{t}(0) = \vec{t}(L)$. The image set $\vec{t}([0, L]) \subset S^1$ is a compact and connected subset of S^1, hence an arc. This arc contains at least one pair $\pm z_0$ of opposite points. Otherwise, there would exist a number $z_0^* \in S^1$ such that neither $+z_0^*$ nor $-z_0^*$ would belong to the arc. But then $\frac{1}{z_0^*}\vec{t}(s)$ would have an imaginary part which would have constant sign, and from

$$\int_0^L \mathrm{Im}\left(\frac{1}{z_0^*}\vec{t}(s)\right) \cdot ds = \mathrm{Im}\left(\frac{1}{z_0^*}\int_0^L \vec{t}(s) ds\right) = \mathrm{Im}\left(\frac{1}{z_0^*}\Big(\gamma(L) - \gamma(0)\Big)\right) = 0$$

we would obtain a contradiction. Without loss of generality, we may assume $z_0 = 1$ and $\vec{t}(0) = \vec{t}(L) = 1$. Moreover, there exists a parameter value

$s_1 \in [0, L]$ such that $\vec{t}(s_1) = -1$. Applying the lemma above, we obtain

$$\int_0^{s_1} |k(s)|ds = \int_0^{s_1} |\vec{t}'(\mu)|d\mu \geq \pi, \quad \int_{s_1}^{L} |k(s)|ds = \int_{s_1}^{L} |\vec{t}'(\mu)|d\mu \geq \pi,$$

and, altogether, $\int_0^L |k(s)|ds \geq 2\pi$. The inequality for plane curves is thus proved. In the case of equality, the lemma implies that the plane curvature $k(s)$ of the curve does not change sign on the interval $[0, s_1]$, and

$$\left| \int_0^{s_1} k(s)ds \right| = \pi = \left| \int_{s_1}^{L} k(s)ds \right| .$$

Now we exclude the case that $k(s)$ has different signs on the intervals $[0, s_1]$ and $[s_1, L]$. If this were the case, then

$$k(0) = k(s_1) = k(L) = 0\,,$$

and $k(s)$ would be positive on the interval $[0, s_1]$ and negative on $[s_1, L]$. But then the inequality

$$0 \leq \int_0^s k(\mu)d\mu \leq \pi$$

would hold for all parameters $s \in [0, L]$, and the tangent vector $\vec{t}(s) = \exp\left[i \int_0^s k(\mu)d\mu\right]$ would stay in the upper half-plane $\{z \in \mathbb{C} : \operatorname{Im}|z| \geq 0\}$; hence we derive a contradiction to

$$\int_0^L \vec{t}(s)ds = \gamma(L) - \gamma(0) = 0\,.$$

The plane curvature $k(s)$ of the curve thus has constant sign, $k(s) \geq 0$. In this case, the whole curve completely lies on one side of the tangent to the curve at an arbitrary point. In fact, without loss of generality, we can choose the point as that corresponding to the parameter value $s = 0$. Consider the function $\beta(s)$ measuring the height of the curve point $\gamma(s)$ over the tangent through $\gamma(0)$,

$$\beta(s) := \operatorname{Re}\left((\gamma(s) - \gamma(0)) \cdot \overline{(i \cdot \vec{t}(0))} \right) .$$

The derivative of $\beta(s)$ is easily computed:

$$\beta'(s) = \operatorname{Re}\left(\vec{t}(0) \cdot \exp\left[i \int_0^s k(u)du\right] \cdot \overline{(i \cdot \vec{t}(0))} \right) = \sin\left(\int_0^s k(u)du \right) .$$

The curvature is non-negative, and thus $\beta(s)$ is negative for parameter values close to $s = 0$. If $\beta(s)$ had any negative value in the interval $[0,\ L]$, then there would exist a value $s_{\min}$ such that

$$\beta(s_{\min}) < 0 \quad \text{and} \quad \beta'(s_{\min}) = 0\,.$$

Since $\int_0^L k(u) \cdot du = 2\pi$, this value $s_{\min}$ is determined by the equation

$$\int_0^{s_{\min}} k(u) \cdot du = \pi ,$$

and, for $s \in [0, s_{\min}]$, the derivative of the function $\beta(s)$ would be negative. Therefore, $\beta(s)$ would not be decreasing in $[0, s_{\min}]$, contradicting $\beta(s_{\min}) < 0$. Summarizing, we obtain that the height function is non-negative, and thus the curve lies completely on one side of each of its tangents. Consider now the half-planes determined by all the tangents. Their intersection is a convex domain whose boundary coincides with the curve $\gamma(s)$. □

We generalize the inequality of the preceding theorem to space curves. To do so, we need a preparation.

Lemma 2. *If $\mathcal{D} \subset S^2$ is a closed curve in the sphere of radius one whose length does not exceed 2π, then $\mathcal{D}$ is contained in a hemisphere.*

Proof. Choose two points P and Q on $\mathcal{D}$ which divide the curve into two segments of equal length, $\mathcal{D} = \mathcal{D}_1 \cup \mathcal{D}_2$. Let the north pole $N = (0,\ 0,\ 1)$ lie on the shorter segment of the great circle through P and Q on the sphere. Denote by $\mathcal{D}_1^*$ the curve segment obtained from $\mathcal{D}_1$ by rotation around the z-axis through the angle π. The union $\mathcal{D}_1 \cup \mathcal{D}_1^*$ is a closed curve on the sphere whose length coincides with that of the original curve $\mathcal{D}$. If the curve segment $\mathcal{D}_1$ intersects the equator $S^1 = \{(x,\ y,\ 0) : x^2+y^2 = 1\}$ of S^2, then the curve $\mathcal{D}_1 \cup \mathcal{D}_1^*$ contains opposite points on the sphere. Hence its length has to be at least 2π. By assumption this length cannot exceed 2π. Thus $\mathcal{D}_1$ does not intersect the equator S^1 at all, if the length of the curve $\mathcal{D}$ is smaller than 2π. Otherwise, $\mathcal{D}_1$ or $\mathcal{D}_2$ is a semi-circle on the equator, respectively. In both cases, the curve $\mathcal{D}$ is contained in the upper hemisphere. □

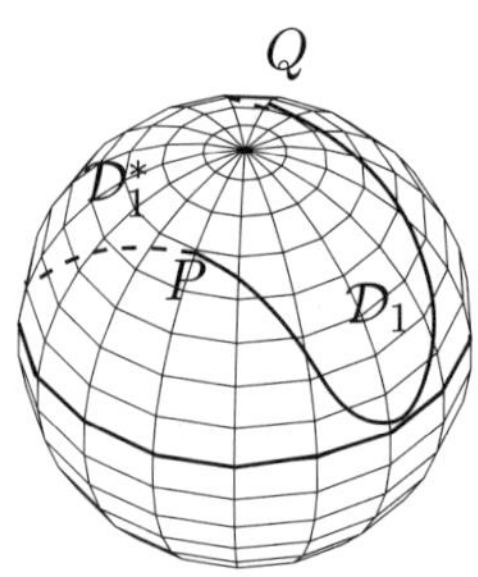

Theorem 9 (Fenchel Inequality). *The total curvature of a closed curve $\mathcal{C} \subset \mathbb{R}^3$ is at least 2π,*

$$\int_{\mathcal{C}} \kappa(s) \cdot ds \ \geq \ 2\pi\,.$$

Equality holds if and only if $\mathcal{C}$ is a plane curve bounding a convex domain.

Proof. Let a non-plane curve be given. Consider its tangent map $\vec{t}$: $[0, L] \to S^2$ as a "curve" in the sphere S^2. The length of the latter coincides with the total curvature of the original curve,

$$L(\vec{t}\,) \ = \ \int_0^L ||\vec{t}\,'(s)|| \cdot ds \ = \ \int_0^L \kappa(s) \cdot ds\,.$$

Note first that the image of $\vec{t}$ cannot be contained completely in a hemisphere of S^2. If this were the case, then there would be a vector $\vec{a} \in S^2$ such that $\langle \vec{a}, \vec{t}(s) \rangle \ \geq \ 0$ for all values of the parameter. But the integral of $\vec{t}(s)$ vanishes, since we started from a closed curve. Hence $\langle \vec{a}, \vec{t}(s) \rangle \ \equiv \ 0$, and the curve $\mathcal{C}$ would be a plane curve. The assertions of the theorem now follow from the previous lemma together with Theorem 8. □

5.2. The Structural Equations of a Surface

A surface is a two-dimensional submanifold of $\mathbb{R}^3$. As in the case of curves, we want to allow for self-intersections and will also consider parametrized surfaces. This is meant to denote a map $F : U \to \mathbb{R}^3$ defined on an open subset U of $\mathbb{R}^2$ (or, more generally, on a two-dimensional manifold U) with the property that the differential $D(F)$ has rank two at each point. As in the case of a two-dimensional submanifold $M^2 \subset \mathbb{R}^3$, we denote by M : $M^2 \to \mathbb{R}^3$ the inclusion into 3-space and view this map as a vector-valued function. Locally, we choose an orthonormal tangent frame e_1, e_2 to the surface and denote by $e_3 := e_1 \times e_2$ the normal vector to the surface. We sometimes consider—without introducing a new notation—e_1, e_2 and e_3 also as vector-valued functions on the manifold M^2. Denote by σ_1 and σ_2 the frame of 1-forms dual to $\{e_1, e_2\}$. The differential dM of the identity map on the surface is the identity map on each tangent space to the surface. Hence the following equation holds:

$$dM \ = \sigma_1 \cdot e_1 \ + \ \sigma_2 \cdot e_2\,,$$

expressing nothing but the decomposition of a tangent vector with respect to the basis e_1, e_2. The exterior product $\sigma_1 \wedge \sigma_2$ is a 2-form on the surface which is independent of the choice of the orthonormal frame e_1, e_2 for a fixed orientation and coincides with the volume form,

$$dM^2 \ = \ \sigma_1 \wedge \sigma_2\,.$$

The scalar products $\langle e_i,\, e_j \rangle$ of the vector-valued functions e_i are constant functions on M^2. By differentiation, we obtain

$$\langle de_i,\, e_j \rangle \;+\; \langle e_i,\, de_j \rangle \;=\; 0$$

and denote the 1-forms by $\omega_{ij} \;:=\; \langle de_i,\, e_j \rangle$. Now combine the differential forms into an antisymmetric (3×3) matrix Ω:

$$\Omega \;=\; \begin{bmatrix} 0 & \omega_{12} & \omega_{13} \\ \omega_{21} & 0 & \omega_{23} \\ \omega_{31} & \omega_{32} & 0 \end{bmatrix}.$$

Using the forms just introduced, the differentials of the vector-valued functions e_i can be represented as

$$de_i \;=\; \sum_{j=1}^{3} \omega_{ij} \cdot e_j \,.$$

Altogether, we constructed five differential forms of degree one, the forms σ_i and ω_{ij}. These are, however, not independent; the so-called *structural equations* of a surface express the relations among them. The fact that the form ω_{12} is completely determined by σ_1 and σ_2 will turn out to be of special importance. Later on, this will imply the fact that the Gaussian curvature belongs to the inner geometry of a surface (Theorema Egregium, Gauss, 1827).

Theorem 10 (Structural Equations of a Surface).

(1) $d\sigma_1 \;=\; \omega_{12} \wedge \sigma_2, \quad d\sigma_2 \;=\; \omega_{21} \wedge \sigma_1 \,.$

The form ω_{12} is completely determined by the forms σ_1 and σ_2 .

(2) $\sigma_1 \wedge \omega_{13} \;+\; \sigma_2 \wedge \omega_{23} \;=\; 0 \,.$

(3) $d\Omega \;=\; \Omega \wedge \Omega \,.$

Proof. Differentiate the function M twice,

$$\begin{aligned} 0 = ddM \;&=\; d(\sigma_1 \cdot e_1 \;+\; \sigma_2 \cdot e_2) \;=\; \big(d\sigma_1 - \omega_{12} \wedge \sigma_2\big) \cdot e_1 \\ &+\; \big(d\sigma_2 - \omega_{21} \wedge \sigma_1\big) \cdot e_2 \;-\; \big(\sigma_1 \wedge \omega_{13} + \sigma_2 \wedge \omega_{23}\big) \cdot e_3 \,. \end{aligned}$$

This yields the equations stated in (1) and (2). The identity for Ω follows using $0 \;=\; dde_i \;=\; \sum_k \big(d\omega_{ik} - \sum_j \omega_{ij} \wedge \omega_{jk}\big) \cdot e_k$. The form ω_{12} can be computed from σ_1 and σ_2 as follows. Represent the differentials

$$d\sigma_1 \;=\; A \cdot \sigma_1 \wedge \sigma_2, \quad d\sigma_2 \;=\; B \cdot \sigma_1 \wedge \sigma_2$$

as multiples of the 2-form $\sigma_1 \wedge \sigma_2$. The first structural equation implies

$$\omega_{12} \;=\; A \cdot \sigma_1 + B \cdot \sigma_2 \,.$$

□

Let a parametrization $F : U \to M^2 \subset \mathbb{R}^3$ of a surface by a a domain $U \subset \mathbb{R}^2$ be given; then the 1-forms σ_i and ω_{ij} can be pulled back to U. The resulting forms will be denoted by $\sigma_i^* := F^*(\sigma_i)$ and $\omega_{ij}^* := F^*(\omega_{ij})$. Then the structural equations of the surface hold as well:

$$d\sigma_i^* = \sum_{j=1}^{2} \omega_{ij}^* \wedge \sigma_j^*, \quad \sigma_1^* \wedge \omega_{13}^* + \sigma_2^* \wedge \omega_{23}^* = 0, \quad d\Omega^* = \Omega^* \wedge \Omega^* .$$

Now we prove a first formulation of the fundamental theorem of surface theory. This allows to determine a surface locally in 3-space from the system of associated 1-forms satisfying the structural equations relating them.

Theorem 11 (Fundamental Theorem of Surface Theory—First Formulation). *On a simply connected, open subset U of $\mathbb{R}^2$ let four differential forms $\overline{\sigma}_1, \overline{\sigma}_2, \overline{\omega}_{13}, \overline{\omega}_{23}$ of degree one be given. Suppose that the forms $\overline{\sigma}_1$ and $\overline{\sigma}_2$ are linearly independent at each point. Define the 1-form $\overline{\omega}_{12}$ by the equations*

$$d\overline{\sigma}_1 = \overline{\omega}_{12} \wedge \overline{\sigma}_2, \quad d\overline{\sigma}_2 = -\overline{\omega}_{12} \wedge \overline{\sigma}_1 .$$

Extending them antisymmetrically, $\overline{\omega}_{ji} := -\overline{\omega}_{ij}$, assume that the given system satisfies the structural equations

$$\overline{\sigma}_1 \wedge \overline{\omega}_{13} + \overline{\sigma}_2 \wedge \overline{\omega}_{23} = 0, \quad d\overline{\Omega} = \overline{\Omega} \wedge \overline{\Omega}.$$

Then there exist a parametrized surface $F : U \to \mathbb{R}^3$ and an orthonormal frame of tangent vector fields such that the induced forms σ_i^* and ω_{ij}^* coincide with the original forms $\overline{\sigma}_i$ and $\overline{\omega}_{ij}$, respectively. The surface together with its orthonormal frame is uniquely determined up to a euclidean motion of $\mathbb{R}^3$.

Proof. The (3×3) matrix $\overline{\Omega}$ is skew-symmetric and satisfies the integrability condition $d\overline{\Omega} = \overline{\Omega} \wedge \overline{\Omega}$. For a given initial condition $A_0 \in \mathrm{SO}(3,\mathbb{R})$, there exists a matrix $\overline{A}$ of functions such that $\overline{\Omega} \cdot \overline{A} = d\overline{A}$ by §4.3, Theorem 6. The rows of this matrix are orthogonal and have length one. Thus they define three vector-valued functions $\overline{e}_1, \overline{e}_2, \overline{e}_3 : U \to \mathbb{R}^3$, and

$$d\overline{e}_i = \sum_{j=1}^{3} \overline{\omega}_{ij} \cdot \overline{e}_j .$$

Consider the 1-form

$$\phi := \overline{\sigma}_1 \cdot \overline{e}_1 + \overline{\sigma}_2 \cdot \overline{e}_2 .$$

Computing the differential of ϕ yields

$$d\phi = (d\overline{\sigma}_1 - \overline{\omega}_{12} \wedge \overline{\sigma}_2) \cdot \overline{e}_1 + (d\overline{\sigma}_2 - \overline{\omega}_{21} \wedge \overline{\sigma}_1) \cdot \overline{e}_2 - (\overline{\sigma}_1 \wedge \overline{\omega}_{13} + \overline{\sigma}_2 \wedge \overline{\omega}_{13}) \cdot \overline{e}_3 = 0 .$$

By Poincarés' lemma (§3.6, Theorem 22) there exists a map $F : U \to \mathbb{R}^3$ such that $dF = \phi = \overline{\sigma}_1 \cdot \overline{e}_1 + \overline{\sigma}_2 \cdot \overline{e}_2$, and hence we obtain the parametrized surface together with the orthonormal frame. Twice in this proof we integrated a differential equation, and thus there remains a certain degree of freedom with respect to the initial condition. First, the solution $\overline{A}$ of the differential equation $\overline{\Omega} \cdot \overline{A} = d\overline{A}$ depends on the prescribed orthogonal matrix $A_0 \in \mathrm{SO}(3,\mathbb{R})$. In the second step, the function $F : U \to \mathbb{R}^3$ is determined by the differential equation $dF = \phi$ only up to a constant vector $a \in \mathbb{R}^3$. Summarizing, we see that the surface is, for prescribed forms $\overline{\sigma}_i$ and $\overline{\omega}_{ij}$, determined up to a euclidean motion. □

Example 3. We discuss the case of a surface of revolution in $\mathbb{R}^3$. Consider a plane curve in the (x, z)-plane with component functions $r(s) > 0$ and $z(s)$. The surface F arises by letting this generating curve revolve around the z-axis, and thus it has the following parametrization defined on $U = \mathbb{R} \times (0, 2\pi)$:

$$F(s,\varphi) = \big(r(s)\sin\varphi,\, r(s)\cos\varphi,\, z(s)\big)\,.$$

For simplicity, we suppose that the generating curve of the surface F is parametrized by arc-length, i.e., at each point the following condition is satisfied:

$$(*) \qquad (r')^2 + (z')^2 = 1\,.$$

(Compare Exercise 15 for the general case.) First we compute two tangent vectors to the surface,

$$\frac{\partial}{\partial s} = \big(r'\sin\varphi,\, r'\cos\varphi,\, z'\big), \qquad \frac{\partial}{\partial\varphi} = \big(r\cos\varphi,\, -r\sin\varphi,\, 0\big),$$

as well as their scalar products,

$$\left\langle \frac{\partial}{\partial s}, \frac{\partial}{\partial s}\right\rangle = (r')^2 + (z')^2 = 1, \quad \left\langle \frac{\partial}{\partial s}, \frac{\partial}{\partial \varphi}\right\rangle = 0, \quad \left\langle \frac{\partial}{\partial \varphi}, \frac{\partial}{\partial \varphi}\right\rangle = r^2\,.$$

Normalizing the second tangent vector correspondingly, we obtain the orthonormal frame

$$e_1 := \frac{\partial}{\partial s}, \qquad e_2 := \frac{1}{r}\frac{\partial}{\partial\varphi}$$

and the basis dual to it,

$$\sigma_1 = ds, \qquad \sigma_2 = r\,d\varphi\,.$$

The unit normal vector to the surface F is the vector product of e_1 and e_2,

$$e_3 := e_1 \times e_2 = \begin{bmatrix} z'\sin\varphi \\ z'\cos\varphi \\ -r' \end{bmatrix}.$$

We compute the differential of $e_1 = \big(r'\sin\varphi, r'\cos\varphi, z'\big)$ componentwise,

$$de_1 = \big(r''\sin\varphi\, ds + r'\cos\varphi\, d\varphi,\, r''\cos\varphi\, ds - r'\sin\varphi\, d\varphi,\, z''\, ds\big),$$

and check that

$$de_1 \;=\; r'\,d\varphi\,e_2 + \frac{r''}{z'}\,ds\,e_3\,.$$

In fact, inserting e_2 and e_3 into this linear combination, we obtain

$$\begin{aligned} &r'\,d\varphi\,e_2 + \frac{r''}{z'}\,ds\,e_3 \\ &= \big(r'\cos\varphi\,d\varphi,\,-r'\sin\varphi\,d\varphi,0\big) + \frac{r''}{z'}\,\big(z'\sin\varphi\,d\varphi,\,z'\cos\varphi\,ds,\,-r'\,ds\big) \\ &= \big(r''\sin\varphi ds + r'\cos\varphi\,d\varphi,\,r''\cos\varphi\,ds - r'\sin\varphi\,d\varphi,\,-r''\,r'/z'\,ds\big)\,. \end{aligned}$$

But, since differenting the relation $(*)$ implies $r'r'' + z'z'' = 0$, this is equal to de_1. Similarly, one proves that

$$de_2 \;=\; -r'\,d\varphi\,e_1 - z'\,d\varphi\,e_3, \quad de_3 \;=\; \frac{-r''}{z'}\,ds\,e_1 + z'\,d\varphi\,e_2\,,$$

and, therefore, we computed the matrix Ω of 1-forms ω_{ij} completely:

$$\Omega \;=\; \begin{bmatrix} 0 & \omega_{12} & \omega_{13} \\ -\omega_{12} & 0 & \omega_{23} \\ -\omega_{13} & -\omega_{23} & 0 \end{bmatrix} \;=\; \begin{bmatrix} 0 & r'\,d\varphi & \frac{r''}{z'}\,ds \\ -r'\,d\varphi & 0 & -z'\,d\varphi \\ \frac{-r''}{z'}\,ds & z'\,d\varphi & 0 \end{bmatrix}.$$

In addition, we show in this example how to compute the form ω_{12} directly from σ_1 and σ_2. Starting from $d\sigma_1 = 0$ and $d\sigma_2 = r'\cdot ds\wedge d\varphi = r'/r\cdot\sigma_1\wedge\sigma_2$, we obtain the formula

$$\omega_{12} \;=\; \frac{r'}{r}\cdot\sigma_2 \;=\; r'\cdot d\varphi$$

using the method described in the proof of Theorem 10. Computing ω_{12} directly from the definition turns out to be even easier:

$$\omega_{12} \;=\; \langle de_1,\,e_2\rangle \;=\; \big\langle r'd\varphi\,e_2 + \frac{r''}{z'}ds\,e_3,\,e_2\big\rangle \;=\; = \; r'\,d\varphi\,.$$

Example 4. Let $h : U \to \mathbb{R}$ be a real function in two variables on the open set $U \subset \mathbb{R}^2$, and denote by F its graph,

$$F(x,y) \;=\; \big(x,\,y,\,h(x,y)\big)\,.$$

We indicate partial derivatives by lower indices. Starting from the two tangent vectors

$$\frac{\partial}{\partial x} \;=\; (1,\,0,\,h_x) \;\text{ and }\; \frac{\partial}{\partial y} \;=\; (0,\,1,\,h_y),$$

we determine an orthonormal frame with the same orientation,

$$e_1 \;=\; \frac{1}{\sqrt{1+h_x^2}}\frac{\partial}{\partial x}, \quad e_2 \;=\; \frac{h_xh_y}{\sqrt{1+h_x^2}\,\sqrt{1+h_x^2+h_y^2}}\left(\frac{1+h_x^2}{h_xh_y}\frac{\partial}{\partial y} - \frac{\partial}{\partial x}\right),$$

as well as the 1-forms dual to these vectors,

$$\sigma_1 = \sqrt{1+h_x^2}\,dx + \frac{h_x h_y}{\sqrt{1+h_x^2}}\,dy, \quad \sigma_2 = \frac{\sqrt{1+h_x^2+h_y^2}}{\sqrt{1+h_x^2}}\,dy\,.$$

The corresponding volume element is

$$\sigma_1 \wedge \sigma_2 = \sqrt{1+h_x^2+h_y^2}\,dx \wedge dy\,.$$

The computation of the whole matrix Ω is more extensive. Because of its geometric relevance, we give the form ω_{12} for later use:

$$\omega_{12} = \langle de_1, e_2 \rangle = \frac{h_y\,dh_x}{(1+h_x^2)\sqrt{1+h_x^2+h_y^2}}\,.$$

In the discussion of curvature properties of surfaces, we will see that particularly interesting graphs arise from holomorphic functions. Therefore, these are discussed next.

Example 5. Let $f : U \to \mathbb{C}$ be a holomorphic function. The *modular surface* of f is the graph of the modulus function $h = |f| = (f\bar{f})^{1/2}$ of f. The relations between the derivatives with respect to the complex parameter z and the (real) partial derivatives are expressed in the following formulas:

$$\frac{\partial}{\partial z} = \frac{1}{2}\left(\frac{\partial}{\partial x} - i\frac{\partial}{\partial y}\right), \quad \frac{\partial}{\partial \bar{z}} = \frac{1}{2}\left(\frac{\partial}{\partial x} + i\frac{\partial}{\partial y}\right).$$

Hence it is possible to express h_x and h_y by the z-derivatives of f and $\bar{f}$. The latter are denoted by $'$. It is easy to see that

$$\frac{\partial h}{\partial z} = \frac{f'\bar{f}}{2h}, \quad \frac{\partial h}{\partial \bar{z}} = \frac{f\bar{f}'}{2h}\,.$$

Thus the derivatives we wanted to compute are

$$h_x = \frac{1}{2h}(f'\bar{f} + f\bar{f}') \quad \text{and} \quad h_y = \frac{i}{2h}(f'\bar{f} - f\bar{f}').$$

Rewriting this yields, e. g. ,

$$1 + h_x^2 + h_y^2 = 1 + |f'|^2\,,$$

and inserting these expressions into the formula of the previous example we immediately obtain an explicit expression for the corresponding form ω_{12} depending on the complex derivatives of the function f. Indicating the lines of constant modulus as well as those of constant argument on its modular surface leads to a clear picture of the behavior of the function in the complex domain. As an example, we consider the function $f(z) = 1/\sin(z)$ outside its poles, shaded according to the argument and showing the level lines of the modulus:

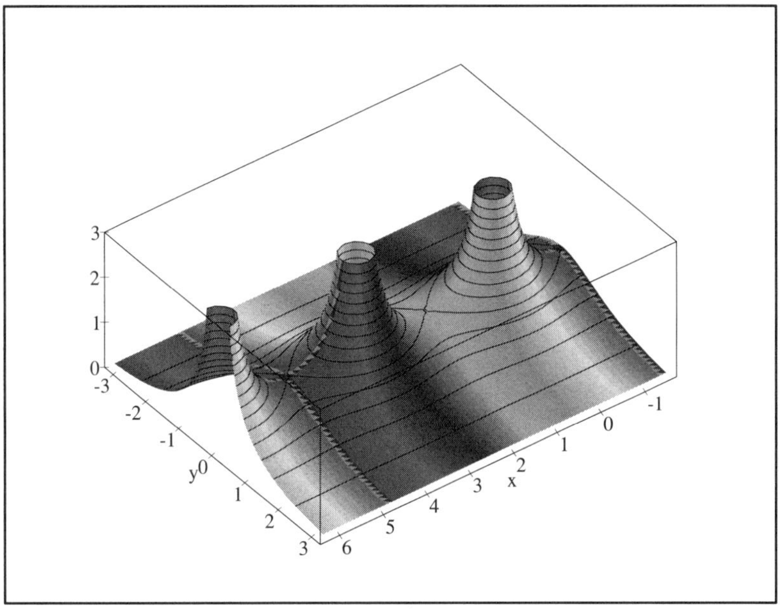

Example 6. Decompose a holomorphic function $f : U \to \mathbb{C}$ into its real and imaginary parts, $f = u + iv$. Taking in Example 4 for h the functions $h = u$ and $h = v$, respectively, we obtain the graphs of the real and the imaginary part of f. Starting from the Cauchy-Riemann equations,

$$u_x = v_y, \quad u_y = -v_x\,,$$

and first using the formula for $\partial/\partial z$ derived in the preceding example, $f' = u_x - iu_y = v_y + iv_x$, we obtain

$$1 + u_x^2 + u_y^2 \;=\; 1 + v_x^2 + v_y^2 \;=\; 1 + |f'|^2.$$

At this point we omit the explicit formula for ω_{12}.

5.3. The First and Second Fundamental Forms of a Surface

In classical differential geometry, it is common to describe a surface by two square matrices, the so-called first and second fundamental forms. The first of these is already familiar from Chapter 3, since the first fundamental form is just the Riemannian metric. The second fundamental form depends on the particular realization of the surface in euclidean space. To be able to define the first fundamental form in the language of differential forms, we introduce the *symmetric product* of 1-forms, in addition to the exteror product we already know.

Definition 8. Let M^n be a manifold, and let σ, μ be two 1-forms on M^n. Define a symmetric bilinear form $\sigma \odot \mu : TM^n \times TM^n \to \mathbb{R}$ by

$$\sigma \odot \mu(\mathcal{V}, \mathcal{W}) := \frac{1}{2}\big(\sigma(\mathcal{V})\mu(\mathcal{W}) + \sigma(\mathcal{W})\mu(\mathcal{V})\big).$$

In the literature, it is common to omit the symbol $\odot$. The symmetric square of a 1-form σ is often denoted by σ^2.

The definition immediately implies that, on a surface with orthonormal frame e_1, e_2 and corresponding dual basis of 1-forms σ_1, σ_2, the following equation holds:

$$\sigma_i \odot \sigma_i(e_j, e_k) = \sigma_i(e_j)\sigma_i(e_k) = \delta_{ij}\delta_{ik}.$$

For the basis e_1, e_2 the bilinear form

$$\mathrm{I} := \sigma_1 \odot \sigma_1 + \sigma_2 \odot \sigma_2$$

is thus precisely the usual euclidean scalar product. It is called the *first fundamental form* of the surface; it coincides with the Riemannian metric defined in Chapter 3. Geometric quantities defined for the surface depending only on the first fundamental form are called quantities belonging to the inner geometry.

Example 7. For the surface of revolution from Example 3, we had $\sigma_1 = ds$ and $\sigma_2 = r\,d\varphi$; thus the first fundamental form is

$$\mathrm{I} = ds \odot ds + r^2 d\varphi \odot d\varphi.$$

This is also written as $\mathrm{I} = ds^2 + r^2 d\varphi^2$.

Example 8. The first fundamental form of a graph (Example 4) is

$$\mathrm{I} = (1 + h_x^2)\,dx^2 + 2\,h_x h_y\,dx\,dy + (1 + h_y^2)\,dy^2.$$

Example 9. On the upper half-plane $\mathcal{H}^2 = \{(x, y) \in \mathbb{R}^2 : y > 0\}$, we consider the 1-forms

$$\sigma_1 = \frac{dx}{y}, \quad \sigma_2 = \frac{dy}{y}.$$

They determine the Riemannian metric

$$\mathrm{I} = \frac{dx^2 + dy^2}{y^2}.$$

The half-plane $\mathcal{H}^2$ endowed with this metric is called the *hyperbolic plane.* The form ω_{12} depending only on σ_1 and σ_2 is equal to σ_1.

If $\mathcal{V}$ and $\mathcal{W}$ are two tangent vector fields to the surface M^2, then, viewing $\mathcal{W}$ as a vector-valued function and differentiating it as such with respect to the vector field $\mathcal{V}$, we denote the resulting vector-valued function by $\mathcal{V}(\mathcal{W})$. In general, $\mathcal{V}(\mathcal{W})$ is not tangent to the surface. However,

Theorem 12. *Let $\mathcal{V}$ and $\mathcal{W}$ be two tangent vector fields on a surface. Then the difference $\mathcal{V}(\mathcal{W}) - \mathcal{W}(\mathcal{V})$ is also tangent to the surface, and it coincides with the commutator of the vector fields,*

$$\mathcal{V}(\mathcal{W}) - \mathcal{W}(\mathcal{V}) \;=\; [\mathcal{V}, \mathcal{W}] .$$

Proof. In the proof we make use of the structural equations of the surface. $\mathcal{V}$ and $\mathcal{W}$ are tangent vector fields, and hence the scalar products $\langle \mathcal{V}, e_3\rangle = \langle \mathcal{W}, e_3\rangle = 0$ vanish identically. This implies

$$\begin{aligned}
&\langle \mathcal{V}(\mathcal{W}) - \mathcal{W}(\mathcal{V}),\, e_3\rangle \\
&= -\langle \mathcal{W},\, de_3(\mathcal{V})\rangle + \langle \mathcal{V},\, de_3(\mathcal{W})\rangle \\
&= -\langle \mathcal{W},\; \omega_{31}(\mathcal{V})e_1 + \omega_{32}(\mathcal{V})e_2\rangle + \langle \mathcal{V},\; \omega_{31}(\mathcal{W})e_1 + \omega_{32}(\mathcal{V})e_2\rangle \\
&= \sigma_1 \wedge \omega_{31}(\mathcal{V}, \mathcal{W}) + \sigma_2 \wedge \omega_{32}(\mathcal{V}, \mathcal{W}).
\end{aligned}$$

Because of the structural equation $\sigma_1 \wedge \omega_{31} + \sigma_2 \wedge \omega_{32} = 0$, the difference $\mathcal{V}(\mathcal{W}) - \mathcal{W}(\mathcal{V})$ is a tangent vector field on the surface. Moreover, from §3.9 and the structural equation, we conclude for $i = 1, 2$ that

$$\begin{aligned}
\sigma_i\big(\mathcal{V}(\mathcal{W}) - \mathcal{W}(\mathcal{V})\big) &= \langle \mathcal{V}(\mathcal{W}) - \mathcal{W}(\mathcal{V}),\, e_i\rangle \\
&= \mathcal{V}\big(\langle \mathcal{W}, e_i\rangle\big) - \langle \mathcal{W}, de_i(\mathcal{V})\rangle - \mathcal{W}\big(\langle \mathcal{V}, e_i\rangle\big) + \langle \mathcal{V}, de_i(\mathcal{W})\rangle \\
&= \mathcal{V}\big(\,\langle \mathcal{W}, e_i\rangle\,\big) - \mathcal{W}\big(\,\langle \mathcal{V}, e_i\rangle\,\big) + \sum_{j=1}^{2} \sigma_j \wedge \omega_{ij}(\mathcal{V}, \mathcal{W}) \\
&= \mathcal{V}\big(\sigma_i(\mathcal{W})\big) - \mathcal{W}\big(\sigma_i(\mathcal{V})\big) - d\sigma_i(\mathcal{V}, \mathcal{W}) \;=\; \sigma_i\big([\mathcal{V}, \mathcal{W}]\big).
\end{aligned}$$

Thus $\mathcal{V}(\mathcal{W}) - \mathcal{W}(\mathcal{V})$ coincides with the commutator $[\mathcal{V}, \mathcal{W}]$. □

We decompose the vector-valued function $\mathcal{V}(\mathcal{W})$ into its tangent and its normal part. The tangent part is denoted by $\nabla_{\mathcal{V}}\mathcal{W}$, whereas the normal part will give rise to the so-called second fundamental form $\mathrm{II}(\mathcal{V}, \mathcal{W})$ of the surface.

Definition 9. Let $\mathcal{V}$ and $\mathcal{W}$ be tangent vector fields on the surface $M^2 \subset \mathbb{R}^3$. The tangent part of $\mathcal{V}(\mathcal{W})$,

$$\nabla_{\mathcal{V}}\mathcal{W} \;:=\; \mathcal{V}(\mathcal{W}) - \langle \mathcal{V}(\mathcal{W}),\, e_3\rangle \cdot e_3 \;=\; \mathcal{V}(\mathcal{W}) + \langle \mathcal{W},\, de_3(\mathcal{V})\rangle \cdot e_3 ,$$

is called the *covariant derivative of the vector field* $\mathcal{W}$ with respect to the vector field $\mathcal{V}$. The normal part of the vector-valued function $\mathcal{V}(\mathcal{W})$ will be denoted by $\mathrm{II}(\mathcal{V}, \mathcal{W})$,

$$\mathrm{II}(\mathcal{V}, \mathcal{W}) \;:=\; \langle \mathcal{W},\, de_3(\mathcal{V})\rangle \;;$$

it is called the *second fundamental form of the surface.*

Remark. The second fundamental form of a surface depends on the orientation of M^2. Changing it reverses the normal vector e_3, and hence II changes sign, too.

First we collect the properties of the covariant derivative ∇.

Theorem 13.

$$\begin{aligned}
&(1)\ \nabla_{\mathcal{V}}(\mathcal{W}_1+\mathcal{W}_2) = \nabla_{\mathcal{V}}(\mathcal{W}_1)+\nabla_{\mathcal{V}}(\mathcal{W}_2);\\
&(2)\ \nabla_{\mathcal{V}_1+\mathcal{V}_2}\mathcal{W} = \nabla_{\mathcal{V}_1}\mathcal{W}+\nabla_{\mathcal{V}_1}\mathcal{W};\\
&(3)\ \textit{If } f : M^2 \to \mathbb{R} \textit{ is a smooth function, then}\\
&\nabla_{\mathcal{V}}(f\cdot\mathcal{W}) = df(\mathcal{V})\cdot\mathcal{W}+f\cdot\nabla_{\mathcal{V}}\mathcal{W} \quad \textit{and} \quad \nabla_{f\cdot\mathcal{V}}\mathcal{W} = f\cdot\nabla_{\mathcal{V}}\mathcal{W};\\
&(4)\ \mathcal{V}\langle\mathcal{W}_1,\ \mathcal{W}_2\rangle = \langle\nabla_{\mathcal{V}}\mathcal{W}_1,\ \mathcal{W}_2\rangle+\langle\mathcal{W}_1,\ \nabla_{\mathcal{V}}\mathcal{W}_2\rangle;\\
&(5)\ \nabla_{\mathcal{V}}\mathcal{W}-\nabla_{\mathcal{W}}\mathcal{V} = [\mathcal{V},\mathcal{W}].
\end{aligned}$$

Proof. The additivity of $\nabla_{\mathcal{V}}\mathcal{W}$ is a straightforward consequence of the definition. The last formula follows from the previous theorem. All the remaining assertions are easily checked, e. g.,

$$\nabla_{\mathcal{V}}(f\cdot\mathcal{W}) = \mathcal{V}(f\cdot\mathcal{W})+\langle f\cdot\mathcal{W},\ de_3(\mathcal{V})\rangle\cdot e_3 = df(\mathcal{V})\cdot\mathcal{W}+f\cdot\nabla_{\mathcal{V}}\mathcal{W}.\ \square$$

Representing the vector field $\mathcal{W}$ in an orthonormal frame e_1, e_2 on the surface, $\mathcal{W} = W^1\cdot e_1 + W^2\cdot e_2$, and taking into account that

$$\begin{aligned}
\nabla_{\mathcal{V}}e_1 &= de_1(\mathcal{V})+\langle e_1, de_3(\mathcal{V})\rangle\ \cdot e_3\\
&= \omega_{12}(\mathcal{V})\cdot e_2+\omega_{13}(\mathcal{V})\cdot e_3+\omega_{31}(\mathcal{V})\cdot e_3 = \omega_{12}(\mathcal{V})\cdot e_2,
\end{aligned}$$

we obtain the important formula

$$\nabla_{\mathcal{V}}\mathcal{W} = (dW^1(\mathcal{V})+\omega_{21}(\mathcal{V})\cdot W^2)\cdot e_1+(dW^2(\mathcal{V})+\omega_{12}(\mathcal{V})\cdot W^1)\cdot e_2\,.$$

The form ω_{12} depends only on σ_1, σ_2, and hence the covariant derivative $\nabla_{\mathcal{V}}\mathcal{W}$ also depends exclusively on quantities belonging to the inner geometry of the surface. The second fundamental form $\mathrm{II}(\mathcal{V},\mathcal{W})$ is a symmetric bilinear form on the tangent bundle of the surface.

Theorem 14. *Let $\mathcal{V}$ and $\mathcal{W}_1, \mathcal{W}_2$ be tangent vector fields on a surface, and let f_1, f_2 be smooth functions. Then*

(1) *the second fundamental form is symmetric,* $\mathrm{II}(\mathcal{V},\mathcal{W}_1) = \mathrm{II}(\mathcal{W}_1,\mathcal{V})$;

(2) $\mathrm{II}(\mathcal{V}, f_1\cdot\mathcal{W}_1+f_2\cdot\mathcal{W}_2) = f_1\cdot\mathrm{II}(\mathcal{V},\mathcal{W}_1)+f_2\cdot\mathrm{II}(\mathcal{V},\mathcal{W}_2)$.

Proof. We prove the symmetry of the second fundamental form; the remaining property immediately follows from the definition. With $\mathcal{V}$ and $\mathcal{W}$, the commutator $[\mathcal{V}, \mathcal{W}] = \mathcal{V}(\mathcal{W}) - \mathcal{W}(\mathcal{V})$ is a tangent vector field as well. Thus

$$\begin{aligned} \langle \mathcal{V},\, de_3(\mathcal{W})\rangle &= -\langle \mathcal{W}(\mathcal{V}),\, e_3\rangle = -\langle \mathcal{V}(\mathcal{W}),\, e_3\rangle + \langle \mathcal{V}(\mathcal{W}) - \mathcal{W}(\mathcal{V}),\, e_3\rangle \\ &= -\langle \mathcal{V}(\mathcal{W}),\, e_3\rangle = \langle \mathcal{W},\, de_3(\mathcal{V})\rangle\,, \end{aligned}$$

i. e. , $\mathrm{II}(\mathcal{V}, \mathcal{W})$ is symmetric. □

Because of the structural equation $de_3 = \omega_{31} \cdot e_1 + \omega_{32} \cdot e_2$, the second fundamental form can be written as a symmetric product of the following 1-forms:

$$\mathrm{II} = \omega_{31} \odot \sigma_1 + \omega_{32} \odot \sigma_2.$$

Hence, in the orthonormal frame e_1, e_2, the second fundamental form is represented as the symmetric matrix

$$\mathrm{II} = \begin{bmatrix} \mathrm{II}(e_1, e_1) & \mathrm{II}(e_1, e_2) \\ \mathrm{II}(e_1, e_2) & \mathrm{II}(e_2, e_2) \end{bmatrix} = \begin{bmatrix} \omega_{31}(e_1) & \omega_{31}(e_2) \\ \omega_{32}(e_1) & \omega_{32}(e_2) \end{bmatrix}.$$

If, on the other hand, $F : U \to M^2 \subset \mathbb{R}^3$ is a parametrization of the surface, and we choose the normal vector

$$e_3 = \frac{\frac{\partial F}{\partial y^1} \times \frac{\partial F}{\partial y^2}}{\|\frac{\partial F}{\partial y^1} \times \frac{\partial F}{\partial y^2}\|} = \frac{1}{\sqrt{g}}\left(\frac{\partial F}{\partial y^1} \times \frac{\partial F}{\partial y^2}\right),$$

then the second fundamental form is given by the symmetric (2×2) matrix

$$b = \begin{bmatrix} b_{11} & b_{12} \\ b_{21} & b_{22} \end{bmatrix}$$

with coefficients

$$b_{ij} = -\left\langle \frac{\partial^2 F}{\partial y^i \partial y^j},\, e_3 \right\rangle = -\frac{1}{\sqrt{g}}\left\langle \frac{\partial^2 F}{\partial y^i \partial y^j},\, \frac{\partial F}{\partial y^1} \times \frac{\partial F}{\partial y^2} \right\rangle.$$

So for each surface $M^2 \subset \mathbb{R}^3$, there exist two symmetric bilinear forms, I and II, of which the first is positive definite at each point. These two fundamental forms are not independent; the structural equations of the surface will lead to a pair of differential equations relating them to each other. In order to formulate these equations, we introduce the curvature tensor of the surface. This is a transformation $\mathcal{R} : TM^2 \times TM^2 \times TM^2 \to TM^2$ associating with every triple of vectors a tangent vector.

Definition 10. Let $\mathcal{U}, \mathcal{V}, \mathcal{W}$ be tangent vector fields on the surface. The *curvature tensor* $\mathcal{R}$ is defined by the following formula:

$$\mathcal{R}(\mathcal{U}, \mathcal{V})\mathcal{W} := \nabla_{\mathcal{U}}\nabla_{\mathcal{V}}\mathcal{W} - \nabla_{\mathcal{V}}\nabla_{\mathcal{U}}\mathcal{W} - \nabla_{[\mathcal{U},\mathcal{V}]}\mathcal{W}.$$

The following relations for the curvature tensor are a formal consequence of the properties of the covariant derivative $\nabla_{\mathcal{V}}\mathcal{W}$ for vector fields stated in Theorem 13. We leave their verification to the reader.

Theorem 15. *Let $\mathcal{U}, \mathcal{V}, \mathcal{W}$ be tangent vector fields, and let f be a function. Then*

$$(1)\ \mathcal{R}(f\cdot\mathcal{U},\mathcal{V})\mathcal{W} = \mathcal{R}(\mathcal{U},f\cdot\mathcal{V})\mathcal{W} = \mathcal{R}(\mathcal{U},\mathcal{V})(f\cdot\mathcal{W}) = f\cdot\mathcal{R}(\mathcal{U},\mathcal{V})\mathcal{W},$$

$$(2)\ \mathcal{R}(\mathcal{U},\mathcal{V})\mathcal{W} = -\mathcal{R}(\mathcal{V},\mathcal{U})\mathcal{W},$$

$$(3)\ \mathrm{I}(\mathcal{R}(\mathcal{U},\mathcal{V})\mathcal{W}_1,\mathcal{W}_2) = -\mathrm{I}(\mathcal{R}(\mathcal{U},\mathcal{V})\mathcal{W}_2,\mathcal{W}_1),$$

$$(4)\ \mathrm{I}(\mathcal{R}(\mathcal{U},\mathcal{V})\mathcal{W}_1,\mathcal{W}_2) = \mathrm{I}(\mathcal{R}(\mathcal{W}_1,\mathcal{W}_2)\mathcal{U},\mathcal{V}),$$

$$(5)\ \mathcal{R}(\mathcal{U},\mathcal{V})\mathcal{W}+\mathcal{R}(\mathcal{V},\mathcal{W})\mathcal{U}+\mathcal{R}(\mathcal{W},\mathcal{U})\mathcal{V} = 0.$$

Recalling that the covariant derivative $\nabla_{\mathcal{V}}\mathcal{W}$ of two tangent vectors only depends on the forms σ_1, σ_2 belonging to the inner geometry, the same also holds for the curvature tensor. We prove a local formula for $\mathcal{R}(\mathcal{U},\mathcal{V})\mathcal{W}$ specifying this fact.

Theorem 16. *Let e_1, e_2 be an orthonormal frame on M^2. Then*

$$\mathcal{R}(\mathcal{U},\mathcal{V})\mathcal{W} = d\omega_{12}(\mathcal{U},\mathcal{V})\cdot\big(\sigma_1(\mathcal{W})\cdot e_2 - \sigma_2(\mathcal{W})\cdot e_1\big)\,.$$

Proof. Because of the symmetry properties of the curvature tensor stated in the preceding theorem, it suffices to prove this formula for the basis fields $\mathcal{U} = e_1,\ \mathcal{V} = e_2,\ \mathcal{W} = e_1$. In this case,

$$\begin{aligned}
\mathcal{R}(e_1,e_2)e_1 &= \nabla_{e_1}(\nabla_{e_2}e_1) - \nabla_{e_2}(\nabla_{e_1}e_1) - \nabla_{[e_1,e_2]}e_1\\
&= \nabla_{e_1}(\omega_{12}(e_2)\cdot e_2) - \nabla_{e_2}(\omega_{12}(e_1)\cdot e_2) - \omega_{12}([e_1,e_2])\cdot e_2\\
&= \big(e_1(\omega_{12}(e_2)) - e_2(\omega_{12}(e_1)) - \omega_{12}([e_1,e_2])\big)\cdot e_2\\
&\quad + \big(\omega_{12}(e_2)\omega_{21}(e_1) - \omega_{12}(e_1)\omega_{21}(e_2)\big)\cdot e_1\\
&= d\omega_{12}(e_1,e_2)\cdot e_2\,,
\end{aligned}$$

and this last expression coincides with the right-hand side of the equation to be proved. □

The first fundamental form is positive definite. Hence the symmetric second fundamental form can be represented by a symmetric endomorphism $S : TM^2 \longrightarrow TM^2$ defined by

$$\mathrm{II}(\mathcal{V},\mathcal{W}) = \mathrm{I}(\mathcal{V},S(\mathcal{W}))\,,$$

the so-called *Weingarten map* of the surface (see §5.4). The equation $\mathrm{II}(\mathcal{V},\mathcal{W}) = \langle\mathcal{V},\, de_3(\mathcal{W})\rangle$ implies that S can be rewritten as

$$S = de_3 = \omega_{31}\cdot e_1 + \omega_{32}\cdot e_2\,.$$

The covariant derivative of an arbitrary endomorphism $E : TM^2 \to TM^2$ of the tangent bundle is a 2-form $\nabla E : TM^2 \times TM^2 \to TM^2$ with values in TM^2. This is defined by the equation

$$\nabla E(\mathcal{V}, \mathcal{W}) \;:=\; \nabla_{\mathcal{V}}(E(\mathcal{W})) - \nabla_{\mathcal{W}}(E(\mathcal{V})) - E([\mathcal{V}, \mathcal{W}])\,.$$

Theorem 17 (Gauss and Codazzi-Mainardi Equations).

(1) $\nabla S = 0$ *(Codazzi-Mainardi equation);*

(2) $\mathcal{R}(\mathcal{U}, \mathcal{V})\mathcal{W} \;=\; \mathrm{II}(\mathcal{V}, \mathcal{W}) \cdot S(\mathcal{U}) - \mathrm{II}(\mathcal{U}, \mathcal{W}) \cdot S(\mathcal{V})$ *(Gauss equation).*

Proof. Using the formula for the exterior derivative of a 1-form, the equation $dS = dde_3 = 0$ first implies

$$0 \;=\; \mathcal{V}(S(\mathcal{W})) - \mathcal{W}(S(\mathcal{V})) - S([\mathcal{V}, \mathcal{W}])\,.$$

Looking at the tangent parts and taking into account that $S([\mathcal{V}, \mathcal{W}])$ is already tangent, we obtain

$$0 \;=\; \nabla_{\mathcal{V}}(S(\mathcal{W})) - \nabla_{\mathcal{W}}(S(\mathcal{V})) - S([\mathcal{V}, \mathcal{W}])\,.$$

Hence $\nabla S = 0$ is proved. Recalling that the second fundamental form is $\mathrm{II}(\mathcal{V}, \mathcal{W}) = \langle \mathcal{V}, de_3(\mathcal{W})\rangle$, we obtain $S(\mathcal{U}) = de_3(\mathcal{U})$. Using now $de_3 = \omega_{31} \cdot e_1 + \omega_{32} \cdot e_2$ and then applying the preceding theorem proves the Gauss equation:

$$\begin{aligned} \mathrm{II}(\mathcal{V}, \mathcal{W})\, S(\mathcal{U}) - \mathrm{II}(\mathcal{U}, \mathcal{W})\, S(\mathcal{V}) &= \omega_{13} \wedge \omega_{32}(\mathcal{U}, \mathcal{V})\, (\sigma_1(\mathcal{W})\, e_2 - \sigma_2(\mathcal{W})\, e_1) \\ &= d\omega_{12}(\mathcal{U}, \mathcal{V})\, (\sigma_1(\mathcal{W})\, e_2 - \sigma_2(\mathcal{W})\, e_1) \\ &= \mathcal{R}(\mathcal{U}, \mathcal{V})\mathcal{W}\,. \end{aligned}$$

□

Next we will prove that prescribing a positive definite first fundamental form as well as a second fundamental form uniquely determines a surface up to a euclidean motion in $\mathbb{R}^3$, provided that both fundamental forms satisfy the Gauss and the Codazzi-Mainardi equations.

Theorem 18 (Fundamental Theorem of Surface Theory—Second Formulation). *Let two symmetric bilinear forms* $\bar{\mathrm{I}}, \overline{\mathrm{II}} : TU \times TU \to \mathbb{R}$ *be given on a simply connected, open subset* U *of* $\mathbb{R}^2$*, and suppose that the first,* $\bar{\mathrm{I}}$*, is positive definite at each point. Starting from* $\bar{\mathrm{I}}$*, the equation*

$$\begin{aligned} 2\bar{\mathrm{I}}(\overline{\nabla}_{\mathcal{U}}\mathcal{V}, \mathcal{W}) &= \mathcal{U}(\bar{\mathrm{I}}(\mathcal{V}, \mathcal{W})) + \mathcal{V}(\bar{\mathrm{I}}(\mathcal{U}, \mathcal{W})) - \mathcal{W}(\bar{\mathrm{I}}(\mathcal{U}, \mathcal{V})) + \bar{\mathrm{I}}([\mathcal{U}, \mathcal{V}], \mathcal{W}) \\ &\quad + \bar{\mathrm{I}}(\mathcal{V}, [\mathcal{W}, \mathcal{U}]) - \bar{\mathrm{I}}(\mathcal{U}, [\mathcal{V}, \mathcal{W}]) \end{aligned}$$

defines a covariant derivative $\overline{\nabla}$ *for vector fields with respect to vector fields, and the curvature transformation is determined by*

$$\overline{\mathcal{R}}(\mathcal{U}, \mathcal{V})\mathcal{W} \;:=\; \overline{\nabla}_{\mathcal{U}}\overline{\nabla}_{\mathcal{V}}\mathcal{W} - \overline{\nabla}_{\mathcal{V}}\overline{\nabla}_{\mathcal{U}}\mathcal{W} - \overline{\nabla}_{[\mathcal{U},\mathcal{V}]}\mathcal{W}\,.$$

If the symmetric endomorphism $\overline{S} : TU \to TU$ *defined from* $\overline{\mathrm{II}}$ *by* $\overline{\mathrm{II}}(\mathcal{U}, \mathcal{V}) = \bar{\mathrm{I}}(\mathcal{U}, \overline{S}(\mathcal{V}))$ *satisfies*

(1) $\overline{\nabla}\,\overline{S} \;=\; 0,$

(2) $\overline{\mathcal{R}}(\mathcal{U},\mathcal{V})\mathcal{W} \;=\; \overline{\mathrm{II}}(\mathcal{V},\mathcal{W})\cdot\overline{S}(\mathcal{U}) - \overline{\mathrm{II}}(\mathcal{U},\mathcal{W})\cdot\overline{S}(\mathcal{V}),$

then there exists a parametrized surface $F : U \to \mathbb{R}^3$ such that the induced first and second fundamental forms of this surface coincide with $\overline{\mathrm{I}}$ and $\overline{\mathrm{II}}$,

$$\overline{\mathrm{I}} \;=\; F^*(\mathrm{I}), \quad \overline{\mathrm{II}} \;=\; F^*(\mathrm{II})\,.$$

This parametrized surface is uniquely determined up to a euclidean motion of $\mathbb{R}^3$.

Proof. The proof is based on a straightforward application of the fundamental theorem of surface theory in its first formulation (Theorem 11, §5.2). To apply this, we choose a frame $\{\overline{e}_1,\overline{e}_2\}$ of vector fields on the open set $U \subset \mathbb{R}^2$ which is orthonormal with respect to the bilinear form $\overline{\mathrm{I}}$, and denote by $\overline{\sigma}_1,\overline{\sigma}_2$ the corresponding dual frame of 1-forms. Define additional 1-forms by the equations

$$\begin{aligned}
\overline{\omega}_{12}(\mathcal{V}) &:= \overline{\mathrm{I}}(\overline{\nabla}_{\mathcal{V}}\overline{e}_1,\overline{e}_2),\\
\overline{\omega}_{13}(\mathcal{V}) &:= \overline{\mathrm{II}}(\mathcal{V},\overline{e}_1) \;=\; \overline{\mathrm{I}}(\overline{S}(\mathcal{V}),\overline{e}_1),\\
\overline{\omega}_{23}(\mathcal{V}) &:= \overline{\mathrm{II}}(\mathcal{V},\overline{e}_2) \;=\; \overline{\mathrm{I}}(\overline{S}(\mathcal{V}),\overline{e}_2).
\end{aligned}$$

This leads to a system $\{\overline{\sigma}_1,\overline{\sigma}_2,\overline{\omega}_{12},\overline{\omega}_{13},\overline{\omega}_{23}\}$ of 1-forms, which is extended by requiring antisymmetry, $\overline{\omega}_{ji} := -\overline{\omega}_{ij}$. The equations we supposed to hold imply that the integrability conditions of the fundamental theorem of surface theory in its first formulation are satisfied. To see this, note first that the equation defining the covariant derivative $\overline{\nabla}$ immediately yields the formulas

$$\mathcal{U}(\overline{\mathrm{I}}(\mathcal{V},\mathcal{W})) = \overline{\mathrm{I}}(\overline{\nabla}_{\mathcal{U}}\mathcal{V},\mathcal{W}) + \overline{\mathrm{I}}(\mathcal{V},\overline{\nabla}_{\mathcal{U}}\mathcal{W}), \quad \overline{\nabla}_{\mathcal{U}}\mathcal{V} - \overline{\nabla}_{\mathcal{V}}\mathcal{U} \;=\; [\mathcal{U},\mathcal{V}]\,.$$

From these, we obtain

$$\begin{aligned}
d\overline{\sigma}_1(\mathcal{U},\mathcal{V}) &= \mathcal{U}(\overline{\sigma}_1(\mathcal{V})) - \mathcal{V}(\overline{\sigma}_1(\mathcal{U})) - \overline{\sigma}_1([\mathcal{U},\mathcal{V}])\\
&= \mathcal{U}(\overline{\mathrm{I}}(\mathcal{V},\overline{e}_1)) - \mathcal{V}(\overline{\mathrm{I}}(\mathcal{U},\overline{e}_1)) - \overline{\mathrm{I}}([\mathcal{U},\mathcal{V}],\overline{e}_1)\\
&= \overline{\mathrm{I}}(\overline{\nabla}_{\mathcal{U}}\mathcal{V} - \overline{\nabla}_{\mathcal{V}}\mathcal{U} - [\mathcal{U},\mathcal{V}],\ \overline{e}_1) + \overline{\mathrm{I}}(\mathcal{V},\ \overline{\nabla}_{\mathcal{U}}\overline{e}_1) - \overline{\mathrm{I}}(\mathcal{U},\ \overline{\nabla}_{\mathcal{V}}\overline{e}_1)\\
&= \overline{\omega}_{12}(\mathcal{U})\cdot\overline{\sigma}_2(\mathcal{V}) - \overline{\omega}_{12}(\mathcal{V})\cdot\overline{\sigma}_2(\mathcal{U}) \;=\; \overline{\omega}_{12}\wedge\overline{\sigma}_2(\mathcal{U},\mathcal{V})\,.
\end{aligned}$$

Hence $d\overline{\sigma}_1 = \overline{\omega}_{12}\wedge\overline{\sigma}_2$, and similarly one verifies that $d\overline{\sigma}_2 = \overline{\omega}_{21}\wedge\overline{\sigma}_1$. In the next step, we show that $\overline{\sigma}_1\wedge\overline{\omega}_{13} + \overline{\sigma}_2\wedge\overline{\omega}_{23} = 0$. This follows from

$$\overline{\sigma}_1\wedge\overline{\omega}_{13}(\mathcal{U},\mathcal{V}) + \overline{\sigma}_2\wedge\overline{\omega}_{23}(\mathcal{U},\mathcal{V}) \;=\; \overline{\mathrm{II}}(\mathcal{V},\mathcal{U}) - \overline{\mathrm{II}}(\mathcal{U},\mathcal{V}) \;=\; 0\,,$$

since the second bilinear form is symmetric. For the computations to follow, note that $\overline{\mathrm{I}}(\overline{e}_1,\overline{e}_1) \equiv 1$ implies $\overline{\mathrm{I}}(\overline{\nabla}_{\mathcal{U}}\overline{e}_1,\overline{e}_1) = 0$, and hence $\overline{\nabla}_{\mathcal{U}}\overline{e}_1$ is parallel to $\overline{e}_2$. Thus for arbitrary vector fields $\mathcal{U},\mathcal{V}$ we always have

$$\overline{\mathrm{I}}(\overline{\nabla}_{\mathcal{U}}\overline{e}_1,\overline{\nabla}_{\mathcal{V}}\overline{e}_2) \;=\; 0\,.$$

Using this equation and the Gauss equation, we compute the form $d\overline{\omega}_{12}$:

$$\begin{aligned} d\overline{\omega}_{12}(\mathcal{U},\mathcal{V}) &= \mathcal{U}\big(\bar{\mathrm{I}}(\overline{\nabla}_{\mathcal{V}}\bar{e}_1,\bar{e}_2)\big) - \mathcal{V}\big(\bar{\mathrm{I}}(\overline{\nabla}_{\mathcal{U}}\bar{e}_1,\bar{e}_2)\big) - \bar{\mathrm{I}}(\overline{\nabla}_{[\mathcal{U},\mathcal{V}]}e_1,e_2) \\ &= \bar{\mathrm{I}}(\overline{\mathcal{R}}(\mathcal{U},\mathcal{V})\bar{e}_1,\bar{e}_2) \;=\; \overline{\mathrm{II}}(\mathcal{V},\bar{e}_1)\cdot\overline{\mathrm{II}}(\mathcal{U},\bar{e}_2) - \overline{\mathrm{II}}(\mathcal{U},\bar{e}_1)\cdot\overline{\mathrm{II}}(\mathcal{V},e_2) \\ &= \overline{\omega}_{13}\wedge\overline{\omega}_{32}(\mathcal{U},\mathcal{V})\,. \end{aligned}$$

From this equation we get the structural equation to be satisfied, $d\overline{\omega}_{12} = \overline{\omega}_{13}\wedge\overline{\omega}_{32}$. The relation for $d\overline{\omega}_{13}$ is derived in a similar way using the Codazzi-Mainardi equation:

$$\begin{aligned} d\overline{\omega}_{13}(\mathcal{U},\mathcal{V}) &= \mathcal{U}\big(\bar{\mathrm{I}}(\overline{S}(\mathcal{V}),\bar{e}_1)\big) - \mathcal{V}\big(\bar{\mathrm{I}}(\overline{S}(\mathcal{U}),\bar{e}_1)\big) - \bar{\mathrm{I}}(\overline{S}([\mathcal{U},\mathcal{V}]),\bar{e}_1) \\ &= \bar{\mathrm{I}}(\overline{\nabla}\;\overline{S}(\mathcal{U},\mathcal{V}),\bar{e}_1) + \bar{\mathrm{I}}(\overline{S}(\mathcal{V}),\overline{\nabla}_{\mathcal{U}}\bar{e}_1) - \bar{\mathrm{I}}(\overline{S}(\mathcal{U}),\overline{\nabla}_{\mathcal{V}}\bar{e}_1) \\ &= \bar{\mathrm{I}}(\overline{S}(\mathcal{V}),\bar{e}_2)\cdot\bar{\mathrm{I}}(\bar{e}_2,\overline{\nabla}_{\mathcal{U}}\bar{e}_1) - \bar{\mathrm{I}}(\overline{S}(\mathcal{U}),\bar{e}_2)\cdot\bar{\mathrm{I}}(\bar{e}_2,\overline{\nabla}_{\mathcal{V}}\bar{e}_1) \\ &= \overline{\omega}_{12}\wedge\overline{\omega}_{23}(\mathcal{U},\mathcal{V})\,. \end{aligned}$$

Thus $d\overline{\omega}_{13} = \overline{\omega}_{12}\wedge\overline{\omega}_{23}$, and an analogous computation yields $d\overline{\omega}_{23} = \overline{\omega}_{21}\wedge\overline{\omega}_{13}$. Together, the system of forms $\{\overline{\sigma}_1,\overline{\sigma}_2,\overline{\omega}_{12},\overline{\omega}_{13},\overline{\omega}_{23}\}$ satisfies the integrability conditions of Theorem 11, §5.2, and hence the existence of a parametrized surface $F: U \to \mathbb{R}^3$ such that $F^*(\mathrm{I}) = \bar{\mathrm{I}}$ and $F^*(\mathrm{II}) = \overline{\mathrm{II}}$ is proved. The uniqueness statement is a consequence of the corresponding uniqueness result of that theorem. □

5.4. Gaussian and Mean Curvature

Consider a surface $M^2 \subset \mathbb{R}^3$ and its first fundamental form, the induced Riemannian metric. Choosing a local orthonormal frame $\{e_1, e_2\}$ with dual frame σ_1, σ_2, we know that the 1-form $\omega_{12} = \langle de_1, e_2\rangle = \langle \nabla e_1, e_2\rangle$ is completely determined by the forms σ_1, σ_2. Since its exterior differential $d\omega_{12}$ is a 2-form, there exists a function G on the surface such that

$$d\omega_{12} \;=\; -\,G\cdot\sigma_1\wedge\sigma_2\,.$$

The function G is independent of the choice of the local frame of vector fields. Every other frame $\{e_1^*, e_2^*\}$ can be represented as

$$\begin{aligned} e_1^* &= \mu\cos\varphi\cdot e_1 + \sin\varphi\cdot e_2\,, \\ e_2^* &= -\mu\sin\varphi\cdot e_1 + \cos\varphi\cdot e_2\,, \end{aligned}$$

where $\mu = \pm 1$ is constant, and φ is a function on that part of M^2 where both orthonormal frames are defined. The dual frames are thus related by

$$\begin{aligned} \sigma_1^* &= \mu\cdot\cos\varphi\cdot\sigma_1 + \sin\varphi\cdot\sigma_2\,, \\ \sigma_2^* &= -\mu\cdot\sin\varphi\cdot\sigma_2 + \cos\varphi\cdot,\sigma_2, \end{aligned}$$

and we obtain $\sigma_1^*\wedge\sigma_2^* = \mu\cdot\sigma_1\wedge\sigma_2$. On the other hand, the form ω_{12}^* is equal to $\omega_{12}^* = \langle de_1^*, e_2^*\rangle = \mu\cdot\omega_{12} + d\varphi$, leading to the formula $d\omega_{12}^* = \mu\cdot d\omega_{12}$. This observation shows that the function G is uniquely defined.

Definition 11. The *Gaussian curvature* $G : M^2 \to \mathbb{R}$ is the function defined by the equation

$$d\omega_{12} \;=\; -\, G \cdot \sigma_1 \wedge \sigma_2 \,.$$

Remark. The Gaussian curvature depends exclusively on the first fundamental form of the surface. If this is given, we can choose an orthonormal frame of 1-forms σ_1, σ_2. From the equations

$$d\sigma_1 \;=\; A \cdot \sigma_1 \wedge \sigma_2 \quad \text{and} \quad d\sigma_2 \;=\; B \cdot \sigma_1 \wedge \sigma_2,$$

we determine two functions A, B and then introduce the 1-form

$$\omega_{12} \;=\; A \cdot \sigma_1 + B \cdot \sigma_2 \,.$$

Finally, the defining equation for the Gaussian curvature is

$$d\omega_{12} \;=\; -\, G \cdot \sigma_1 \wedge \sigma_2.$$

This way, the Gaussian curvature can be computed in practice.

The formula for the curvature tensor of the surface stated earlier can be simplified by means of the Gaussian curvature:

$$\mathcal{R}(\mathcal{U}, \mathcal{V})\mathcal{W} \;=\; G \cdot (\sigma_2(\mathcal{W}) \cdot e_1 - \sigma_1(\mathcal{W}) \cdot e_2) \cdot dM^2(\mathcal{U}, \mathcal{V}),$$

and, conversely, G can be expressed by the curvature tensor.

Theorem 19. *The Gaussian curvature of a surface is*

$$G = \langle \mathcal{R}(e_1, e_2)e_2, e_1 \rangle \,,$$

where $\mathcal{R}$ is the curvature tensor, and $\{e_1, e_2\}$ is an orthonormal frame of vector fields.

Now we turn to the invariants of the second fundamental form, which we view, by means of the Riemannian metric, as a symmetric endomorphism

$$S : TM^2 \to TM^2,$$

the *Weingarten map* of the surface introduced in the previous section. Its normalized trace is the mean curvature of the surface.

Definition 12. The *mean curvature* H of a surface is the function

$$H \;:=\; \mathrm{tr}\,(S)/2 \,.$$

The mean curvature of a surface can also be determined as the divergence (in $\mathbb{R}^3$) of the normal vector field.

Theorem 20. *Let M^2 be a surface, and let $\mathcal{N}$ be a vector field of length one defined on an open neighborhood of M^2 in $\mathbb{R}^3$. Assume that the restriction $\mathcal{N}\big|_{M^2}$ of $\mathcal{N}$ to M^2 is perpendicular to the surface. Then*

$$H \;=\; \frac{1}{2}\mathrm{div}(\mathcal{N}) \,,$$

where the divergence is taken with respect to the euclidean metric on $\mathbb{R}^3$.

Proof. From $\langle \mathcal{N}, d\mathcal{N}\rangle \equiv 0$, we obtain

$$2H = \langle e_1, d\mathcal{N}(e_1)\rangle + \langle e_2, d\mathcal{N}(e_2)\rangle + \langle \mathcal{N}, d\mathcal{N}(\mathcal{N})\rangle .$$

The last sum, however, is independent of the particular choice of the orthonormal basis in $\mathbb{R}^3$. This implies

$$2H = \left\langle \frac{\partial}{\partial x}, \frac{\partial \mathcal{N}}{\partial x} \right\rangle + \left\langle \frac{\partial}{\partial y}, \frac{\partial \mathcal{N}}{\partial y} \right\rangle + \left\langle \frac{\partial}{\partial z}, \frac{\partial \mathcal{N}}{\partial z} \right\rangle = \operatorname{div}(\mathcal{N}) . \qquad \square$$

Apart from the trace, the second fundamental form S has one more invariant, the determinant. But this does not lead to a new geometric quantity, as we shall now see.

Theorem 21. *The determinant of the second fundamental form is the Gaussian curvature,*

$$\det(S) = G .$$

Proof. Let $x \in M^2$ be an arbitrary point of the surface, and choose an orthonormal basis $\{e_1, e_2\}$ consisting of eigenvectors of the symmetric endomorphism in the tangent space T_xM^2. If κ_1, κ_2 are the eigenvalues, then, with respect to this basis, S has the matrix

$$S = \begin{bmatrix} \kappa_1 & 0 \\ 0 & \kappa_2 \end{bmatrix} .$$

The assertion now follows directly from the Gauss equation,

$$\begin{aligned} G &= \langle \mathcal{R}(e_1, e_2)e_2, e_1\rangle = \langle \mathrm{II}(e_2, e_2)S(e_1), e_1\rangle - \langle \mathrm{II}(e_1, e_2)S(e_2), e_1\rangle \\ &= \kappa_1 \cdot \kappa_2 = \det(S). \end{aligned} \qquad \square$$

Remark. The preceding theorem allows to compute the Gaussian curvature, belonging to the inner geometry of a surface, from the second fundamental form. Take an arbitrary (not necessarily orthonormal) basis v_1, v_2 in the tangent space T_xM^2 at a point of the surface, and denote the matrices of the first and second fundamental forms by g and b, respectively,

$$g = \begin{bmatrix} g_{11} & g_{12} \\ g_{21} & g_{22} \end{bmatrix}, \quad g_{ij} = \mathrm{I}(v_i, v_j), \quad b = \begin{bmatrix} b_{11} & b_{12} \\ b_{21} & b_{22} \end{bmatrix}, \quad b_{ij} = \mathrm{II}(v_i, v_j) .$$

The Gaussian and the mean curvature can then be computed using the formulas

$$G = \frac{\det(b)}{\det(g)} \quad \text{and} \quad H = \frac{1}{2}\operatorname{tr}(b \cdot g^{-1}) .$$

Their proof is a consequence of a more general observation. A symmetric bilinear form B on a real vector space V without any additional structure has only a single invariant, the signature. If, however, a non-degenerate scalar product G is given in addition, then this induces an isomorphism

$G : V^* \to V$ between the vector space V and its dual space V^*. Thus B can be viewed as a G-symmetric endomorphism, $S := B \circ G^{-1} : V \to V$ in the vector space V, and all coefficients of its characteristic polynomial are invariants of B with respect to G. These, in turn, can be computed by means of an arbitrary basis, and the result does not depend on the particular basis. Two of these invariants of the characteristic polynomial are the trace and the determinant.

Example 10. For the hyperbolic plane $\mathcal{H}^2$ from Example 9, a straightforward computation shows that $d\omega_{12} = \sigma_1 \wedge \sigma_2$. Therefore, the Gaussian curvature is -1. A complete realization of $\mathcal{H}^2$ in 3-space does not exist (Hilbert, 1901). Certain open subsets of the hyperbolic plane, however, can be realized as surfaces. The related mean curvature is not uniquely determined; it depends on the particular realization as a surface in $\mathbb{R}^3$. One of them will be presented at the end of the discussion concerning surfaces of revolution to follow.

Example 11. For the surface of revolution from Example 3, we had

$$de_3 \;=\; -\frac{r''}{z'}\, ds \cdot e_1 + z'\, d\varphi \cdot e_2\,.$$

This enables us to compute the Weingarten map explicitly:

$$S\left(\frac{\partial}{\partial s}\right) \;=\; -\frac{r''}{z'} \cdot e_1 \;=\; -\frac{r''}{z'} \cdot \frac{\partial}{\partial s}, \quad S\left(\frac{\partial}{\partial \varphi}\right) \;=\; z' \cdot e_2 \;=\; \frac{z'}{r} \cdot \frac{\partial}{\partial \varphi}\,.$$

Thus, in the basis $\partial/\partial s$, $\partial/\partial\varphi$, the Weingarten map is described by the matrix

$$S \;=\; \begin{bmatrix} -r''/z' & 0 \\ 0 & z'/r \end{bmatrix}.$$

From this, the Gaussian and the mean curvature are immediately computed:

$$G \;=\; -\frac{r''}{r}, \quad H \;=\; \frac{1}{2}\left(\frac{z'}{r} - \frac{r''}{z'}\right).$$

Examples for surfaces of revolution with constant Gaussian curvature are the cylinder ($G = 0$, $r = 1$, $z = s$), the sphere ($G = 1$, $r = \sin s$, $z = \cos s$), or the *pseudo-sphere* ($G = -1$): The latter is an example of a (necessarily non-compact) surface of constant negative curvature, and hence a 2-dimensional model of hyperbolic geometry. It is defined as the surface obtained as the surface of revolution generated by the *tractrix* with the parametrization

$$\left(e^{-s}, \int_0^s \sqrt{1 - e^{-2t}}dt\right) \;=\; \left(e^{-s}, -\sqrt{1 - e^{-2s}} + \operatorname{arctanh}\bigl(\sqrt{1 - e^{-2s}}\bigr)\right).$$

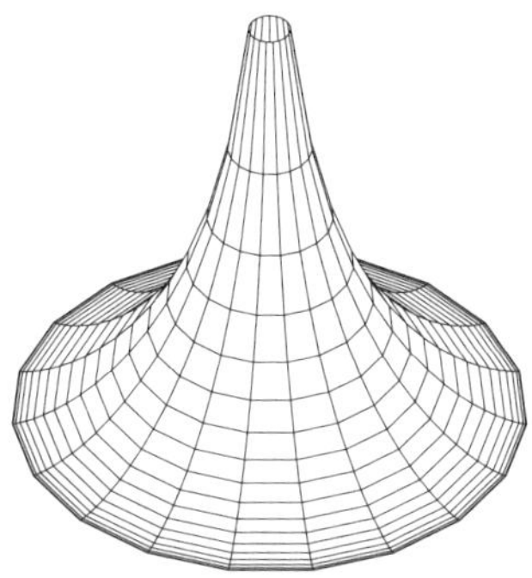

Example 12. The Gaussian curvature of the graph of the function $h(x, y)$ (Example 4) is preferably computed using the differential of the form ω_{12}. The resulting expression is

$$G = \frac{h_{xx}\, h_{yy} - h_{xy}^2}{(1 + h_x^2 + h_y^2)^2}.$$

To compute the mean curvature, it is convenient to describe the graph as the zero set of the function $g : U \to \mathbb{R}$, $g(x, y, z) := h(x, y) - z$. From Theorem 5, §3.2, we then know that the normalized gradient vector field,

$$\frac{\mathrm{grad}(g)}{||\mathrm{grad}(g)||} = \frac{1}{\sqrt{1 + h_x^2 + h_y^2}}\big(h_x,\, h_y,\, -1\big),$$

is perpendicular to the surface at each point. Hence the mean curvature can be computed as the divergence of this vector field (Theorem 20), and we obtain

$$H = \frac{(1 + h_y^2)h_{xx} - 2h_x h_y h_{xy} + (1 + h_x^2)h_{yy}}{2(1 + h_x^2 + h_y^2)^{3/2}}.$$

Example 13. In order to apply this formula to the computation of the Gaussian curvature for a modular surface (Example 5), we still have to determine the second derivatives of $h = |f|$. From

$$h_x = \frac{\partial h}{\partial z} + \frac{\partial h}{\partial \bar{z}} \quad \text{and} \quad h_y = i\left(\frac{\partial h}{\partial z} - \frac{\partial h}{\partial \bar{z}}\right)$$

we first conclude formally that

$$h_{xx} = \frac{\partial^2 h}{\partial z^2} + 2\,\frac{\partial^2 h}{\partial z \partial \bar{z}} + \frac{\partial^2 h}{\partial \bar{z}^2},$$

$$h_{yy} = -\frac{\partial^2 h}{\partial z^2} + 2\,\frac{\partial^2 h}{\partial z \partial \bar{z}} - \frac{\partial^2 h}{\partial \bar{z}^2}, \qquad h_{xy} = i\left(\frac{\partial^2 h}{\partial z^2} - \frac{\partial^2 h}{\partial \bar{z}^2}\right).$$

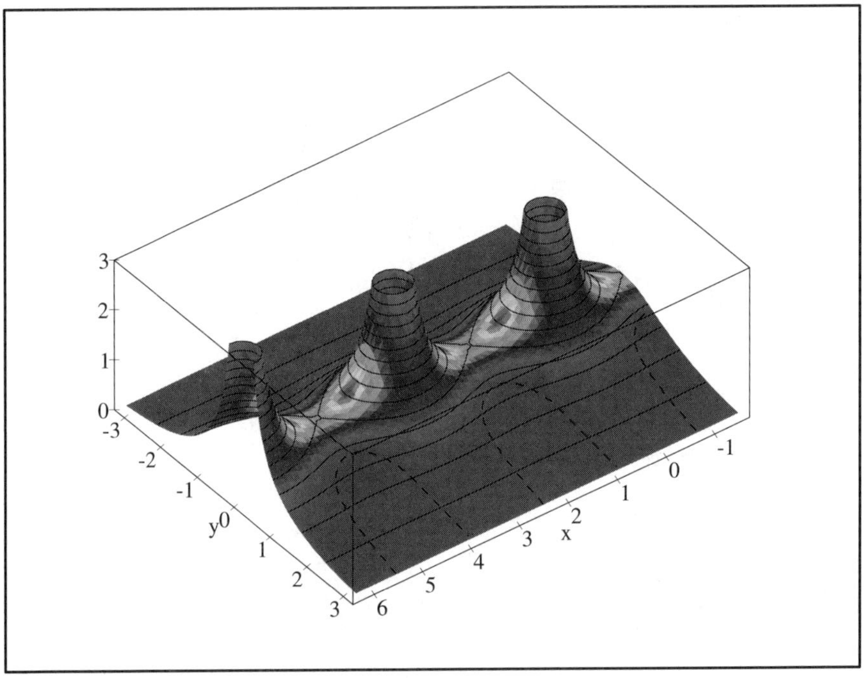

After expanding, the numerator of the Gaussian curvature turns out to be

$$h_{xx}\,h_{yy}-h_{xy}^2 \;=\; 4\left[\left(\frac{\partial^2 h}{\partial z\partial\bar z}\right)^2-\frac{\partial^2 h}{\partial\bar z^2}\cdot\frac{\partial^2 h}{\partial\bar z^2}\right].$$

We compute the derivatives:

$$\frac{\partial^2 h}{\partial z^2}=\frac{f''\bar f}{2\,h}-\frac{(f'\bar f)^2}{4\,h^3},\quad \frac{\partial^2 h}{\partial\bar z^2}=\frac{f\bar f''}{2\,h}-\frac{(f\bar f')^2}{4\,h^3},\quad \frac{\partial^2 h}{\partial z\partial\bar z}=\frac{f'\bar f'}{4\,h},$$

from which we deduce the formula for the Gaussian curvature:

$$G\;=\;\frac{|f''|^2}{(1+|f'|^2)^2}\left(\mathrm{Re}\frac{f'^2}{f\,f''}-1\right).$$

In the example discussed before, $f(z)=1/\sin z$, the Gaussian curvature has the form

$$G\;=\;\frac{\left|2\,\dfrac{\cos^2 z}{\sin^3 z}+\dfrac{1}{\sin z}\right|^2\left[\mathrm{Re}\,\dfrac{\cos^2 z}{1+\cos^2 z}-1\right]}{\left[1+\left|\dfrac{\cos z}{\sin^2 z}\right|^2\right]^2}.$$

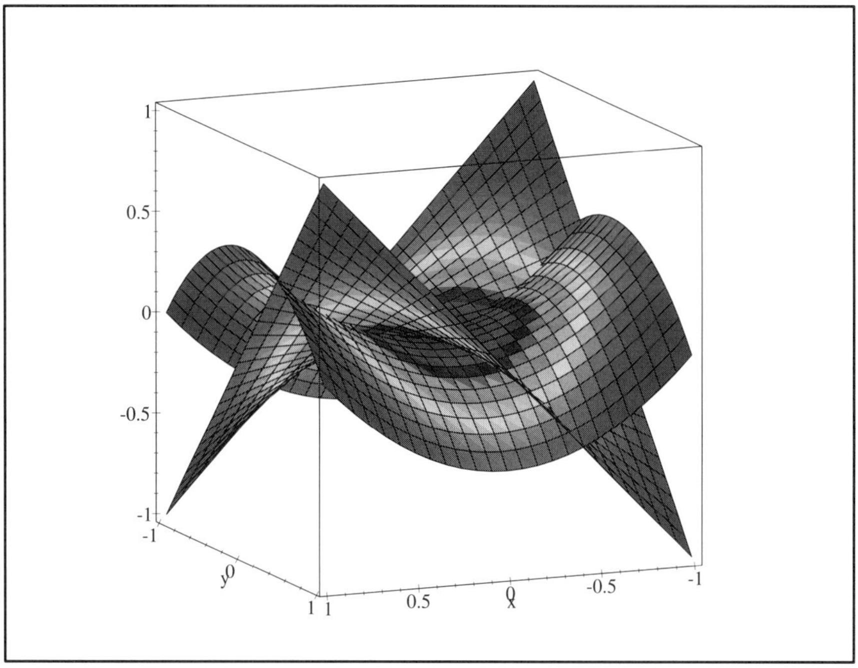

Of particular interest is the curve on the surface separating the points of positive Gaussian curvature from those of negative Gaussian curvature. Here, it is implicitly described by the equation

$$\operatorname{Re} \frac{\cos^2 z}{1+\cos^2 z} \;=\; 1,$$

and is indicated as a dashed line in the figure on the previous page showing the modular surface shaded according to the Gaussian curvature.

Example 14. We now continue the discussion of the surfaces defined by the real and the imaginary part of a holomorphic function, $f = u + iv : U \to \mathbb{C}$ (Example 6). There we already derived from the Cauchy-Riemann equations the relation

$$1 + u_x^2 + u_y^2 \;=\; 1 + v_x^2 + v_y^2 \;=\; 1 + |f'|^2.$$

A straightforward calculation yields the second derivative of f:

$$f'' \;=\; u_{xx} - iu_{xy} \;=\; v_{xy} + iv_{xx}.$$

The formula for the Gaussian curvature of a graph thus leads to the remarkable fact that the surfaces defined by the real and the imaginary parts have the same Gaussian curvature,

$$G = \frac{-|f''|^2}{(1+|f'|^2)^2}.$$

The picture on the previous page illustrates this fact for the function $f(z) = z^2$, again shaded according to the values of the Gaussian curvature. The surface becoming pointed at the corners belongs to $\operatorname{Im} z^2$, and the saddle-like surface corresponds to $\operatorname{Re} z^2$. The Gaussian curvature is

$$G = \frac{-4}{(1+4|z|^2)^2}.$$

On the other hand, the mean curvatures of these surfaces differ:

$$H(\operatorname{Re} z^2) = \frac{4\,(y^2 - x^2)}{(1+4x^2+4y^2)^{3/2}}, \quad H(\operatorname{Im} z^2) = \frac{-8\,xy}{(1+4x^2+4y^2)^{3/2}}.$$

Now we want to interpret the formula for the Gaussian curvature expressing it as the determinant of the second fundamental form in a differential-geometric way. To this end, consider the normal vector e_3 as a map on the surface M^2 with values in the unit sphere $S^2 \subset \mathbb{R}^3$. The volume form dS^2 of the sphere evaluated on two vectors $\mathcal{V}, \mathcal{W} \in T_x S^2$ at the point $x \in S^2$ is the vector product

$$dS^2(\mathcal{V}, \mathcal{W}) = \langle \mathcal{V} \times \mathcal{W},\, x \rangle .$$

This formula is a special case of the general one for the volume form of the sphere $S^{n-1} \subset \mathbb{R}^n$ stated in §3.4, Example 28. From this, the 2-form $e_3^*(dS^2)$ induced on M^2 is easily computed:

$$\begin{aligned} e_3^*(dS^2)(\mathcal{V}, \mathcal{W}) &= \langle de_3(\mathcal{V}) \times de_3(\mathcal{W}),\, e_3 \rangle \\ &= \langle (\omega_{31}(\mathcal{V})\, e_1 + \omega_{32}(\mathcal{V})\, e_2) \times (\omega_{31}(\mathcal{W})\, e_1 + \omega_{32}(\mathcal{W})\, e_2), e_3 \rangle \\ &= -\,\omega_{13} \wedge \omega_{32}(\mathcal{V}, \mathcal{W}) = -d\omega_{12}(\mathcal{V}, \mathcal{W}) = G \cdot dM^2(\mathcal{V}, \mathcal{W}). \end{aligned}$$

We summarize the result in

Theorem 22. *Let $e_3 : M^2 \to S^2$ be the unit normal map of the surface, and let G be its Gaussian curvature. Then the induced 2-form $e_3^*(dS^2)$ is equal to the volume form multiplied by the Gaussian curvature:*

$$e_3^*(dS^2) = G \cdot dM^2.$$

Fix a point $x \in M^2$ on the surface and denote by $D(x, \varepsilon)$ the set of all points on the surface whose distance from x is less than ε. Then the two-dimensional surface area of the set $D^*(x, \varepsilon) := e_3(D(x, \varepsilon)) \subset S^2$ can be computed using the transformation formula for the integral as follows:

$$\operatorname{vol}_{S^2}(D^*(x,\varepsilon)) = \int\limits_{D^*(x,\varepsilon)} dS^2 = \pm \int\limits_{D(x,\varepsilon)} e_3^*(dS^2) = \pm \int\limits_{D(x,\varepsilon)} G \cdot dM^2.$$

Dividing both sides by the surface area, $\mathrm{vol}_{M^2}(D(x,\varepsilon))$, and applying the mean value theorem of integral calculus for continuous functions, we obtain the formula

$$|G(x)| \;=\; \lim_{\varepsilon\to 0} \frac{\mathrm{vol}_{S^2}(D^*(x,\varepsilon))}{\mathrm{vol}_{M^2}(D(x,\varepsilon))}\,.$$

This was the original geometric definition of the curvature for a surface in 3-space. It is—similarly to the curvature of a curve—the infinitesimal volume distortion of the normal map of the surface. For this approach to curvature, it is at first not obvious at all that the Gaussian curvature depends only on the first fundamental form of the surface. This fact forms the contents of the Theorema Egregium (Gauss, 1827). Proceeding, however, like we did here, this becomes immediate.

The total Gaussian curvature of a compact, oriented surface without boundary depends only on the topology of the manifold. We prove the so-called Gauss-Bonnet formula, expressing this fact explicitly. To do so, we need a preparation.

Lemma 3. *Let $\mathcal{V}$ be a vector field of length one on the oriented surface M^2, and let $\{e_1,e_2\}$ be an orthonormal frame in this orientation. Representing $\mathcal{V}$ as a linear combination, $\mathcal{V} = \cos\tau\cdot e_1+\sin\tau\cdot e_2$, the 1-form*

$$\chi_{\mathcal{V}} \;:=\; \omega_{12}+d\tau$$

is independent of the chosen frame. The exterior derivative of $\chi_{\mathcal{V}}$ is

$$d\chi_{\mathcal{V}} \;=\; -\,G\cdot dM^2\,.$$

Proof. Every other frame e_1^*,e_2^* can be written as

$$\begin{aligned} e_1^* &= \cos\varphi\cdot e_1+\sin\varphi\cdot e_2\,,\\ e_2^* &= -\sin\varphi\cdot e_1+\cos\varphi\cdot e_2\,, \end{aligned}$$

and we already proved the formula $\omega_{12}^*=\omega_{12}+d\varphi$. Moreover, $\tau^*=\tau-\varphi$, and hence $\omega_{12}^*+d\tau^*=\omega_{12}+d\tau$. □

Stokes' theorem now immediately implies

Theorem 23. *Let M^2 be an oriented, compact manifold without boundary, and let $\mathcal{V}$ be a vector field of length one defined outside finitely many points $x_1,\ldots,x_k\in M^2$. If $D(x_i,\varepsilon)\subset M^2$ are disk neighborhoods of the points x_i of radius $\varepsilon>0$, then*

$$\int_{M^2} G\cdot dM^2 \;=\; \lim_{\varepsilon\to 0}\left(\sum_{i=1}^{k}\int_{\partial D(x_i,\varepsilon)}\chi_{\mathcal{V}}\right).$$

Hence we see that the integral of the Gaussian curvature is independent of the geometry of the surface. We evaluate the last formula explicitly by choosing special vector fields. To this end, consider a smooth function $f : M^2 \to \mathbb{R}$ all of whose critical points are non-degenerate, and denote the number of its critical points with corresponding index by

$$\begin{aligned} m_0(f) &:= \text{ the number of minima of } f, \\ m_1(f) &:= \text{ the number of saddle points of } f, \\ m_2(f) &:= \text{ the number of maxima of } f. \end{aligned}$$

The vector field $\mathcal{V} = \mathrm{grad}(f)/\|\mathrm{grad}(f)\|$ is defined outside the set of critical points. Computing the limits on the right-hand side of the formula, we obtain

$$\lim_{\varepsilon \to 0} \int_{\partial D(x_i,\varepsilon)} \chi_{\mathcal{V}} \;=\; \begin{cases} 2\pi & \text{if } x_i \text{ is a minimum or maximum;} \\ -2\pi & \text{if } x_i \text{ is a saddle point.} \end{cases}$$

To prove this, choose coordinates x, y with respect to which the function f has normal form, $f(x, y) = \frac{1}{2}(\pm x^2 \pm y^2)$. We are only interested in the limit of these integrals for $\varepsilon \to 0$, and hence we may, without loss of generality, take the metric to be flat. From

$$\frac{\mathrm{grad} f}{\|\mathrm{grad}(f)\|} \;=\; \pm\frac{x}{\sqrt{x^2+y^2}}\frac{\partial}{\partial x} \pm \frac{y}{\sqrt{x^2+y^2}}\frac{\partial}{\partial y}$$

we obtain

$$\cos\tau \;=\; \pm\frac{x}{\sqrt{x^2+y^2}}, \qquad \sin\tau \;=\; \pm\frac{y}{\sqrt{x^2+y^2}}.$$

Computing the differential $d\tau$ leads to the formula

$$d\tau \;=\; \pm\frac{-y\cdot dx + x\cdot dy}{x^2+y^2}.$$

Here we have to take the positive sign in the cases of a minimum or a maximum, and the negative sign for a saddle point. Since

$$\int_{S^1} d\tau \;=\; \pm 2\pi,$$

we obtain the limit we wanted to compute, and hence we have proved the Gauss-Bonnet theorem.

Theorem 24 (Gauss-Bonnet Formula). *Let M^2 be a compact, oriented surface without boundary, and let $f : M^2 \to \mathbb{R}$ be a smooth function all of whose critical points are non-degenerate. Then*

$$\int_{M^2} G\cdot dM^2 \;=\; 2\pi\cdot(m_0(f) - m_1(f) + m_2(f)),$$

where $m_0(f), m_1(f), m_2(f)$ denote the number of minima, saddle points, and maxima of f, respectively.

Corollary 1. *Let M^2 be a compact, oriented surface without boundary.*

(1) *The integral of the Gaussian curvature depends only on the topology of the surface, but does not depend on its fundamental forms.*

(2) *The alternating sum $m_0(f) - m_1(f) + m_2(f)$ is independent of the choice of the function f all of whose critical points are non degenerate.*

Definition 13. The *Euler characteristic* $\chi(M^2)$ of a compact surface without boundary is the number

$$\chi(M^2) \ := \ m_0(f) - m_1(f) + m_2(f) .$$

Using this number, the Gauss-Bonnet formula for compact, oriented surfaces without boundary can be written as

$$\frac{1}{2\pi} \int_{M^2} G \cdot dM^2 \ = \ \chi(M^2) .$$

Example 15. The Euler characteristic of the sphere S^2 is 2. In fact, the restriction of the height function $h(x, y, z) = z$ to $S^2 = \{(x, y, z) \in \mathbb{R}^3 : x^2 + y^2 + z^2 = 1\}$ has one minimum, one maximum, and no saddle points. Thus

$$\frac{1}{2\pi} \int_{M^2} G \cdot dM^2 \ = \ 2$$

for every surface M^2 which is diffeomorphic (but not necessarily isometric) to S^2.

Example 16. The Euler characteristic of the torus T^2 is zero. It is not difficult to find, e.g., on the torus of revolution (see §3.1), a function with one minimum, one maximum, and two saddle points. Thus

$$\int_{M^2} G \cdot dM^2 \ = \ 0$$

for every surface which is diffeomorphic to the torus.

The Gauss-Bonnet formula in its general form comprises an additional relation for every vector field $\mathcal{V}$ of length one with finitely many singularities. First we note that the limit

$$\lim_{\varepsilon \to 0} \int_{D(x_i, \varepsilon)} \chi_{\mathcal{V}}$$

can also be computed, along the same lines as for gradient fields, for an arbitrary vector field. Start by choosing coordinates y^1, y^2 around the point $x_i \in M^2$ such that this point corresponds to $y^1 = 0,\ y^2 = 0$. Then represent the vector field $\mathcal{V}$ in these coordinates,

$$\mathcal{V} \ = \ V^1 \frac{\partial}{\partial y^1} + V^2 \frac{\partial}{\partial y^2} .$$

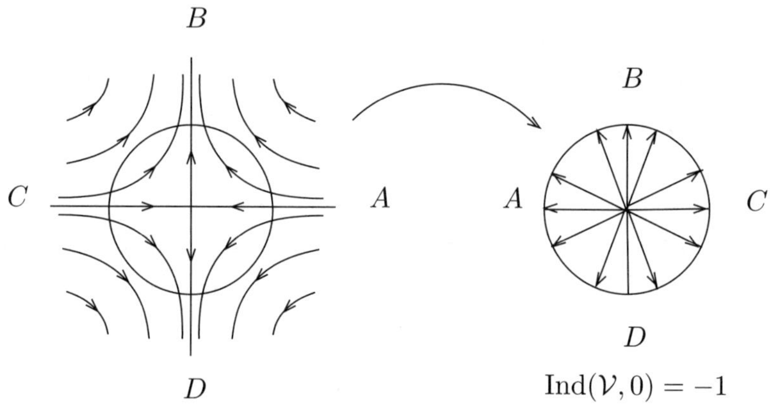

As before, we may assume, without loss of generality, that the metric is flat in this neighborhood of the point, since we do not want to compute the integral itself, but only its limit. Then $(V^1)^2+(V^2)^2 \equiv 1$, and hence $\cos\tau = V^1$, $\sin\tau = V^2$. From this we obtain $-\sin\tau\cdot d\tau = dV^1$, $\cos\tau\cdot d\tau = dV^2$, and thus the form $\chi_{\mathcal{V}} = d\tau$ is given by

$$\chi_{\mathcal{V}} \;=\; d\tau \;=\; V^1\cdot dV^2 - V^2\cdot dV^1 .$$

Therefore,

$$\lim_{\varepsilon\to 0}\int_{D(x_i,\varepsilon)}\chi_{\mathcal{V}} \;=\; \int_{S^1}\left(V^1\cdot dV^2 - V^2\cdot dV^1\right) \;:=\; 2\pi\,\mathrm{Ind}(\mathcal{V},x_i)\,.$$

This number is called the *index of the vector field* $\mathcal{V}$ at the singular point $x_i \in M^2$. Notice that integrating $V^1\,dV^1 + V^2\,dV^2$ along S^1 yields the same result. The geometric meaning of the index is illustrated in the figure above. The picture on the left shows some vector field (the circle is only drawn for greater clarity and has no direct relation to the vector field itself). Starting at A, move around the circle once, and assign to each point on the circle the normalized vector in S^1 pointing into the direction of the vector field at that point. In the case shown in this figure, one full turn in the positive direction on the left corresponds to one full turn on the right, but in the opposite direction. Hence the index of this vector field at the indicated point is -1.

From the Gauss-Bonnet theorem, we obtain

Theorem 25 (Hopf-Poincaré). *Let M^2 be a compact, oriented surface without boundary, and let $\mathcal{V}$ be a vector field of length one defined outside $\{x_1,\ldots,x_k\}$. Then the sum*

$$\sum_{i=1}^{k}\mathrm{Ind}(\mathcal{V},x_i) \;=\; \chi(M^2)$$

is independent of the choice of the vector field.

A further integral formula deals with the integral of the mean curvature. It goes back to Steiner (1840) and Minkowski (1900), and plays an important role in the theory of ovaloids.

Theorem 26 (Minkowski-Steiner). *Let $M^2 \subset \mathbb{R}^3$ be a compact, oriented surface without boundary, and let e_3 be a unit normal field. Then*

$$\int_{M^2} H(x) \cdot dM^2(x) \;=\; \int_{M^2} \langle x,\, e_3(x)\rangle \; G(x) \cdot dM^2(x) \,.$$

Proof. Consider the three functions $p_i(x) = \langle M(x),\, e_i(x)\rangle$ on the surface M^2 and compute their differentials using the structural equations:

$$\begin{aligned} dp_1 &= \sigma_1 + \omega_{12} \cdot p_2 + \omega_{13} \cdot p_3 \,, \\ dp_2 &= \sigma_2 + \omega_{21} \cdot p_1 + \omega_{23} \cdot p_3 \,, \\ dp_3 &= \omega_{31} \cdot p_1 + \omega_{32} \cdot p_2 \,. \end{aligned}$$

Note also that the 1-form

$$\omega \;:=\; p_1 \cdot \omega_{23} + p_2 \cdot \omega_{31}$$

does not depend on the choice of the tangent frame e_1, e_2. Thus ω is a globally defined form on M^2. We compute the differential of ω, making use of the structural equations as well as the formula for the differentials dp_i. This yields

$$d\omega = \sigma_1 \wedge \omega_{23} + \sigma_2 \wedge \omega_{31} + 2 \cdot p_3 \cdot \omega_{13} \wedge \omega_{23} = -2H \cdot dM^2 + 2p_3 \cdot G \cdot dM^2 \,.$$

Stokes' theorem now implies the Minkowski-Steiner integral formula. □

The eigenvalues κ_1, κ_2 of the second fundamental form are called the *principal curvatures* of the surface at the point. Their arithmetic mean is the mean curvature of the surface, whereas their product coincides with the Gaussian curvature,

$$H \;=\; \frac{1}{2}\big(\kappa_1 + \kappa_2\big), \quad G \;=\; \kappa_1 \cdot \kappa_2 \,.$$

The points of a surface are divided into several types, depending on the signs of their principal curvatures. A point of the surface is called

(1) *elliptic* if $\kappa_1 \cdot \kappa_2 > 0$;

(2) *hyperbolic* if $\kappa_1 \cdot \kappa_2 < 0$;

(3) *parabolic* if $\kappa_1 \cdot \kappa_2 = 0$ and $\kappa_1^2 + \kappa_2^2 > 0$;

(4) *flat* if $\kappa_1 = \kappa_2 = 0$.

Umbilic points are the points of the surface for which the principal curvatures coincide, $\kappa_1 = \kappa_2$. An umbilic point is either elliptic or flat.

Example 17. The line of points with vanishing Gaussian curvature indicated in the modular surface of the function $f(z) = 1/\sin z$ (see p. 160) consists of parabolic points. For this reason, it is called the *parabolic curve.* By definition, it separates the hyperbolic from the elliptic points.

Theorem 27. *If the surface $M^2 \subset \mathbb{R}^3$ consists exclusively of umbilic points, it is part of a plane or a sphere.*

Proof. By assumption, the Weingarten map is a multiple of the identity, $S = \kappa \cdot \mathrm{Id}$. The Codazzi equation,

$$0 = \nabla S = \nabla(\kappa \cdot \mathrm{Id}) = \nabla(\kappa) \cdot \mathrm{Id},$$

implies that κ is constant. In case $\kappa = 0$, the normal vector e_3 is also constant. In fact, $de_3 = \omega_{31} \cdot e_1 + \omega_{32} \cdot e_2 = 0$. Thus M^2 is part of a plane. If $\kappa \neq 0$, we obtain from $\kappa \cdot dM = \kappa \cdot \mathrm{Id} = S = de_3$ the equation $d(e_3 - \kappa M) = 0$. Hence $e_3 - \kappa M$ is a constant vector a. Thus,

$$\|a + \kappa M\|^2 \equiv 1,$$

i. e., M^2 is part of a sphere. □

Theorem 28. *A surface $M^2 \subset \mathbb{R}^3$ with vanishing Gaussian curvature, $G \equiv 0$, and constant mean curvature, $H \neq 0$, is part of a cylinder.*

Proof. The first principal curvature vanishes, $\kappa_1 = 0$, and the second principal curvature is constant, $\kappa_2 \neq 0$. In an orthonormal frame e_1, e_2 consisting of eigenvectors of the second fundamental form, we obtain the formulas

$$\omega_{13} = 0 \quad \text{and} \quad \omega_{23} = \kappa_2 \cdot \sigma_2 .$$

Differentiate both equations:

$$0 = d\omega_{13} = \omega_{12} \wedge \omega_{23} = \kappa_2 \cdot \omega_{12} \wedge \sigma_2 ,$$
$$0 = \omega_{21} \wedge \omega_{13} = d\omega_{23} = \kappa_2 \cdot d\sigma_2 .$$

Since $\kappa_2 \neq 0$, on the one hand, ω_{12} is proportional to σ_2, and σ_2 is a closed form, $d\sigma_2 = 0$. But then $d\sigma_1 = \omega_{12} \wedge \sigma_2 = 0$, so that altogether we obtain

$$d\sigma_1 = 0, \quad d\sigma_2 = 0 \quad \text{and} \quad \omega_{12} = 0 .$$

The structural equations imply

$$de_1 = 0, \quad de_2 = \kappa_2 \cdot \sigma_2 \cdot e_3, \quad de_3 = \kappa_2 \cdot \sigma_2 \cdot e_2 .$$

Hence e_1 is constant, and the surface M^2 lies on a cylinder whose axis points in the direction of e_1. □

Theorem 29. *There exists no piece of a surface $M^2 \subset \mathbb{R}^3$ satisfying $G = -1$ and $H = 0$.*

Proof. Let κ_1, κ_2 be the principal curvatures. Then $0 \equiv H = \frac{1}{2}(\kappa_1 + \kappa_2)$ and $-1 = G = \kappa_1\kappa_2$ immediately imply $\kappa_1 = -\kappa_2 = 1$. Thus we obtain, with respect to a tangent orthonormal frame consisting of eigenvectors of the second fundamental form, the equations

$$\omega_{13} = \sigma_1 \quad \text{and} \quad \omega_{23} = -\sigma_2 .$$

Differentiating these equations yields

$$\begin{aligned} d\sigma_1 &= d\omega_{13} = \omega_{12} \wedge \omega_{23} = -\omega_{12} \wedge \sigma_2 , \\ d\sigma_2 &= -d\omega_{23} = -\omega_{21} \wedge \omega_{13} = \omega_{12} \wedge \sigma_1 . \end{aligned}$$

On the other hand, the form $\omega_{12} = A \cdot \sigma_1 + B \cdot \sigma_2$ is a linear combination of σ_1, σ_2 with coefficients A and B still to be determined from

$$d\sigma_1 = A \cdot \sigma_1 \wedge \sigma_2 \quad \text{and} \quad d\sigma_2 = B \cdot \sigma_1 \wedge \sigma_2 .$$

But then

$$\begin{aligned} A \cdot \sigma_1 \wedge \sigma_2 = d\sigma_1 &= -\omega_{12} \wedge \sigma_2 = -A \cdot \sigma_1 \wedge \sigma_2, \\ B \cdot \sigma_1 \wedge \sigma_2 = d\sigma_2 &= \omega_{12} \wedge \sigma_1 = -B \cdot \sigma_1 \wedge \sigma_2 , \end{aligned}$$

and hence $A = B = 0$. The form $\omega_{12} = 0$ vanishes, contradicting $G = -1$. □

Definition 14. A surface $M^2 \subset \mathbb{R}^3$ with vanishing mean curvature, $H \equiv 0$, is called a *minimal surface.*

The preceding theorem states that there does not exist any minimal surface with constant, non-vanishing Gaussian curvature. Consider, more generally, the normal map $e_3 : M^2 \to S^2$ of a minimal surface. From $\kappa_1 = -\kappa_2 = \sqrt{-G}$, we obtain the formulas

$$\omega_{13} = \sqrt{-G}\,\sigma_1, \quad \omega_{23} = -\sqrt{-G}\,\sigma_2 .$$

The relation $de_3 = \omega_{31} \cdot e_1 + \omega_{32} \cdot e_2$ evaluated on tangent vectors $\mathcal{V}, \mathcal{W} \in TM^2$ immediately implies

$$\langle de_3(\mathcal{V}), de_3(\mathcal{W}) \rangle = -G \cdot \langle \mathcal{V}, \mathcal{W} \rangle .$$

If I_{S^2} and I_{M^2} denote the fundamental forms of the sphere S^2 and the surface M^2, respectively, then the induced fundamental form (Riemannian metric) $e_3^*(\mathrm{I}_{S^2})$ is proportional to I_{M^2}:

$$e_3^*(\mathrm{I}_{S^2}) = -G \cdot \mathrm{I}_{M^2} .$$

Thus we have proved the following theorem.

Theorem 30. *The normal map $e_3 : M^2 \to S^2$ of a minimal surface preserves angles, and the Riemannian metric $g^* := -G \cdot \mathrm{I}_{M^2}$ defined on M^2 is a metric of constant positive Gaussian curvature.*

The vector-valued position function $M : M^2 \to \mathbb{R}^3$ of a minimal surface gives rise to harmonic functions.

Theorem 31. *Let $M : M^2 \to \mathbb{R}^3$ be a surface, and let Δ be its Laplacian acting on functions. Then M^2 is a minimal surface if and only if the vector-valued function M is harmonic, $\Delta(M) = 0$.*

Proof. The Laplacian can be expressed by the exterior derivative and the Hodge operator (see §3.11). Using the structural equations, we compute as follows:

$$\begin{aligned}
\Delta(M) \cdot dM^2 &= d * d(M) = d * (\sigma_1 \cdot e_1 + \sigma_2 \cdot e_2) = d(\sigma_2 \cdot e_1 - \sigma_1 \cdot e_2) \\
&= de_1 \wedge \sigma_2 + d\sigma_2 \cdot e_1 - de_2 \wedge \sigma_1 - d\sigma_1 \cdot e_2 \\
&= (\omega_{12} \cdot e_2 + \omega_{13} \cdot e_3) \wedge \sigma_2 + \omega_{21} \wedge \sigma_1 \cdot e_1 \\
&\quad -(\omega_{21} \cdot e_1 + \omega_{23} \cdot e_3) \wedge \sigma_1 - \omega_{12} \wedge \sigma_2 \cdot e_2 \\
&= (\omega_{13} \wedge \sigma_2 - \omega_{23} \wedge \sigma_1) \cdot e_3 \\
&= -(\kappa_1 + \kappa_2) \cdot \sigma_1 \wedge \sigma_2 \cdot e_3 = -2H \cdot e_3 \cdot dM^2 .
\end{aligned}$$

□

Corollary 2. *There does not exist a compact minimal surface $M^2 \subset \mathbb{R}^3$ without boundary.*

Proof. Otherwise, by Hopf's theorem (see §3.8), the components of the position function would have to be constant, a contradiction. □

The normal map $e_3 : M^2 \to S^2$ of a minimal surface with non-vanishing Gaussian curvature is a local conformal diffeomorphism. The stereographic projection from $S^2 \backslash \{\text{north pole}\}$ to $\mathbb{R}^2$ is a conformal diffeomorphism, too. Hence, by inverting the normal map $e_3 : M^2 \to S^2 \to \mathbb{R}^2$, every minimal surface can locally be parametrized by a map $F : U \to M^2$ such that

$$(*) \qquad \left\langle \frac{\partial F}{\partial y^1}, \frac{\partial F}{\partial y^1} \right\rangle = \left\langle \frac{\partial F}{\partial y^2}, \frac{\partial F}{\partial y^2} \right\rangle \quad \text{and} \quad \left\langle \frac{\partial F}{\partial y^1}, \frac{\partial F}{\partial y^2} \right\rangle = 0.$$

Coordinates like these are generally called *isothermic coordinates.* The minimality of the surface expressed in these coordinates is equivalent to the requirement that $F = F(y^1, y^2)$ is harmonic with respect to the Laplacian $\Delta = \partial^2/(\partial y^1)^2 + \partial^2/(\partial y^2)^2$ on $\mathbb{R}^2$,

$$\Delta(F) = 0 .$$

Identifying $\mathbb{R}^2$ with the complex numbers $\mathbb{C}$, every harmonic map $F : U \to \mathbb{R}^3$ can locally be represented as the real part of a holomorphic map $\Phi : U \to \mathbb{C}^3$; this is a direct consequence of the Cauchy-Riemann equations and Poincaré's lemma. The real partial derivatives of Φ can be expressed by the complex derivative ($z = y^1 + iy^2$), since Φ is holomorphic:

$$\frac{\partial \Phi}{\partial y^1} = \frac{\partial \Phi}{\partial z} \quad \text{and} \quad \frac{\partial \Phi}{\partial y^2} = i\frac{\partial \Phi}{\partial z}\,.$$

Denoting the derivative $\partial\Phi/\partial z$ by Ψ and writing $\Psi = (\Psi_1, \Psi_2, \Psi_3)$ with respect to the coordinates in $\mathbb{C}^3$, condition $(*)$ is equivalent to the quadratic equation

$$\Psi_1^2 + \Psi_2^2 + \Psi_3^2 = 0\,.$$

The solutions of this equation are described by two arbitrary holomorphic functions $f(z)$ and $g(z)$:

$$\Psi_1(z) = \frac{f(z)}{2}(1 - g^2(z)),\ \Psi_2(z) = i\frac{f(z)}{2}(1 + g^2(z)),\ \Psi_3(z) = f(z)g(z)\,.$$

Summarizing, we obtain

Theorem 32 (Weierstrass Representation of a Minimal Surface). *Every minimal surface $M^2 \subset \mathbb{R}^3$ with non-vanishing Gaussian curvature can locally be represented in terms of two holomorphic functions $f(z)$ and $g(z)$. A particular parametrization is provided by the formula*

$$F = \mathrm{Re}\Big(\int \frac{f(z)}{2}(1 - g^2(z))dz,\, i\int \frac{f(z)}{2}(1 + g^2(z))dz,\, \int f(z)g(z)dz\Big).$$

Remark. A simply connected minimal surface always occurs as *one* minimal surface within an S^1-parameter family; it can thus be deformed. In fact, if $\Phi := \Phi_R + i\cdot\Phi_I : U \to \mathbb{C}^3$ is holomorphic, then the equation

$$\mathrm{Re}(e^{i\alpha}\Phi) = \cos(\alpha)\cdot\Phi_R - \sin(\alpha)\cdot\Phi_I$$

defines a family of minimal surfaces, since the quadratic equation $\Psi_1^2 + \Psi_2^2 + \Psi_3^2 = 0$ is homogeneous.

Example 18. Very nice non-trivial minimal surfaces already arise by inserting relatively simple functions. Consider, for example, the following triple of holomorphic functions:

$$\Psi = (\Psi_1,\, \Psi_2,\, \Psi_3) = \left(\frac{1 - z^2}{2iz^2},\, \frac{1 + z^2}{2z^2},\, \frac{1}{iz}\right),$$

satisfying the condition $\Psi_1^2 + \Psi_2^2 + \Psi_3^2 = 0$, and their complex primitives,

$$F = (F_1,\, F_2,\, F_3) = \left(-\frac{1}{2i}(z + 1/z),\, \frac{1}{2}(z - 1/z),\, \frac{1}{i}\ln z\right).$$

The real and the imaginary parts of these functions determine a pair of minimal surfaces which are called *adjoint* to one another. Setting $\ln z = \varrho + i\tau$, these are described by the formulas

$$F_R := \mathrm{Re}(F) = \big(-\sinh\varrho\,\sin\tau,\, \sinh\varrho\,\cos\tau,\, \tau\big),$$
$$F_I := \mathrm{Im}\,(F) = \big(-\cosh\varrho\,\cos\tau,\, -\cosh\varrho\,\sin\tau,\, \varrho\big)\,.$$

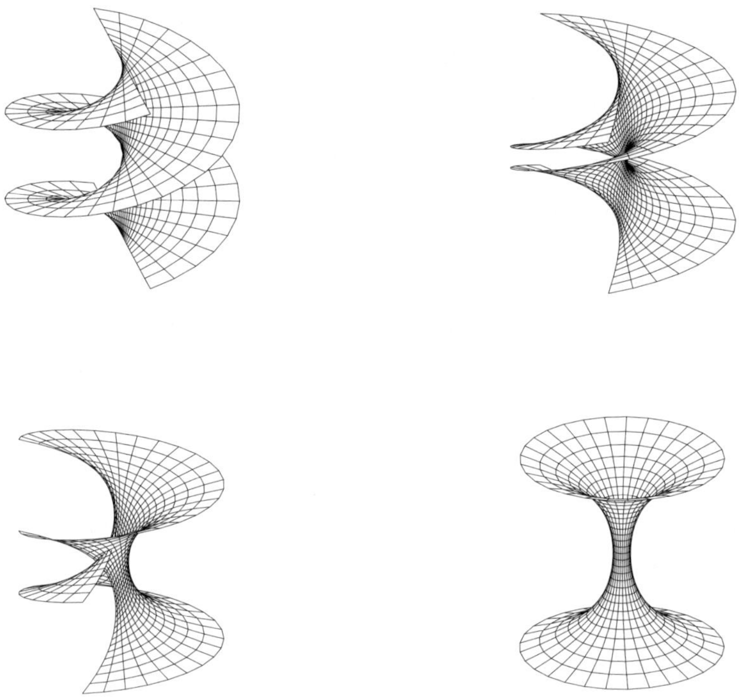

Hence $F_\alpha = \cos\alpha\, F_R + \sin\alpha\, F_I$ determines a 1-parameter family of minimal surfaces joining F_R to F_I. The surface F_R is a helicoid, whereas F_I is the surface of revolution generated by the catenary[2], the catenoid. The sequence of pictures above illustrates the deformation of both these minimal surfaces into one another. Further examples of classical minimal surfaces can be found in the exercises.

5.5. Curves on Surfaces and Geodesic Lines

Consider an oriented surface $M^2 \subset \mathbb{R}^3$ in 3-space, and on it a curve $\gamma : [a, b] \to M^2$, which we assume to be naturally parametrized. The tangent vector $\vec{t}(s)$ to the curve is also tangent to the surface, and will be denoted by $e_1(s)$. If $e_3(s)$ is the normal vector to the surface M^2 along the curve, the equation $e_1(s) \times e_2(s) = e_3(s)$ determines a third vector, which is again tangential to M^2 along the curve. This way, we construct at each point of the curve an orthonormal frame, the so-called *Darboux frame* of the curve

[2]We recall that this is the curve formed by a perfectly flexible inextensible chain hanging from two supports.

γ with respect to the surface M^2. Unlike the Frenet frame, the Darboux frame of a curve in a surface is defined at every point of the curve, even at those points where its curvature vanishes. Now we decompose the curvature vector $\vec{k}(s) := \kappa(s) \cdot \vec{h}(s)$ of the curve γ in $\mathbb{R}^3$ into the part $\vec{k}_g(s)$ which is tangential and the part $\vec{k}_n(s)$ which is perpendicular to the surface.

Definition 15. The vector field $\vec{k}_g(s)$ is called the *geodesic curvature vector*, the vector field $\vec{k}_n(s)$ the *normal curvature vector* of the curve with respect to the surface,

$$\vec{k}(s) \;=\; \kappa(s) \cdot \vec{h}(s) \;=\; \vec{k}_g(s) + \vec{k}_n(s)\,.$$

The vector $\vec{k}_n(s) := \kappa_n(s) \cdot e_3$ is proportional to the normal vector of the surface with a factor $\kappa_n(s)$, called the *normal curvature.* Analogously, $\vec{k}_g(s) := -\kappa_g(s) \cdot e_2$ is proportional to e_2, since the principal normal vector $\vec{h}(s)$ is orthogonal to the tangent vector. From this we obtain the *geodesic curvature* $\kappa_g(s)$ of the curve, and the square of the curvature $\kappa(s)$ of the curve is the sum

$$\kappa^2(s) \;=\; \kappa_g^2(s) + \kappa_n^2(s).$$

Rewriting the geodesic curvature,

$$\kappa_g(s) \;=\; \langle \vec{k}_g, e_2 \rangle \;=\; \langle \vec{k}, e_2 \rangle \;=-\langle de_1(e_1), e_2 \rangle \;=\; -\,\mathrm{I}(\nabla_{e_1} e_1, e_2)\,,$$

we are led to the following theorem.

Theorem 33. *The geodesic curvature of a curve in a surface depends only on the first fundamental form of the surface. In the Darboux frame,*

$$\kappa_g(s) \;=\; -\,\mathrm{I}(\nabla_{e_1} e_1, e_2)\,.$$

Similarly, the normal curvature $\kappa_n(s)$ of a curve can be expressed by the second fundamental form of the surface and its tangent vector:

$$\kappa_n \;=\; \langle \vec{k}_n, e_3 \rangle \;=\; \langle \vec{k}, e_3 \rangle \;=\; \left\langle \frac{d\vec{t}}{ds}, e_3 \right\rangle \;=\; \langle de_1(e_1), e_3 \rangle \;=\; -\,\mathrm{II}(\vec{t}, \vec{t})\,.$$

Theorem 34. *If γ and γ^* are any two curves tangent at a point of the surface M^2, their normal curvatures at this point coincide.*

Now we consider a variation of the curve $\gamma : [a, b] \to M^2$ within the surface M^2 with fixed initial and end points. We may imagine this as a family of curves,

$$\gamma\varepsilon\ (s) := \gamma(s) + \varepsilon \cdot h(s) \cdot e_2(s) + \mathcal{O}(\varepsilon^2),$$

determined by a function h vanishing at the ends of the interval, $h(a) = h(b) = 0$. The length of the curve γ_ε is

$$L(\gamma_\varepsilon(s)) \;=\; \int_a^b \left\| \vec{t}(s) + \varepsilon \frac{dh}{ds}(s) \cdot e_2(s) + \varepsilon \cdot h(s) \cdot \frac{de_2}{ds}(s) + \mathcal{O}(\varepsilon^2) \right\|,$$

and computing the derivative with respect to the parameter ε yields

$$\frac{d}{d\varepsilon}\Big(L(\gamma_\varepsilon(s))\Big)\Big|_{\varepsilon=0} = \int_a^b \langle \vec{t}(s),\, \frac{dh}{ds}(s)\cdot e_2(s) + h(s)\frac{de_2}{ds}(s)\rangle\, ds\,.$$

The formulas $\langle \vec{t}(s),\, e_2(s)\rangle = 0$ and

$$\langle \vec{t}(s),\, \frac{de_2}{ds}(s)\rangle \;=\; \langle e_1,\, \nabla_{e_1} e_2\rangle \;=\; \omega_{21}(e_1) \;=\; \kappa_g(s)$$

then, eventually, imply

$$\frac{d}{d\varepsilon}\Big(L(\gamma_\varepsilon(s))\Big)\Big|_{\varepsilon=0} = \int_a^b h(s)\cdot\kappa_g(s)ds\,.$$

Theorem 35. *Let $\gamma : [a,b] \to M^2$ be a curve in the surface, realizing the shortest distance between the points $\gamma(a)$ and $\gamma(b)$ within the surface. Then the geodesic curvature of the curve vanishes,*

$$\kappa_g(s) \;=\; 0\,.$$

Definition 16. A curve in the surface is called a *geodesic line* if its geodesic curvature vanishes. This condition is equivalent to $\nabla_{\vec{t}}\vec{t} = 0$.

If the geodesic curvature $\kappa_g = 0$ vanishes, the curvature vector

$$\kappa(s)\cdot\vec{h}(s) \;=\; \vec{k}(s) \;=\; \vec{k}_n(s)$$

is proportional to the normal vector e_3 to the surface. Hence a curve with non-vanishing curvature, $\kappa(s) \neq 0$, is a geodesic line in the surface M^2 if and only if the principal normal vector $\vec{h}(s)$ coincides with the normal vector e_3 to the surface,

$$\vec{h}(s) \;=\; \pm e_3(\gamma(s))\,.$$

Geodesic lines realize the shortest distance between two points of the surface only if the points lie sufficiently close to one another. A longer geodesic line does not necessarily have to be the shortest path between its end points within the surface. Examples—on S^2 or on the cylinder—are easily constructed (see Example 21). We represent the curve in local coordinates y^1, y^2 on the surface as $\gamma(s) = (y^1(s), y^2(s))$. If $g_{ij}(y^1,y^2)$ are the coefficients of the Riemannian metric and s is the natural parameter of the curve, then

$$\sum_{i,j=1}^{2} g_{ij}(y^1(s),y^2(s))\cdot\frac{dy^i}{ds}\cdot\frac{dy^j}{ds} \;=\; 1\,.$$

The *Christoffel symbols* Γ^k_{ij} occur in the representation of the covariant derivatives of the basis vector fields $\partial/\partial y^i$ as linear combinations of themselves:

$$\nabla_{\frac{\partial}{\partial y^i}}\frac{\partial}{\partial y^j} \;=:\; \sum_{k=1}^{2}\Gamma^k_{ij}\,\frac{\partial}{\partial y^k}\,.$$

Since

$$\nabla_{\vec{t}}\Big(\frac{\partial}{\partial y^j}\Big) = \sum_{i,k=1}^{2} \frac{dy^i}{ds} \cdot \Gamma_{ij}^k \frac{\partial}{\partial y^k},$$

the following formula holds for the tangent vector $\vec{t}$:

$$\nabla_{\vec{t}}\vec{t} = \sum_{k=1}^{2} \Big(\frac{d^2y^k}{ds^2} + \sum_{i,j=1}^{2} \Gamma_{ij}^k \cdot \frac{dy^i}{ds} \cdot \frac{dy^j}{ds}\Big) \frac{\partial}{\partial y^k} .$$

We summarize this observation in the following theorem:

Theorem 36. *Let $\gamma(s) = (y^1(s), y^2(s))$ be a curve represented in local coordinates. It is a geodesic line if and only if the following system of differential equations is satisfied ($k = 1, 2$):*

$$\frac{d^2y^k}{ds^2} + \sum_{i,j=1}^{2} \Gamma_{ij}^k \frac{dy^i}{ds}\frac{dy^j}{ds} = 0 .$$

This is a second order system of differential equations. The general existence and uniqueness result from the theory of ordinary differential equations ensures the existence of precisely one geodesic line through any fixed point in a given direction.

Corollary 3. *For each point $x \in M^2$ and every tangent vector $\mathcal{V} \in T_xM^2$ of length one on a surface without boundary, there exists exactly one geodesic line $\gamma : (-\varepsilon, \varepsilon) \to M^2$ satisfying the initial conditions $\gamma(0) = x, \gamma'(0) = \mathcal{V}$.*

Example 19. Let $\gamma(s)$ be a geodesic line on the two-dimensional sphere S^2 of radius $R > 0$. Since $\kappa_g = 0$, the normal curvature κ_n and the spatial curvature κ of the curve coincide. Hence

$$\kappa = |\kappa_n| = \mathrm{II}(\vec{t}, \vec{t}) = \frac{1}{R},$$

since the sphere consists exclusively of umbilic points. The normal vector to the sphere is $e_3(x) = \frac{1}{R}x$. Thus, on the one hand, we obtain

$$\begin{aligned} \frac{d^2\vec{t}(s)}{ds^2} &= \frac{d}{ds}(\vec{k}(s)) = \frac{d}{ds}(\vec{k}_n(s)) = -\frac{1}{R}\frac{d}{ds}(e_3(\gamma(s))) \\ &= -\frac{1}{R^2}\frac{d\gamma(s)}{ds} = -\frac{1}{R^2}\vec{t}(s), \end{aligned}$$

and, on the other hand, the Frenet formulas for the curve imply

$$\frac{d^2\vec{t}(s)}{ds^2} = \frac{1}{R}\frac{d\vec{h}(s)}{ds} = -\frac{1}{R^2}\vec{t}(s) + \frac{1}{R}\tau \cdot \vec{b}(s) .$$

Consequently, the torsion τ of the curve vanishes identically. The curve $\gamma(s)$ is thus a plane curve in S^2 with constant curvature $1/R$, i. e., a principal circle in S^2.

Example 20. The relation $\kappa^2 = \kappa_g^2 + \kappa_n^2$ immediately implies that every curve with vanishing curvature is a geodesic line. Formulated differently, if a straight line of $\mathbb{R}^3$ lies in a surface, then it is necessarily a geodesic line in this surface.

Example 21. Consider the cylinder M^2 in $\mathbb{R}^3$ whose axis is parallel to the unit vector $\vec{a}$. The normal vector at a point $x \in M^2$ is proportional to the projection of x onto the plane perpendicular to $\vec{a}$,

$$e_3(x) = \frac{x - \langle x, \vec{a}\rangle \vec{a}}{\|x - \langle x, \vec{a}\rangle \vec{a}\|}.$$

For a geodesic line $\gamma(s)$ on the cylinder, we immediately obtain from $\vec{h}(s) = e_3(\gamma(s))$ the equations $\langle \vec{h}(s), \vec{a}\rangle = 0$ and

$$C := \langle \vec{h}(s), \gamma(s)\rangle = \langle e_3(\gamma(s)), \gamma(s)\rangle = \|\gamma(s) - \langle \gamma(s), \vec{a}\rangle \vec{a}\|.$$

Here C is positive and constant, since $\gamma(s)$ lies on the cylinder. As $d\vec{t}(s)/ds = \kappa(s) \cdot \vec{h}(s)$ and $d\vec{b}(s)/ds = -\tau(s) \cdot \vec{h}(s)$, the equation $\langle \vec{h}(s), \vec{a}\rangle = 0$ implies that

$$A := \langle \vec{t}(s), \vec{a}\rangle \quad \text{and} \quad B := \langle \vec{b}(s), \vec{a}\rangle$$

are constant. In case $A = 0$, the curve lies in an affine plane perpendicular to $\vec{a}$ and on the cylinder; hence it is a circle. For $A \neq 0$, we differentiate the equation $\langle \vec{h}(s), \vec{a}\rangle = 0$ once, and then the Frenet formulas imply the relation

$$-A \cdot \kappa(s) + B \cdot \tau(s) = 0.$$

Differentiating the constant C^2 twice, we deduce that

$$C \cdot \kappa(s) + 1 = A^2.$$

The curvature $\kappa(s)$ as well as the torsion $\tau(s)$ are thus constant, and hence the curve $\gamma(s)$ is a helix (see Exercise 6). In summary, the geodesic lines on the cylinder are helices, circles, and the straight lines parallel to the $\vec{a}$-axis. Two points on the cylinder may be joined by infinitely many geodesic lines, only one of which is the shortest path between them:

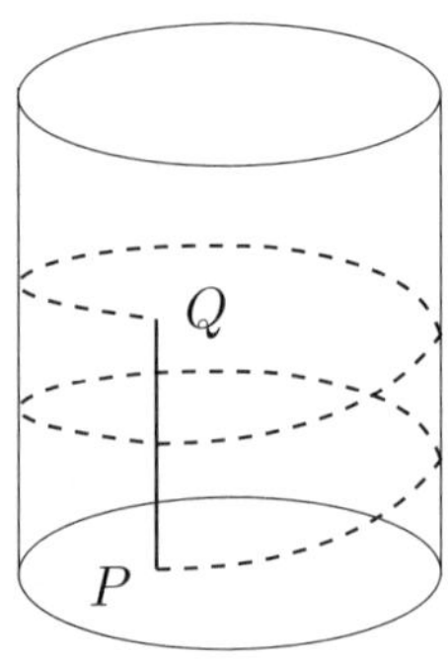

The equations characterizing geodesics are non-linear differential equations of second order, and hence, in general, difficult to solve. Things become a little easier if, because of the symmetry of the surface, one can find sufficiently many functions which are constant along geodesics:

Definition 17. A function $f : TM \to \mathbb{R}$ is called a *first integral* of the geodesic flow if f is constant on all tangent curves $\gamma'(t)$ of geodesics.

The length of γ' itself is always one non-trivial first integral, which is interpreted as energy. It expresses nothing but the fact that geodesic lines are parametrized by arc length. Further first integrals are obtained by *Noether's theorem* starting from isometries of the surface.

Theorem 37 (Noether's Theorem). *Let $\mathcal{V}$ be a tangent vector field on the surface M^2 whose flow $\Phi_t : M^2 \to M^2$ consists only of isometries. Then the function*

$$f_{\mathcal{V}} : TM \longrightarrow \mathbb{R}, \quad f_{\mathcal{V}}(\mathcal{W}) \; = \; \mathrm{I}(\mathcal{W}, \mathcal{V}),$$

is a first integral of the geodesic flow.

Proof. Consider a geodesic line $\gamma(s)$ on M^2 in its natural parametrization. Setting $\mathcal{W} = \gamma'$, the function defined in the theorem has the following derivative with respect to s:

$$\frac{d}{ds}\left\langle \gamma'(s), \mathcal{V}(\gamma'(s))\right\rangle \; = \; \left\langle \gamma''(s), \mathcal{V}(\gamma'(s))\right\rangle + \left\langle \gamma'(s), d\mathcal{V}(\gamma'(s))\right\rangle .$$

The vector γ'' is nothing but the curvature vector of the curve in $\mathbb{R}^3$, and for a geodesic line this is always perpendicular to the surface. Hence the first summand vanishes. On the other hand, the flow Φ_t was supposed to consist of isometries,

$$\langle \mathcal{W}_1, \mathcal{W}_2 \rangle \; = \; \langle d\Phi_t(\mathcal{W}_1), d\Phi_t(\mathcal{W}_2)\rangle .$$

Differentiating this identity and evaluating the result at $t = 0$, we obtain

$$\langle d\mathcal{V}(\mathcal{W}_1), \mathcal{W}_2\rangle + \langle d\mathcal{V}(\mathcal{W}_2), \mathcal{W}_1\rangle \; = \; 0 ,$$

and the second term vanishes, too, since it has to be antisymmetric in $\gamma'(s)$. □

Example 22. We want to find all geodesic lines on a surface of revolution. For simplicity, assume as before that the generating curve is given in natural parametrization. With respect to the coordinates s and φ on the surface, a geodesic line is parametrized as $\gamma(t) = (s(t), \varphi(t))$ and has the first integral

$$E \; := \; ||\gamma'(t)||^2 \; = \; s'^2 + r(s)^2\varphi'^2 .$$

Since the rotation through the angle φ around the z-axis is an isometry of the surface, the vector field $\partial/\partial\varphi$ satisfies the assumptions of the Noether theorem and gives rise to a further first integral,

$$M \;:=\; \langle \gamma', \partial/\partial\varphi\rangle \;=\; r^2\varphi' .$$

The existence of this second invariant is known as *Clairaut's theorem* (1731) and has the following geometric interpretation. Consider a geodesic line through a point of the surface, and also the meridian $\eta(t) = (s_0, \varphi(t))$ through this same point. The angle α formed by the geodesic and the meridian is computed as

$$\cos(\alpha) \;=\; \frac{\langle\gamma', \eta'\rangle}{||\gamma'||\cdot||\eta'||} \;=\; \frac{r^2\varphi'^2}{\sqrt{E}\cdot\sqrt{r^2\varphi'^2}} \;=\; \frac{r\,\varphi'}{\sqrt{E}} .$$

Therefore, M is constant if and only if $r(s)\cdot\cos(\alpha)$ is constant. For the qualitative discussion of the geodesic lines, we take $E = 1$. Inserting $\varphi' = M/r^2$ into $E = 1$, we obtain

$$s'^2 r^2 \;=\; r^2 - M^2, \quad \text{thus } 0 \le M^2 \le r^2 .$$

If M vanishes identically, then $\varphi' = 0$, and φ has to be constant. This corresponds precisely to the generating curve of the surface of revolution; each profile curve of F is thus a geodesic. In the other extremal case, $M^2 = r^2$, we have $s' = 0$, i. e., s has to be constant. This curve is a meridian on F; note, however, that in this case differentiating $M^2 = r^2$ immediately leads to $r\,r' = 0$, which, in turn, implies $r' = 0$. A meridian of F is a geodesic if and only if the radius function r has an extremal point there. All that remains is the generic case, $0 < M^2 < r^2$. The condition $E = 1$ is equivalent to

$$s' \;=\; \pm\frac{1}{r}\sqrt{r^2 - M^2} .$$

In order to eliminate the curve parameter t, we view s as a function of φ; hence $ds/d\varphi = s'/\varphi'$,

$$\frac{ds}{d\varphi} \;=\; \pm\frac{r}{M}\sqrt{r^2 - M^2} ,$$

and thus

$$M\int_{s_0}^{s(\varphi)} \frac{1}{r(\tilde{s})\sqrt{r(\tilde{s})^2 - M^2}}\, d\tilde{s} \;=\; \int_0^{\varphi} d\tilde{\varphi} \;=\; \pm\varphi .$$

This provides a parametrization of the geodesic lines in the case of surfaces for which the integral on the left-hand side is elementarily integrable and can be solved for s.

Example 23. Let $\mathcal{H}^2$ be the hyperbolic plane (see Example 9). We again parametrize the geodesic γ as $\gamma(s) = (x(s),\, y(s))$. Then the condition that γ is parametrized by arc length takes the form

$$\frac{x'^2}{y^2} + \frac{y'^2}{y^2} \;=\; 1\,.$$

Since the metric does not depend on x, translations in the x-direction are isometries; the associated vector field is $\partial/\partial x$. Hence

$$M \;:=\; \langle \gamma',\, \partial/\partial x \rangle \;=\; x'/y^2$$

is a first integral. If M vanishes, x has to be constant, and the geodesic lines are precisely the open straight half-lines parallel to the y-axis. In the general case, rewrite the first condition, inserting M. A short calculation leads to

$$y' \;=\; \pm y\sqrt{1 - M^2 y^2}.$$

Again, we view y as a function of x and set $r := 1/|M|$,

$$\frac{dy}{dx} \;=\; \frac{y'}{x'} \;=\; \pm\frac{1}{y}\sqrt{r^2 - y^2}\,.$$

Directly integrating this equation and denoting the integration constant by a yields

$$-\sqrt{r^2 - y^2} \;=\; \pm(x-a), \quad \text{hence} \;\; (x-a)^2 + y^2 \;=\; r^2\,.$$

The trace of the geodesic line is thus a semi-circle whose center lies on the x-axis.

Once again we now turn to the Gauss-Bonnet formula. Let M^2 be an oriented, compact surface with boundary ∂M^2, and let $\mathcal{V}$ be a vector field of length one defined on M^2 except at finitely many points $x_1, \ldots, x_k$ in the interior of M^2. Assume, moreover, that $\mathcal{V}$ is perpendicular to ∂M^2 at the boundary ∂M^2 and outwardly directed. Consider the 1-form $\chi_{\mathcal{V}}$ introduced in §5.4. From the formulas proved there, we deduce

$$\int_{M^2} d\chi_{\mathcal{V}} \;=\; \int_{\partial M^2} \chi_{\mathcal{V}} - 2\pi \sum_{i=1}^{k} \mathrm{Ind}\,(\mathcal{V}, x_i)\,.$$

Near the boundary ∂M^2, we choose an orthonormal frame e_1, e_2 such that $e_1 = \mathcal{V}$. Then $\chi_{\mathcal{V}} = \omega_{12}$, and e_2 is the unit tangent vector to the boundary. Since $\omega_{12}(e_2) \;=\; \mathrm{I}(\nabla_{e_2} e_1, e_2) \;=\; -\mathrm{I}(e_1, \nabla_{e_2} e_2) \;=\; \kappa_g$, we obtain the Gauss-Bonnet formula for manifolds with boundary.

Theorem 38 (Gauss-Bonnet Formula). *Let M^2 be a compact, oriented surface with boundary, and let $\mathcal{V}$ be a vector field of length one defined on M^2 except at finitely many points $x_1, \ldots, x_k$ in the interior of M^2. Assume,* *moreover, that at the boundary $\mathcal{V}$ is perpendicular to the boundary and outwardly directed. Then*

$$\int_{M^2} G \cdot dM^2 + \int_{\partial M^2} \kappa_g \cdot d(\partial M^2) \;=\; 2\pi \cdot \sum_{i=1}^{k} \mathrm{Ind}(\mathcal{V}, x_i)\,.$$

This has similar consequences as in the case of surfaces without boundary. On the one hand, the sum of the integrals

$$\int_{M^2} G \cdot dM^2 + \int_{\partial M^2} \kappa_g \cdot d(\partial M^2)$$

is independent of the first fundamental form. On the other hand, the sum of the indices of a vector field does not depend on the vector field.

5.6. Maps between Surfaces

The length $\|\mathcal{V}\| = \sqrt{\mathrm{I}(\mathcal{V},\mathcal{V})}$ of a tangent vector can be determined by measuring the length of curves. To see this, choose a curve γ on the surface M^2 in arbitrary parametrization whose tangent at $t = 0$ coincides with $\mathcal{V}$, $\dot\gamma(0) = \mathcal{V}$. The length of the curve segment $\gamma([0,t])$ is defined as

$$L(t) \;=\; \int_0^t \sqrt{\mathrm{I}(\dot\gamma(\mu), \dot\gamma(\mu))}d\mu\,,$$

and computing the derivative at $t = 0$ yields

$$\left.\frac{dL(t)}{dt}\right|_{t=0} \;=\; \sqrt{\mathrm{I}(\mathcal{V},\mathcal{V})}\,.$$

The first fundamental form of the surface is thus completely determined by the lengths of curve segments, and vice versa. Hence each map $f : M^2 \to \overline{M}^2$ between two surfaces which does not alter the lengths of curve segments has a differential $df : TM^2 \to T\overline{M}^2$ preserving the first fundamental forms $\mathrm{I}_{\overline{M}^2}$ and I_{M^2},

$$f^*(\mathrm{I}_{\overline{M}^2}) \;=\; \mathrm{I}_{M^2}\,.$$

Definition 18. An *isometry* $f : M^2 \to \overline{M}^2$ between two surfaces is a length-preserving smooth map.

The Gaussian curvature of a surface is a geometric quantity depending on the first fundamental form. This immediately implies

Theorem 39. *Let $f : M^2 \to \overline{M}^2$ be an isometry. Then the Gaussian curvatures $G_{\overline{M}^2}, G_{M^2}$ at corresponding points coincide:*

$$G_{M^2}(x) \;=\; G_{\overline{M}^2}(f(x))\,.$$

Corollary 4. *There does not exist an isometric map from an open subset of S^2 onto an open subset of the euclidean space $\mathbb{R}^2$.*

For this reason, cartography was forced to look for other suitable maps. Especially important are conformal (angle-preserving) and volume-preserving maps. Let us first turn to the former ones. A map $f : M^2 \to \overline{M}^2$ is called *conformal* (or angle-preserving) if the angle between two intersecting curves γ_1, γ_2 in M^2 is the same as that of the image curves $f \circ \gamma_1, f \circ \gamma_2$ in $\overline{M}^2$. These maps can be characterized using the first fundamental form.

Theorem 40. *A map $f : M^2 \to \overline{M}^2$ is conformal if and only if there exists a positive function $h : M^2 \to (0, \infty)$ such that $f^*(\mathrm{I}_{\overline{M}^2}) = h \cdot \mathrm{I}_{M^2}$.*

Proof. This is an easily proved fact from linear algebra. Let $\langle\ ,\ \rangle_1$ and $\langle\ ,\ \rangle_2$ be two positive definite scalar products on a two-dimensional vector space satisfying the equation

$$\frac{\langle \mathcal{V}, \mathcal{W}\rangle_1}{\|\mathcal{V}\|_1 \|\mathcal{W}\|_1} = \frac{\langle \mathcal{V}, \mathcal{W}\rangle_2}{\|\mathcal{V}\|_2 \|\mathcal{W}\|_2}$$

for all vectors $\mathcal{V}, \mathcal{W}$. Then there exists a positive number h such that

$$\langle \mathcal{V}, \mathcal{W}\rangle_2 = h \cdot \langle \mathcal{V}, \mathcal{W}\rangle_1 . \qquad \square$$

We already encountered conformal maps. The normal map $e_3 : M^2 \to S^2$ of a minimal surface in $\mathbb{R}^3$ is conformal (Theorem 30). Complex analytic functions are conformal maps from the plane $\mathbb{R}^2$ to itself. In fact, if $f = (u, v) = u + i \cdot v$ is a map $f : \mathbb{R}^2 \to \mathbb{R}^2$, then the following formula holds for the first fundamental form $\mathrm{I}_{\mathbb{R}^2} = dx^2 + dy^2$:

$$\begin{aligned} f^*(\mathrm{I}_{\mathbb{R}^2}) &= (u_x\, dx + u_y\, dy)^2 + (v_x\, dx + v_y\, dy)^2 \\ &= (u_x^2 + v_x^2)dx^2 + 2(u_x u_y + v_x v_y)dx \odot dy + (v_x^2 + v_y^2)dy^2 . \end{aligned}$$

The Cauchy-Riemann equations, $u_x = v_y, u_y = -v_x$, imply that, for the induced form,

$$f^*(\mathrm{I}_{\mathbb{R}^2}) = |f'(z)|^2 \cdot \mathrm{I}_{\mathbb{R}^2} .$$

Hence every complex-differentiable function $f = u + iv$ is a conformal map, and the square of the absolute value of its derivative, $|f'(z)|^2$, appears as the conformal factor in the sense of two-dimensional geometry. The best-known conformal map is the *stereographic projection* from the two-dimensional sphere to the plane. This shows that conformal maps from $\mathbb{R}^2$ to S^2 exist, and so angle-preserving geographic maps can be drawn.

Example 24. In coordinates, the stereographic projection $f : \mathbb{R}^2 \to S^2$ is the map

$$f(x, y) = \left(\frac{2x}{1 + x^2 + y^2}, \frac{2y}{1 + x^2 + y^2}, \frac{1 - x^2 - y^2}{1 + x^2 + y^2}\right).$$

The partial derivative vectors of f are

$$\frac{\partial f}{\partial x} = \frac{2}{(1+x^2+y^2)^2}\left(1-x^2+y^2,\,-2xy,\,-2x\right),$$
$$\frac{\partial f}{\partial y} = \frac{2}{(1+x^2+y^2)^2}\left(-2xy,\,1+x^2-y^2,\,-2y\right),$$

and hence

$$\left\langle \frac{\partial f}{\partial x}, \frac{\partial f}{\partial x}\right\rangle = \frac{4}{(1+x^2+y^2)^2} = \left\langle \frac{\partial f}{\partial y}, \frac{\partial f}{\partial y}\right\rangle \quad \text{and} \quad \left\langle \frac{\partial f}{\partial x}, \frac{\partial f}{\partial y}\right\rangle = 0\,.$$

These formulas show that the stereographic projection is conformal.

Example 25 (Mercator projection)**.** Let (φ, ψ) be spherical coordinates on the sphere $S^2\backslash\{N, S\}$ outside the north and the south pole (see §3.2, Example 14). Define a map from the sphere S^2 to $\mathbb{R}^2$ by the formula

$$f(\varphi,\psi) = \left(\varphi, \log\left(\tan\left(\frac{\pi}{4}+\frac{\psi}{2}\right)\right)\right).$$

This turns out to be a conformal map with an additional interesting property: curves of constant azimuth (i. e. forming constant angles with the meridians on S^2) are mapped to straight lines under f. For this reason, the Mercator projection (1569) has been one of the most widely used methods for the creation of nautical maps for centuries. The reader may find an extensive discussion of further important map projections in the book by K. Strubecker (Volume 2) cited in the bibliography.

Another interesting class of maps between surfaces are those preserving the two-dimensional volume (surface area) of measurable subsets of the surface. Such a map $f : M^2 \to \overline{M}^2$ is characterized by the property that the volume form is preserved,

$$f^*(d\overline{M}^2) = dM^2\,.$$

We will call such an f an *area-preserving map.* Isometries are the maps which are both conformal and area-preserving.

Theorem 41. *A map $f : M^2 \to \overline{M}^2$ is an isometry if and only if f is conformal and area-preserving.*

Proof. The relation $f^*(\mathrm{I}_{\overline{M}^2}) = h \cdot \mathrm{I}_{M^2}$ implies $f^*(d\overline{M}^2) = h^2 dM^2$, and hence also the assertion. □

The sphere S^2 (or pieces of it) thus cannot be mapped to the plane in an angle- and area-preserving way. There exist, however, angle-preserving maps, as the stereographic projection shows. We now want to describe an example for an area-preserving map from S^2 to $\mathbb{R}^2$.

Example 26. Delete the north and the south pole from the sphere S^2, join them by a straight line, and consider the cylinder whose axis this line is:

$$Z = \{(x, y, z) \in \mathbb{R}^3 : x^2 + y^2 = 1\}.$$

To each point different from the north and the south pole, $P \neq N, S$, there corresponds one straight line containing the point P which is perpendicular to the fixed axis. This line intersects the cylinder in two points, and we denote the one closer to P by $f(P)$. This defines the so-called *Lambert projection*,

$$f : S^2 \backslash \{N, S\} \longrightarrow Z .$$

Composing this with an isometry from the cylinder to the plane, we obtain an area-preserving map from S^2 to $\mathbb{R}^2$.

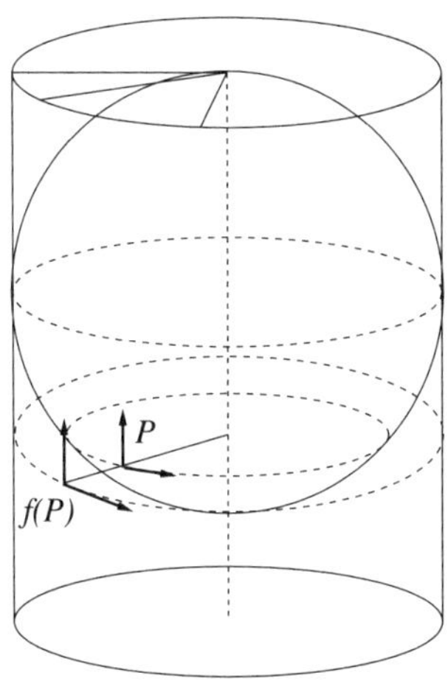

To see this, let e_1, e_2 and e_1^*, e_2^* be the orthonormal frames on S^2 and Z, respectively, indicated in the above picture. Then an elementary geometric argument immediately leads to the relation

$$df(e_1) = re_1^*, \quad df(e_2) = \frac{1}{r} e_2^* .$$

Here r is the radius of the meridian of S^2 tangent to e_2. These formulas prove that f is an area-preserving map.

5.7. Higher-Dimensional Riemannian Manifolds

Following §3.11, we will conclude this chapter by a brief survey of higher-dimensional Riemannian manifolds. Only the basic facts are discussed, and no comprehensive introduction into Riemannian geometry is intended. As in the case of surfaces, every pseudo-Riemannian manifold (M^m, g) can be endowed with a covariant derivative ∇ of vector fields with respect to vector fields which is uniquely determined by two conditions. This formalizes the notion of parallel displacement of vectors on a manifold. In fact, Levi-Civita's idea (1916) that every Riemannian space admits a unique parallel displacement preserving the metric turned out to be fundamental for the further development of geometry.

Theorem 42. *Every pseudo-Riemannian manifold has precisely one covariant derivative $\nabla_{\mathcal{V}}\mathcal{W}$ of vector fields with the following properties:*

$$(1)\ \nabla_{\mathcal{V}}(\mathcal{W}_1+\mathcal{W}_2) = \nabla_{\mathcal{V}}(\mathcal{W}_1)+\nabla_{\mathcal{V}}(\mathcal{W}_2)\ ;$$

$$(2)\ \nabla_{\mathcal{V}_1+\mathcal{V}_2}\mathcal{W} = \nabla_{\mathcal{V}_1}\mathcal{W}+\nabla_{\mathcal{V}_1}\mathcal{W}\ ;$$

(3) *if f is a smooth function, then*

$$\nabla_{\mathcal{V}}(f\cdot\mathcal{W}) = df(\mathcal{V})\cdot\mathcal{W}+f\cdot\nabla_{\mathcal{V}}\mathcal{W} \quad \text{and} \quad \nabla_{f\cdot\mathcal{V}}\mathcal{W} = f\cdot\nabla_{\mathcal{V}}\mathcal{W};$$

$$(4)\ \mathcal{V}g(\mathcal{W}_1,\ \mathcal{W}_2) = g(\nabla_{\mathcal{V}}\mathcal{W}_1,\ \mathcal{W}_2)+g(\mathcal{W}_1,\ \nabla_{\mathcal{V}}\mathcal{W}_2)\ ;$$

$$(5)\ \nabla_{\mathcal{V}}\mathcal{W}-\nabla_{\mathcal{W}}\mathcal{V} = [\mathcal{V},\mathcal{W}].$$

The vector field $\nabla_{\mathcal{V}}\mathcal{W}$ is determined by the expression

$$\begin{aligned}2g(\nabla_{\mathcal{U}}\mathcal{V},\ \mathcal{W}) &= \mathcal{U}(g(\mathcal{V},\ \mathcal{W}))+\mathcal{V}(g(\mathcal{U},\ \mathcal{W}))-\mathcal{W}(g(\mathcal{U},\mathcal{V}))+g([\mathcal{U},\mathcal{V}],\ \mathcal{W})\\ &\quad +g(\mathcal{V},\ [\mathcal{W},\mathcal{U}])-g(\mathcal{U},\ [\mathcal{V},\mathcal{W}])\,.\end{aligned}$$

Proof. We prove uniqueness of the covariant derivative and derive in particular the stated formula for $\nabla_{\mathcal{V}}\mathcal{W}$. By assumption, we have

$$\begin{aligned}\mathcal{U}(g(\mathcal{V},\ \mathcal{W})) &= g(\nabla_{\mathcal{U}}\mathcal{V},\ \mathcal{W})+g(\mathcal{V},\ \nabla_{\mathcal{U}}\mathcal{W})\,,\\ \mathcal{V}(g(\mathcal{U},\ \mathcal{W})) &= g(\nabla_{\mathcal{V}}\mathcal{U},\ \mathcal{W})+g(\mathcal{U},\ \nabla_{\mathcal{V}}\mathcal{W})\,,\\ \mathcal{W}(g(\mathcal{U},\ \mathcal{V})) &= g(\nabla_{\mathcal{W}}\mathcal{U},\ \mathcal{V})+g(\mathcal{U},\ \nabla_{\mathcal{W}}\mathcal{V})\,.\end{aligned}$$

Because of the properties of ∇, we may introduce the commutators of the vector fields into these equations:

$$\begin{aligned}\mathcal{W}(g(\mathcal{U},\ \mathcal{V})) &= g(\nabla_{\mathcal{U}}\mathcal{W},\ \mathcal{V})+g(\mathcal{U},\ \nabla_{\mathcal{V}}\mathcal{W})+g([\mathcal{W},\mathcal{U}],\ \mathcal{V})-g(\mathcal{U},\ [\mathcal{V},\mathcal{W}])\,,\\ \mathcal{V}(g(\mathcal{U},\ \mathcal{W})) &= g(\nabla_{\mathcal{U}}\mathcal{V},\ \mathcal{W})+g(\mathcal{U},\ \nabla_{\mathcal{V}}\mathcal{W})-g(\mathcal{W},\ [\mathcal{U},\mathcal{V}])\,.\end{aligned}$$

From the first, fourth, and fifth equation we obtain the formula for the scalar product $g(\nabla_{\mathcal{U}}\mathcal{V},\ \mathcal{W})$. Hence there exists at most one covariant derivative with the required properties. Conversely, the formula determines the vector $\nabla_{\mathcal{V}}\mathcal{W}$, and an analogous computation shows that the operation thus defined is a covariant derivative with all the properties needed. □

Definition 19. The covariant derivative $\nabla_{\mathcal{V}}\mathcal{W}$ is called the *Levi-Civita connection* of the pseudo-Riemannian manifold.

Let $\mathcal{V}_1, \ldots, \mathcal{V}_m$ be an arbitrary frame of vector fields defined on an open subset of M^m. Representing the covariant derivative of the basis vector fields as

$$\nabla_{\mathcal{V}_i}\mathcal{V}_j \;:=\; \sum_{k=1}^{m} \Gamma_{ij}^k \cdot \mathcal{V}_k\,,$$

we introduce the *Christoffel symbols* of the Levi-Civita connection. The symmetry properties of the functions Γ_{ij}^k strongly depend on the chosen frame. If the vector fields $\mathcal{V}_i$ mutually commute, $[\mathcal{V}_i, \mathcal{V}_j] = 0$, and if $g_{ij} = g(\mathcal{V}_i, \mathcal{V}_j)$ are the coefficients of the Riemannian metric, then formula (5) from Theorem 42 simplifies, and we obtain the expression

$$\Gamma_{ij}^k \;=\; \frac{1}{2}\sum_{\alpha=1}^{m} g^{k\alpha}\big(\mathcal{V}_i(g_{j\alpha}) + \mathcal{V}_j(g_{i\alpha}) - \mathcal{V}_\alpha(g_{ij})\big)\,.$$

In this case, the Christoffel symbols are symmetric in the lower indices,

$$\Gamma_{ij}^k \;=\; \Gamma_{ji}^k \qquad \text{(for commuting frames)}\,.$$

If the frame $\mathcal{V}_1 := e_1, \ldots, \mathcal{V}_m := e_m$ consists of mutually orthogonal vector fields of constant length ± 1, then the equality

$$0 \;=\; \mathcal{V}_i\big(g(\mathcal{V}_j, \mathcal{V}_k)\big) \;=\; g(\nabla_{\mathcal{V}_i}\mathcal{V}_j, \mathcal{V}_k) + g(\mathcal{V}_j, \nabla_{\mathcal{V}_i}\mathcal{V}_k)$$

implies the following symmetry property of the Christoffel symbols:

$$\Gamma_{ij}^k \;=\; -\Gamma_{ik}^j \qquad \text{(for orthonormal frames)}\,.$$

Example 27. The first fundamental form of a surface of revolution can be written (see Example 3) as

$$\mathrm{I} \;=\; ds^2 + r(s)^2\, d\varphi^2\,.$$

With respect to the commuting frame $\mathcal{V}_1 = \partial/\partial s$, $\mathcal{V}_2 = \partial/\partial\varphi$ we obtain, for the "classical" Christoffel symbol Γ_{22}^1,

$$\Gamma_{22}^1 \;=\; \frac{1}{2}\sum_{\alpha=s,\varphi} g^{s\alpha}\left[\frac{\partial}{\partial\varphi}g_{\varphi\alpha} + \frac{\partial}{\partial\varphi}g_{\varphi\alpha} - \frac{\partial}{\partial s}g_{\varphi\varphi}\right] \;=\; \frac{1}{2}\left[-\frac{\partial\, r(s)^2}{\partial s}\right] \;=\; -r\cdot r',$$

and similarly $\Gamma_{12}^2 \;=\; \Gamma_{21}^2 \;=\; r'/r$. The other Christoffel symbols vanish. From Example 3 we take the formula $\omega_{12} = (r'/r)\,\sigma_2$ with respect to the orthonormal frame $e_1 = \partial/\partial s$, $e_2 = (1/r)\cdot\partial/\partial\varphi$ and its dual frame σ_1, σ_2 considered there. Together with the definition of the "Cartan" Christoffel symbols,

$$\omega_{12} \;=\; \Gamma_{11}^2\,\sigma_1 + \Gamma_{21}^2\,\sigma_2 \;=\; 0\cdot\sigma_1 + \frac{r'}{r}\sigma_2,$$

this yields $\Gamma_{11}^2 = 0$ and $\Gamma_{21}^2 = r'/r$. Similarly we have

$$\omega_{21} \;=\; \Gamma_{12}^1\,\sigma_1 + \Gamma_{22}^1\,\sigma_2 \;=\; 0\cdot\sigma_1 - \frac{r'}{r}\sigma_2\,,$$

implying $\Gamma^1_{12} = 0$, $\Gamma^1_{22} = -\Gamma^2_{21}$.

Again denoting by $\sigma_1, \ldots, \sigma_m$ the 1-forms dual to the frame $e_1, \ldots, e_m$, we introduce the *local connection forms*

$$\omega_{ij} := \varepsilon_j \cdot g(\nabla e_i, e_j) = \sigma_j(\nabla e_i) \quad \text{with} \quad \varepsilon_j := g(e_j, e_j).$$

By definition, they satisfy

$$\omega_{ij} = \sum_{\alpha=1}^{m} \Gamma^j_{\alpha i} \cdot \sigma_\alpha \quad \text{and} \quad \varepsilon_i \cdot \omega_{ij} + \varepsilon_j \cdot \omega_{ji} = 0.$$

We compute the exterior derivative $d\sigma_i$ using the properties of the Levi-Civita connection:

$$\begin{aligned} d\sigma_i(\mathcal{V}, \mathcal{W}) &= \mathcal{V}\big(\sigma_i(\mathcal{W})\big) - \mathcal{W}\big(\sigma_i(\mathcal{V})\big) - \sigma_i([\mathcal{V}, \mathcal{W}]) \\ &= \sum_{\alpha=1}^{m} \varepsilon_i \cdot \{g(\nabla_{\mathcal{V}} e_i, e_\alpha) \cdot \sigma_\alpha(\mathcal{W}) - g(\nabla_{\mathcal{W}} e_i, e_\alpha) \cdot \sigma_\alpha(\mathcal{V})\} \\ &= \sum_{\alpha=1}^{m} \sigma_\alpha \wedge \omega_{\alpha i}(\mathcal{V}, \mathcal{W}). \end{aligned}$$

This leads to the first structural equation of a pseudo-Riemannian space. In order to formulate the second structural equation, denote by $\Omega := (\omega_{ij})$ the antisymmetric $(m \times m)$ matrix composed of the 1-forms ω_{ij}. This additional structural equation computes the matrix $d\Omega - \Omega \wedge \Omega$ of 2-forms, and to formulate it we need the curvature of the space. The *curvature tensor* $\mathcal{R}(\mathcal{U}, \mathcal{V})\mathcal{W} := \nabla_{\mathcal{U}}\nabla_{\mathcal{V}}\mathcal{W} - \nabla_{\mathcal{V}}\nabla_{\mathcal{U}}\mathcal{W} - \nabla_{[\mathcal{U},\mathcal{V}]}\mathcal{W}$ of a pseudo-Riemannian manifold is introduced generalizing the case of surfaces, and all the identities of Theorem 15 in §5.3 remain valid. In particular, the *first Bianchi identity* holds:

$$\mathcal{R}(\mathcal{U}, \mathcal{V})\mathcal{W} + \mathcal{R}(\mathcal{V}, \mathcal{W})\mathcal{U} + \mathcal{R}(\mathcal{W}, \mathcal{U})\mathcal{V} = 0.$$

However, in the higher-dimensional case the curvature tensor does not reduce to a single function. By definition, it is a $(3, 1)$-tensor. This can be transformed into a $(4, 0)$-tensor, the *Riemannian curvature tensor*:

$$\mathcal{R}(\mathcal{U}, \mathcal{V}, \mathcal{W}, \mathcal{W}_1) := g(\mathcal{R}(\mathcal{U}, \mathcal{V})\mathcal{W}, \mathcal{W}_1).$$

Referring to an orthonormal frame $e_1, \ldots, e_m$ of basis vector fields of lengths ± 1, we obtain from

$$\mathcal{R}(e_i, e_j)e_k := \sum_{l=1}^{m} R^l_{ijk}\, e_l$$

the components R^l_{ijk} of the curvature tensor. These can be computed by the formula

$$R^l_{ijk} = \varepsilon_l \cdot \mathcal{R}(e_i, e_j, e_k, e_l)$$

from the Riemannian curvature tensor.

Theorem 43 (Structural Equations for Pseudo-Riemannian Manifolds). *Let $e_1, \dots, e_m$ be a local orthonormal frame on a pseudo-Riemannian manifold, and let $\sigma_1, \dots, \sigma_m$ be the dual frame. Then the following equations hold:*

(1) $d\sigma_i = \sum\limits_{\alpha=1}^{m} \sigma_\alpha \wedge \omega_{\alpha i}$;

(2) $d\omega_{ij} = \sum\limits_{\alpha=1}^{m} \omega_{i\alpha} \wedge \omega_{\alpha j} + \dfrac{1}{2} \sum\limits_{\alpha,\beta=1}^{m} R^{j}_{\alpha\beta i}\, \sigma_\alpha \wedge \sigma_\beta$.

Proof. We already derived the first equation. To prove the second, we use the properties of the Levi-Civita connection and obtain, after several elementary steps,

$$d\omega_{ij}(\mathcal{V}, \mathcal{W}) = \sigma_j(\mathcal{R}(\mathcal{V}, \mathcal{W})e_i) + \varepsilon_j \cdot \big(g(\nabla_{\mathcal{W}} e_i, \nabla_{\mathcal{V}} e_j) - g(\nabla_{\mathcal{V}} e_i, \nabla_{\mathcal{W}} e_j)\big)$$
$$= \sigma_j(\mathcal{R}(\mathcal{V}, \mathcal{W})e_i) + \sum_{\alpha=1}^{m} \varepsilon_j \varepsilon_\alpha \big(\sigma_\alpha(\nabla_{\mathcal{W}} e_i)\sigma_\alpha(\nabla_{\mathcal{V}} e_j) - \sigma_\alpha(\nabla_{\mathcal{W}} e_j)\sigma_\alpha(\nabla_{\mathcal{V}} e_i)\big).$$

Moreover,

$$\varepsilon_j \cdot \varepsilon_\alpha \cdot \sigma_\alpha(\nabla_{\mathcal{V}} e_j) = \varepsilon_j \cdot g(\nabla_{\mathcal{V}} e_j, e_\alpha) = -\,\varepsilon_j \cdot g(e_j, \nabla_{\mathcal{V}} e_\alpha) = -\,\omega_{\alpha j}(\mathcal{V})$$

and, similarly, $\varepsilon_j \cdot \varepsilon_\alpha \cdot \sigma_\alpha(\nabla_{\mathcal{W}} e_j) = -\,\omega_{\alpha j}(\mathcal{W})$. Altogether, this completes the proof,

$$d\omega_{ij}(\mathcal{V}, \mathcal{W}) = \sum_{\alpha=1}^{m} \omega_{i\alpha} \wedge \omega_{\alpha j}(\mathcal{V}, \mathcal{W}) + \sigma_j(\mathcal{R}(\mathcal{V}, \mathcal{W})e_i)$$
$$= \sum_{\alpha=1}^{m} \omega_{i\alpha} \wedge \omega_{\alpha j}(\mathcal{V}, \mathcal{W}) + R^{j}_{\mathcal{V}\mathcal{W}i}\,. \qquad \square$$

Corollary 5. *If the curvature tensor of a pseudo-Riemannian manifold M^m vanishes, then in a neighborhood of each point there exists a chart in which the coefficients of the Riemannian metric are constant,*

$$g = \mathrm{diag}(\pm 1, \dots, \pm 1)\,.$$

Proof. First choose a local orthonormal frame $e_1, \dots e_m$ and consider the matrix $\Omega = (\omega_{ij})$ of connection forms. Then

$$d\Omega = \Omega \wedge \Omega\,,$$

since the curvature tensor vanishes. By Theorem 6 in §4.3, there exists a (pseudo-)orthogonal matrix A of functions such that

$$\Omega = dA \cdot A^{-1} \quad \text{and} \quad A(0) = \mathrm{diag}(\pm 1, \dots, \pm 1)\,.$$

Denote by $\tau := (\sigma_1 \ldots \sigma_m) \cdot A$ a special system of 1-forms determined by $\sigma_1, \ldots, \sigma_m$ and the matrix A. The structural equations imply that this form is closed,

$$d\tau = d\sigma \cdot A - \sigma \wedge dA = \sigma \wedge \Omega \cdot A - \sigma \wedge \Omega \cdot A = 0 .$$

Hence, by Poincaré's lemma, each 1-form τ_i can locally be represented as the differential of a function f_i. This way, we construct local coordinates on the manifold, and for the pseudo-Riemannian metric g we obtain the expression

$$g = \pm\sigma_1^2 + \ldots \pm \sigma_m^2 = \pm\tau_1^2 + \ldots \pm \tau_m^2 = \pm(df_1)^2 + \ldots \pm (df_m)^2 . \qquad \square$$

Definition 20. A pseudo-Riemannian manifold with vanishing curvature tensor is called *flat*.

The preceding corollary thus states that every flat pseudo-Riemannian space is locally isometric to the standard space $\mathbb{R}^{m,k}$ with the following metric of index k:

$$dx^1 \odot dx^1 + \ldots + dx^{m-k} \odot dx^{m-k} - dx^{m-k+1} \odot dx^{m-k+1} - \ldots - dx^m \odot dx^m .$$

Consider a curve $\gamma(t)$ in the pseudo-Riemannian manifold (M^m, g) and a vector field $\mathcal{W}(t)$ along this curve. In addition, we may suppose that locally the latter is the restriction of a vector field $\mathcal{W}$ defined on M^m. In the neighborhood of a point on the curve we choose the coordinates $y^1, \ldots, y^m$ and represent the vector field $\mathcal{W}$ and the tangent vector $\dot{\gamma}(t)$ in these coordinates:

$$\mathcal{W} = \sum_{i=1}^{m} W^i \frac{\partial}{\partial y^i}, \quad \dot{\gamma}(t) = \sum_{i=1}^{m} \dot{\gamma}^i(t) \frac{\partial}{\partial y^i},$$

$$\frac{d}{dt} W^k(\gamma(t)) = \sum_{i=1}^{m} \dot{\gamma}^i(t) \frac{\partial W^k}{\partial y^i}(\gamma(t)) .$$

Then

$$(\nabla_{\dot{\gamma}(t)} \mathcal{W})(\gamma(t)) = \sum_{k=1}^{m} \left[\frac{dW^k(\gamma(t))}{dt} + \sum_{i,j=1}^{m} \dot{\gamma}^i(t) \cdot W^j(\gamma(t)) \cdot \Gamma_{ij}^k(\gamma(t)) \right] \frac{\partial}{\partial y^k} .$$

Along the curve $\gamma(t)$, this vector field is independent of the particular (local) extension of the vector field $\mathcal{W}(t)$, originally defined only on $\gamma(t)$, to a vector field $\mathcal{W}$. The formula thus uniquely defines a vector field $\nabla \mathcal{W}/dt$, the *covariant derivative* of $\mathcal{W}(t)$.

Definition 21. A vector field $\mathcal{W}(t)$ is called *parallel* along $\gamma(t)$ if

$$\frac{\nabla \mathcal{W}}{dt} = 0 .$$

In coordinates, this condition is equivalent to the following system of ordinary differential equations:

$$\frac{dW^k}{dt} + \sum_{i,j=1}^{m} \dot{\gamma}^i \cdot W^j \cdot \Gamma_{ij}^k = 0 .$$

The existence and uniqueness theorem for ordinary differential equations ensures the existence of a unique parallel vector field $W(t)$ for every initial vector $W_a \in TM^m_{\gamma(a)}$ at the initial point $\gamma(a)$ on the curve,

$$\frac{\nabla W}{dt} = 0, \quad W(a) = W_a .$$

The value $W(b)$ at the end point $\gamma(b)$ of the curve is called the *parallel displacement* P_γ of the vector W_a along the curve $\gamma(t)$,

$$P_\gamma(W_a) := W(b) .$$

Because of the linearity of the differential equation and the property

$$\mathcal{U}(g(\mathcal{V}, \mathcal{W})) = g(\nabla_{\mathcal{U}}\mathcal{V}, \mathcal{W}) + g(\mathcal{V}, \nabla_{\mathcal{U}}\mathcal{W})$$

of the Levi-Civita connection, the parallel displacement

$$P_\gamma : T_{\gamma(a)}M^m \longrightarrow T_{\gamma(b)}M^m$$

is a linear isometric map between tangent spaces. The parallel displacement $P_{\gamma_1 * \gamma_2}$ along a curve formed by combining two curves γ_1 and γ_2, for which the end point of the first coincides with the initial point of the second, into a single curve $\gamma_1 * \gamma_2$ leads to the same result as the superposition of the single parallel displacements,

$$P_{\gamma_2} \circ P_{\gamma_1} = P_{\gamma_1 * \gamma_2} .$$

Consider a surface (M^2, g) with a positive definite Riemannian metric and a closed curve γ in M^2 bounding a domain $N^2 \subset M^2$. Parallel displacement along γ then is a rotation in the tangent space $T_{\gamma(0)}M^2$. The corresponding rotation angle can be computed from the Gaussian curvature.

Theorem 44. *If the closed curve γ bounds a domain N^2 in a surface M^2 with Gaussian curvature G, the parallel displacement along the curve is the rotation through the angle*

$$\phi = \int_{N^2} G \cdot dM^2 = 2\pi \cdot \chi(N^2) - \int_\gamma k_g \cdot ds .$$

Proof. Choose an orthonormal frame e_1, e_2 along γ in which e_1 is tangent to the curve. In any parametrization $\gamma : [0, L] \to M^2$ of the curve, a vector field $\mathcal{V} = V^1(s) \cdot e_1(s) + V^2(s) \cdot e_2(s)$ is parallel if

$$\frac{dV^1}{ds} = \omega_{12}(e_1) \cdot V^2 \quad \text{and} \quad \frac{dV^2}{ds} = -\omega_{12}(e_1) \cdot V^1 .$$

Consider the complex-valued function $z : [0, L] \to \mathbb{C}$ defined by $z(s) := V^1(s) + i \cdot V^2(s)$. Then we can write this system of ordinary differential equations in the form

$$\frac{dz}{ds} = - i \cdot \omega_{12}(e_1) \cdot z \, .$$

The solution of this is

$$z(s) = z(0) \cdot \exp \left[- i \cdot \int_0^s \omega_{12}(e_1) \cdot ds \right] ,$$

and, in particular, for the rotation angle of the parallel displacement we obtain

$$\phi = - \int_\gamma \omega_{12} = - \int_{N^2} d\omega_{12} = \int_{N^2} G \cdot dM^2 \, . \qquad \square$$

Example 28. It is a peculiar property of the hyperbolic space $\mathcal{H}^2$ that the hyperbolic angle coincides with the usual euclidean angle. In general, the angle does not change under parallel displacement. Hence the parallel displacement of a vector along a *geodesic line* in a surface is the vector at the end point of the line forming the same angle with the tangent vector as the original vector did at the initial point. The points PQR in the picture below are the vertices of a geodesic triangle in the hyperbolic plane (see Example 23). Start at P with any tangent vector (1), translate it to the point Q (2), then along the upper geodesic arc to R (3), and finally along the straight segment RP back to P (4). The resulting vector obviously does not coincide with the original one.

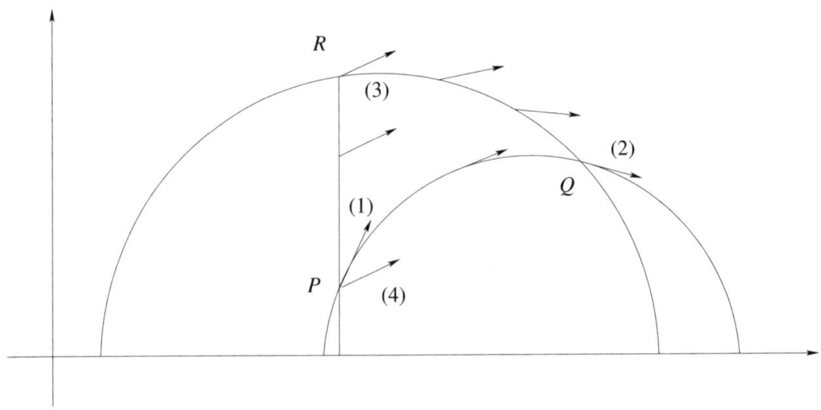

Parallel displacement allows one to generalize the *covariant derivative* $\nabla_{\mathcal{V}} T$ with respect to vector fields to arbitrary tensors T. We explain this for tensors of type $(k, 0)$. Assume that at a point $x \in M^m$, the vectors $\mathcal{W}_1, \ldots \mathcal{W}_k \in$

T_xM^m are given, and let $\gamma(t)$ be the integral curve of the vector field $\mathcal{V}$ passing at time $t = 0$ through the point $x \in M^m$. Then we define

$$(\nabla_{\mathcal{V}}T)(\mathcal{W}_1, \ldots, \mathcal{W}_k)(x) \;:=\; \frac{d}{dt}(T_{\gamma(t)}(P_\gamma(\mathcal{W}_1), \ldots, P_\gamma(\mathcal{W}_k))\big|_{t=0}.$$

Here P_γ denotes parallel displacement along the curve γ from the point $x = \gamma(0)$ to the curve point $\gamma(t)$. If, for example, T is a $(1,0)$-tensor and $\mathcal{W}$ a vector field, we obtain the formula

$$(\nabla_{\mathcal{V}}T)(\mathcal{W})(x) \;=\; \mathcal{V}(T(\mathcal{W}))(x) - T(\nabla_{\mathcal{V}}\mathcal{W})(x)\,.$$

Theorem 45. *The covariant derivative $\nabla_{\mathcal{V}}T$ of a $(k,0)$-tensor is given by*

$$\begin{aligned}(\nabla_{\mathcal{V}}T)(\mathcal{W}_1, \ldots, \mathcal{W}_k) \;=\; & \mathcal{V}\big(T(\mathcal{W}_1, \ldots, \mathcal{W}_k)\big) \;-\; T\big(\nabla_{\mathcal{V}}\mathcal{W}_1, \mathcal{W}_2, \ldots, \mathcal{W}_k\big) \\ & -\; \ldots \;-\; T\big(\mathcal{W}_1, \ldots, \mathcal{W}_{k-1}, \nabla_{\mathcal{V}}\mathcal{W}_k\big)\,.\end{aligned}$$

Remark. The metric g, viewed as a $(2,0)$-tensor, is parallel:

$$(\nabla_{\mathcal{V}}g)(\mathcal{W}_1, \mathcal{W}_2) \;=\; \mathcal{V}(g(\mathcal{W}_1, \mathcal{W}_2)) - g(\nabla_{\mathcal{V}}\mathcal{W}_1, \mathcal{W}_2) - g(\mathcal{W}_1, \nabla_{\mathcal{V}}\mathcal{W}_2) \;=\; 0\,.$$

A completely elementary but somewhat lengthy computation, using the properties of the Levi-Civita connection and the definition of the curvature tensor, leads to an identity for the covariant derivative $\nabla\mathcal{R}$ of the curvature tensor, the so-called *second Bianchi identity*. We omit the simple proof and formulate only the result.

Theorem 46 (Second Bianchi Identity)**.** *Let $\mathcal{U}, \mathcal{V}, \mathcal{W}, \mathcal{W}_1, \mathcal{W}_2$ be five vector fields on the pseudo-Riemannian manifold (M^m, g). Then*

$$(\nabla_{\mathcal{U}}\mathcal{R})(\mathcal{V}, \mathcal{W}, \mathcal{W}_1, \mathcal{W}_2) + (\nabla_{\mathcal{V}}\mathcal{R})(\mathcal{W}, \mathcal{U}, \mathcal{W}_1, \mathcal{W}_2) + (\nabla_{\mathcal{W}}\mathcal{R})(\mathcal{U}, \mathcal{V}, \mathcal{W}_1, \mathcal{W}_2) = 0.$$

Define the length of a curve $\gamma : [a,b] \to M^m$ in a pseudo-Riemannian space by the integral

$$L(\gamma) \;:=\; \int_a^b \sqrt{|g(\dot{\gamma}(t), \dot{\gamma}(t))|}\,dt\,.$$

For a variation $\gamma_\varepsilon : [a,b] \to M^m$ of the original curve $\gamma(t)$ with fixed initial and end points, $\gamma_\varepsilon(a) = \gamma(a), \gamma_\varepsilon(b) = \gamma(b)$, the field of the variation,

$$\mathcal{W}(t) \;:=\; \frac{d}{d\varepsilon}(\gamma_\varepsilon(t))\big|_{\varepsilon=0},$$

is a vector field defined along the curve $\gamma(t)$ vanishing at the end points. Its covariant derivative $\nabla\mathcal{W}/dt$ along the curve $\gamma(t)$ coincides with the covariant derivative $\nabla\dot{\gamma}_\varepsilon(t)/d\varepsilon$ of the vector field $\dot{\gamma}_\varepsilon(t)$ defined along the curve $\eta(\varepsilon) := \gamma_\varepsilon(t)$,

$$\frac{\nabla\mathcal{W}}{dt}(t) \;=\; \frac{\nabla\dot{\gamma}_\varepsilon(t)}{d\varepsilon}\bigg|_{\varepsilon=0}.$$

This formula is an immediate consequence of the local expressions for the two covariant derivatives. Parametrizing the curve $\gamma(t)$ by arc length, we obtain the following formula for the variation of the length of the curve:

$$\begin{aligned}\frac{d}{d\varepsilon}(L(\gamma_\varepsilon))\big|_{\varepsilon=0} &= \int_a^b g\Big(\frac{\nabla\dot\gamma_\varepsilon(t)}{d\varepsilon}_{|\varepsilon=0}, \dot\gamma(t)\Big)dt \\ &= \int_a^b g\Big(\frac{\nabla\mathcal{W}}{dt}(t), \dot\gamma(t)\Big)dt \;=\; -\int_a^b g\Big(\mathcal{W}(t), \frac{\nabla\dot\gamma}{dt}(t)\Big)dt\,.\end{aligned}$$

Theorem 47. *If the curve $\gamma : [a,b] \to M^m$ is a critical point of the length functional defined on the set of all curves with fixed initial and end points, then the tangent vector $\dot\gamma(t)$ is parallel along the curve $\gamma(t)$,*

$$\frac{\nabla\dot\gamma}{dt} \;=\; 0\,.$$

Definition 22. A curve $\gamma(t)$ in the pseudo-Riemannian space (M^m, g) whose tangent vector field is parallel is called a *geodesic line.*

Remark. If the curve $\gamma(t) = (\gamma^1(t), \ldots, \gamma^m(t))$ is given in local coordinates, the equation for it to be a geodesic line is

$$\frac{d^2\gamma^k}{dt^2} + \sum_{i,j=1}^{m} \frac{d\gamma^i}{dt}\frac{d\gamma^j}{dt}\Gamma_{ij}^k \;=\; 0\,.$$

This is an ordinary differential equation of second order, and Corollary 3 from §5.5 holds correspondingly:

Corollary 6. *To each point $x \in M^m$ and every vector $\mathcal{V} \in T_xM^m$ on a pseudo-Riemannian manifold without boundary there exists precisely one geodesic line $\gamma_{\mathcal{V}} : (-\varepsilon, \varepsilon) \to M^m$ such that $\gamma_{\mathcal{V}}(0) = x$ and $\dot\gamma_{\mathcal{V}}(0) = \mathcal{V}$.*

The geodesic line $\gamma_{\mathcal{V}}(t)$ thereby determined is not necessarily defined for all values of the parameter $t \in \mathbb{R}$. The subset $\Sigma_x \subset T_xM^m$ of all vectors $\mathcal{V} \in T_xM^m$ for which $\gamma_{\mathcal{V}}(t)$ exists for $t = 1$ is an open neighborhood of the zero vector in T_xM^m. The *exponential map* at the point x of the pseudo-Riemannian manifold maps Σ_x to the base space,

$$\mathrm{Exp}_x : \Sigma_x \longrightarrow M^m, \quad \mathrm{Exp}_x(\mathcal{V}) \;=\; \gamma_{\mathcal{V}}(1)\,.$$

The exponential map is smooth, and its differential $d(\mathrm{Exp}_x) : T_0(T_xM^m) = T_xM^m \to T_xM^m$ is the identity,

$$d\,\mathrm{Exp}_x(\mathcal{W}) = \frac{d}{dt}(\mathrm{Exp}(t\mathcal{W}))\big|_{t=0} = \frac{d}{dt}(\gamma_{t\cdot\mathcal{W}}(1))\big|_{t=0} = \frac{d}{dt}(\gamma_{\mathcal{W}}(t))\big|_{t=0} = \mathcal{W}\,.$$

Hence the exponential map Exp_x is a diffeomorphism from an open neighborhood $0 \in U_x \subset T_xM^m$ of the zero vector in the tangent space onto an open neighborhood $x \in V_x \subset M^m$ of the point in the base space.

Simpler curvature invariants of the pseudo-Riemannian manifold arise from the curvature tensor by contraction (taking the trace), the most important ones being the Ricci tensor and the scalar curvature. The sectional curvature appears as the Gaussian curvature of subplanes of the tangent space. We shall now discuss these curvature quantities.

Definition 23. The *Ricci tensor* Ric is the first contraction of the curvature tensor,

$$\mathrm{Ric}(\mathcal{U},\mathcal{V}) := \sum_{i,j=1}^{m} g^{ij}\mathcal{R}(e_i,\mathcal{U},\mathcal{V},e_j)\,.$$

The algebraic identities of the curvature tensor imply that Ric is a symmetric $(2,0)$-tensor,

$$\mathrm{Ric}(\mathcal{U},\mathcal{V}) = \mathrm{Ric}(\mathcal{V},\mathcal{U})\,.$$

Taking the trace of the Ricci tensor with respect to the pseudo-Riemannian metric g (§5.4) leads to the *scalar curvature* τ of the space.

Definition 24. The *scalar curvature* of a Riemannian manifold is the function

$$\tau := \sum_{i,j,k,l=1}^{m} g^{ij}g^{kl}\mathcal{R}(e_i,e_k,e_l,e_j)\,.$$

Contrary to the surface case ($m=2$), the curvature tensor is no longer determined by the scalar curvature for manifolds of dimension at least three. Thus the Ricci tensor and the scalar curvature do not contain all the curvature information about a manifold. The last of the curvature quantities is the sectional curvature. Apart from the point of the manifold it depends, in addition, on a non-degenerate 2-plane in the tangent space. This is a two-dimensional subspace $E^2 \subset T_xM^m$ whose orthogonal complement,

$$(E^2)^{\perp} = \{\mathcal{W} \in T_xM^m : g(\mathcal{V},\mathcal{W}) = 0 \quad \text{for all } \mathcal{V} \in E^2\},$$

is complementary to E^2, $E^2 \oplus (E^2)^{\perp} = T_xM^m$. Intersecting E^2 with an open set $0 \in U_x \subset T_xM^m$ on which the exponential map is defined and a diffeomorphism, we obtain a surface in M^m,

$$M^2(E^2) := \mathrm{Exp}_x(E^2 \cap U_x)\,.$$

Its tangent plane at the point x is E^2, and, by restriction, the pseudo-Riemannian metric g of M^m induces a metric on $M^2(E^2)$. The Gaussian curvature $G(x)$ of the latter at the point $x \in M^2(E^2)$ is called the *sectional curvature* $K(x,E^2)$ of the pseudo-Riemannian manifold at the point x in the direction of the 2-plane E^2. It can be computed using the curvature tensor of M^m.

Theorem 48. *Let $\mathcal{V}_1, \mathcal{V}_2$ be a basis of the non-degenerate plane E^2. Then*

$$K(x, E^2) \;=\; \frac{\mathcal{R}(\mathcal{V}_1, \mathcal{V}_2, \mathcal{V}_2, \mathcal{V}_1)}{g(\mathcal{V}_1, \mathcal{V}_1)\, g(\mathcal{V}_2, \mathcal{V}_2) - g^2(\mathcal{V}_1, \mathcal{V}_2)}\,.$$

Proof. For the proof, we assume that the metric of M^m is positive definite. This is no essential restriction, but some of the formulas simplify. Note first that the right-hand side of the formula for $K(x, E^2)$ does not depend on the choice of the basis $\mathcal{V}_1, \mathcal{V}_2$; thus it suffices to prove this formula for an orthonormal basis in E^2. Fix an orthonormal basis $e_1(x), e_2(x), \ldots, e_m(x)$ at the point $x \in M^2(E^2) \subset M^m$ such that the first two vectors belong to E^2. We extend this basis to a local orthonormal frame by parallel displacement along the geodesic straight segments $\{\mathrm{Exp}_x(t \cdot \mathcal{W}) : \mathcal{W} \in T_x M^m\}$. At the point x this frame is parallel,

$$\nabla e_i(x) \;=\; 0\,.$$

Furthermore, on $M^2(E^2)$ the vector fields e_1, e_2 are tangent to this surface, since $M^2(E^2)$ consists of geodesic straight segments with initial vectors in E^2. Let σ_α, ω_{ij} be the forms dual to the frame $e_1, \ldots, e_m$ and the connection forms of the Riemannian manifold M^m, respectively. The embedding $f : M^2(E^2) \to M^m$ allows us to restrict these forms to the surface,

$$\mu_\alpha \;:=\; f^*(\sigma_\alpha), \quad \xi_{ij} \;:=\; f^*(\omega_{ij})\,.$$

Then $\mu_3 = \mu_4 = \ldots = \mu_m = 0$, and the structural equations of M^m imply, after restriction to $M^2(E^2)$, that

$$d\mu_1 = \mu_2 \wedge \xi_{21}, \quad d\mu_2 = \mu_1 \wedge \xi_{12}, \quad d\xi_{12} = \sum_\alpha \xi_{1\alpha} \wedge \xi_{\alpha 2} + R^2_{121}\, \mu_1 \wedge \mu_2.$$

The pair μ_1, μ_2 is, within the surface $M^2(E^2)$, the frame dual to e_1, e_2, and the last equations show that the connection form ω^*_{12} of $M^2(E^2)$ coincides with ξ_{12}, $\omega^*_{12} \;=\; \xi_{12}$. The Gaussian curvature G of $M^2(E^2)$ is determined by

$$-\, G \cdot \mu_1 \wedge \mu_2 \;=\; d\omega^*_{12} \;=\; d\xi_{12} \;=\; \sum_\alpha \xi_{1\alpha} \wedge \xi_{\alpha 2} + R^2_{121}\, \mu_1 \wedge \mu_2\,.$$

At the point $x \in M^2(E^2)$, however, we have $\xi_\alpha(e_1) = g(\nabla_{e_1} e_1, e_\alpha) = 0$, and hence we obtain

$$-\, G(x) \;=\; R^2_{121} \;=\; g(\mathcal{R}(e_1, e_2)\, e_1, e_2) \;=\; -\, g(\mathcal{R}(e_1, e_2)\, e_2, e_1)\,. \qquad \square$$

Special Riemannian manifolds are the so-called Einstein spaces and spaces of constant sectional curvature.

Definition 25. A pseudo-Riemannian manifold is called an *Einstein space* if the Ricci tensor is proportional to the metric, $\mathrm{Ric} \;=\; h \cdot g$.

Remark. Contracting the equation $\mathrm{Ric} = h \cdot g$ leads to $\tau = m \cdot h$, i. e., the factor of proportionality is uniquely determined by the scalar curvature of the manifold,

$$\mathrm{Ric} \;=\; \frac{\tau}{m} \cdot g\,.$$

We prove that this factor is constant.

Theorem 49. *If (M^m, g) is a connected Einstein space of dimension $m \geq 3$, then the scalar curvature τ is constant.*

Proof. Choose an orthonormal frame $e_1, \ldots, e_m$ and an arbitrary vector field $\mathcal{U}$. After we insert the vector fields $\mathcal{V} = \mathcal{W}_2 := e_i$, $\mathcal{W} := e_j$, and $\mathcal{W}_1 := e_j$ into the second Bianchi identity and sum over all indices $1 \leq i, j \leq m$, an elementary computation leads to the equation

$$\sum_j \varepsilon_j \Big((\nabla_{\mathcal{U}})\mathrm{Ric}(e_j, e_j) - 2(\nabla_{e_j}\mathrm{Ric})(\mathcal{U}, e_j) \Big) \;=\; 0\,.$$

If the Ricci tensor is proportional to the metric, $\mathrm{Ric} = h \cdot g$, the relation $\nabla g = 0$ implies $\nabla_{\mathcal{U}}\mathrm{Ric} = \mathcal{U}(h) \cdot g$, and the last equation becomes

$$m \cdot \mathcal{U}(h) - 2 \sum \varepsilon_j e_j(h)\, g(\mathcal{U}, e_j) \;=\; 0\,.$$

Considering the vector fields $\mathcal{U} := e_k$, this yields $(m-2)\; dh(e_k) = 0$, and, in case $m > 2$, the differential dh of the function h vanishes. □

Example 29. In spherical coordinates, the standard metric of the two-dimensional sphere is the quadratic form

$$g_{|S^2} \;=\; \cos^2(\psi) \cdot d\varphi^2 + d\psi^2\,.$$

Choose a pseudo-Riemannian metric h in the product $\mathbb{R}_+ \times \mathbb{R}$ and a function $f : \mathbb{R}_+ \times \mathbb{R} \to \mathbb{R}$. On the four-dimensional manifold $(\mathbb{R}^3 - 0) \times \mathbb{R} = S^2 \times \mathbb{R}_+ \times \mathbb{R}$ we can then consider the rotationally symmetric metric

$$g \;:=\; h \oplus e^{2f(r,t)} \cdot g_{|S^2}\,.$$

A longer computation reveals that g is a four-dimensional Einstein metric with vanishing scalar curvature if and only if the following pair of differential equations is satisfied:

$$e^{-2f} \;=\; \Delta(f) + 2|\mathrm{grad}(f)|^2 \quad \text{and} \quad G \cdot h \;=\; 2 \cdot \mathrm{Hess}^h(f) + 2 \cdot df \odot df.$$

Here, Δ and grad are the Laplacian and the gradient of the two-dimensional pseudo-Riemannian manifold $(\mathbb{R}_+ \times \mathbb{R},\; h)$, G is its Gaussian curvature, and

$$\mathrm{Hess}^h(f)(\mathcal{V}, \mathcal{W}) \;:=\; h(\nabla_{\mathcal{V}}(\mathrm{grad}(f)), \mathcal{W})$$

is the symmetric *Hessian form* of the function f with respect to the metric h. Choosing, for example, the metric

$$h = \left[1 + \frac{2M}{r}\right] dr^2 + c\,\frac{4M}{r}\, dr \odot dt - c^2 \left[1 - \frac{2M}{r}\right] dt^2$$

and the function $f(r,t) = \log(r)$ gives rise to the *Schwarzschild-Eddington metric* on $(\mathbb{R}^3 - 0) \times \mathbb{R}$, which, in Einstein's general theory of relativity, models the gravitational field generated by a mass concentrated at the point $x = 0$. Further solutions of this system of differential equations describe the gravitational field of a rotating mass (*Kerr metric*).

Definition 26. A pseudo-Riemannian manifold (M^m, g) is called a *space of constant sectional curvature* (or *space form*) if there exists a function $K^* : M^m \to \mathbb{R}$ such that for every 2-plane $E^2 \subset T_x M^m$

$$K(x, E^2) = K^*(x)\,.$$

Theorem 50. *The sectional curvature $K = K^*$ of every space form M^m of dimension $m \geq 3$ is constant, and the curvature tensor is algebraically determined by the metric,*

$$\mathcal{R}(\mathcal{U},\mathcal{V})\mathcal{W} = K^*(g(\mathcal{V},\mathcal{W}) \cdot \mathcal{U} - g(\mathcal{U},\mathcal{W}) \cdot \mathcal{V})\,.$$

Proof. Consider the tensor

$$\begin{aligned} &\mathcal{S}(\mathcal{U},\mathcal{V},\mathcal{W}_1,\mathcal{W}_2) \\ &\quad := \mathcal{R}(\mathcal{U},\mathcal{V},\mathcal{W}_1,\mathcal{W}_2) - K^*\big(g(\mathcal{V},\mathcal{W}_1)g(\mathcal{U},\mathcal{W}_2) - g(\mathcal{U},\mathcal{W}_1)g(\mathcal{V},\mathcal{W}_2)\big)\,, \end{aligned}$$

and note that $\mathcal{S}$ has the same symmetry properties as the curvature tensor:

(1) $\mathcal{S}(\mathcal{U},\mathcal{V},\mathcal{W}_1,\mathcal{W}_2) = -\mathcal{S}(\mathcal{V},\mathcal{U},\mathcal{W}_1,\mathcal{W}_2)\,.$

(2) $\mathcal{S}(\mathcal{U},\mathcal{V},\mathcal{W}_1,\mathcal{W}_2) = -\mathcal{S}(\mathcal{V},\mathcal{U},\mathcal{W}_2,\mathcal{W}_1)\,.$

(3) $\mathcal{S}(\mathcal{U},\mathcal{V},\mathcal{W}_1,\mathcal{W}_2) + \mathcal{S}(\mathcal{V},\mathcal{W}_1,\mathcal{U},\mathcal{W}_2) + \mathcal{S}(\mathcal{W}_1,\mathcal{U},\mathcal{V},\mathcal{W}_2) = 0\,.$

(4) $\mathcal{S}(\mathcal{U},\mathcal{V},\mathcal{W}_1,\mathcal{W}_2) = \mathcal{S}(\mathcal{W}_1,\mathcal{W}_2,\mathcal{U},\mathcal{V})\,.$

An additional property of the tensor $\mathcal{S}$ results from the formula for the sectional curvature:

(5) $\mathcal{S}(\mathcal{U},\mathcal{V},\mathcal{V},\mathcal{W}) = 0\,.$

A calculation only making use of equations (1) – (5) shows that this tensor has to vanish, $\mathcal{S} \equiv 0$. In fact, from

$$\begin{aligned} 0 &= \mathcal{S}(\mathcal{U},\mathcal{V}+\mathcal{W}_2,\mathcal{V}+\mathcal{W}_2,\mathcal{U}) \\ &= \mathcal{S}(\mathcal{U},\mathcal{V},\mathcal{V},\mathcal{U}) + \mathcal{S}(\mathcal{U},\mathcal{V},\mathcal{W}_2,\mathcal{U}) + \mathcal{S}(\mathcal{U},\mathcal{W}_2,\mathcal{V},\mathcal{U}) + \mathcal{S}(\mathcal{U},\mathcal{W}_2,\mathcal{W}_2,\mathcal{U}) \\ &= 2 \cdot \mathcal{S}(\mathcal{U},\mathcal{V},\mathcal{W}_2,\mathcal{U}) \end{aligned}$$

we first conclude that $S(\mathcal{U}, \mathcal{V}, \mathcal{W}_2, \mathcal{U}) = 0$. Hence we have succeeded in replacing the third argument $\mathcal{V}$ in the identity (5) by an arbitrary vector $\mathcal{W}_2$. We repeat this step once more, starting from $S(\mathcal{U}, \mathcal{V}, \mathcal{W}_2, \mathcal{U}) = 0$. Thus, as the fourth entry another independent vector $\mathcal{W}_1$ can be substituted for the vector $\mathcal{U}$, that is, S vanishes, and the curvature tensor of a space form is determined by

$$\mathcal{R}(\mathcal{U}, \mathcal{V})\mathcal{W} = K^*(g(\mathcal{V}, \mathcal{W}) \cdot \mathcal{U} - g(\mathcal{U}, \mathcal{W}) \cdot \mathcal{V}) .$$

Contracting the curvature tensor, the resulting Ricci tensor is proportional to the metric,

$$\mathrm{Ric} = (m-1)\, K^* \cdot g .$$

Hence M^m is an Einstein space, and K^* is a constant function. □

The components $R^j_{\alpha\beta i}$ of the curvature tensor of a space form with positive definite metric (for simplicity) are

$$R^j_{\alpha\beta i} = K^* \cdot (\delta_{\beta i}\delta_{\alpha j} - \delta_{\beta j}\delta_{\alpha i}),$$

and the structural equations simplify,

$$d\sigma_i = \sum_{\alpha=1}^{m} \sigma_\alpha \wedge \omega_{\alpha i}, \quad d\omega_{ij} = \sum_{\alpha=1}^{m} \omega_{i\alpha} \wedge \omega_{\alpha j} - K^* \cdot \sigma_i \wedge \sigma_j .$$

In the middle of the 19th century, Riemann gave the complete local description of these spaces.

Theorem 51 (Riemann, 1854). *Each space form is locally isometric to the spherical space S^m, the euclidian space $\mathbb{R}^m$, or the hyperbolic space $\mathcal{H}^m$.*

Exercises

1. Let $\gamma(s)$ be a curve in its natural parametrization, and let $\mu(t)$ be this same curve in an arbitrary parametrization, which hence is related to γ via $\mu(t) = \gamma(s(t))$.

a) Prove the following formula for the second derivative of μ:

$$\ddot{\mu}(t) = \frac{d^2\mu(t)}{dt^2} = \frac{d^2\gamma}{ds^2}\left(\frac{ds}{dt}\right)^2 + \frac{d\gamma}{ds}\frac{d^2 s}{dt^2} = \kappa(t)\,(\dot{s})^2\cdot\vec{h} + \ddot{s}\cdot\vec{t}.$$

b) Deduce from this the formula for the curvature in a general parametrization:

$$\kappa(t) = \frac{\|\dot{\mu}(t)\times\ddot{\mu}(t)\|}{\|\dot{\mu}(t)\|^3}.$$

c) Prove that the torsion can be written in the Frenet frame as

$$\tau(t) = -\left\langle \vec{t}\times\vec{h}',\, \vec{h}\right\rangle = \det\left(\vec{t},\, \vec{h},\, \vec{h}'\right) = \frac{\det\left(\gamma',\, \gamma'',\, \gamma'''\right)}{\kappa(t)^2}.$$

d) Derive in an analogous way to a) a formula for $\dddot{\mu}(t)$, and conclude that the torsion is given in an arbitrary parametrization by

$$\tau(t) = \frac{\det\left(\dot{\mu}(t),\, \ddot{\mu}(t),\, \dddot{\mu}(t)\right)}{\|\dot{\mu}(t)\times\ddot{\mu}(t)\|^2}.$$

2. Let $\mathcal{C}\subset\mathbb{R}^3$ be a curve all of whose tangents pass through one and the same point. Prove that $\mathcal{C}$ is part of a straight line.

3. Let $\mathcal{C}\subset\mathbb{R}^3$ be a curve all of whose tangents are parallel to one and the same straight line. Prove that $\mathcal{C}$ is part of a straight line.

4. Show that the tractrix (Example 11) is the curve passing through the point $(1,0)$ on the horizontal axis with the property that the length of the segment of the tangent line from any point on the curve to the vertical axis is constant. The following intuitive interpretation gave it its German name "Schleppkurve" (tow curve): It is the path that a dog pulling on a leash to the west is constrained to take when his master walks along a north-south path.

5. Compute the curvature and the torsion of the following curves:

a) $\gamma(t) = \left(t,\, a(e^{t/a} + e^{-t/a})/2,\, 0\right)$ (catenary);

b) $\gamma(t) = \big(a\cos t,\, a\sin t,\, b\,t\big)$ (helix);

c) $\gamma(t) = \big(a(t - \sin t),\, a(1 - \cos t),\, b\,t\big)$;

d) $\gamma(t) = \big(t,\, t^2,\, t^3\big)$.

6. Prove that $\gamma(t) = \big(a\,t,\, b\,t^2,\, t^3\big)$ with $a, b > 0$ is a slope line if and only if $2b^2 = 3a$.

7. Prove that a C^3-curve $\mathcal{C}$ is a helix if and only if its curvature $\kappa > 0$ and its torsion τ are constant.

8. Determine all plane curves $\mathcal{C} \subset \mathbb{R}^2$ satisfying

a) $\kappa(s) = \text{const}$;

b) $\kappa(s) = 1/\sqrt{s}$;

c) $\kappa(s) = 1/\sqrt{1 - s^2}$.

9. The *Darboux vector* of a curve with non-vanishing curvature is the vector $\vec{d} := \tau\cdot\vec{t} + \kappa\cdot\vec{b}$. Prove that the Frenet formulas can be written in the following form:

$$\frac{d\vec{t}}{ds} = \vec{d} \times \vec{t}, \quad \frac{d\vec{h}}{ds} = \vec{d} \times \vec{h}, \quad \frac{d\vec{b}}{ds} = \vec{d} \times \vec{b}.$$

10. Consider the spherical curve $\gamma^*(s) := \vec{t}(s) : [0, L] \to S^2 \subset \mathbb{R}^3$ and suppose that the curvature $\kappa(s)$ of the original curve $\gamma(s)$ does not vanish. Prove that $\gamma^*(s)$ is a regular curve, and compute its curvature κ^* and its torsion τ^*. Which curve $\gamma^*(s)$ arises in the case of the helix $\gamma(s)$?

11. A slope line with slope angle α lies on a sphere of radius R if and only if its curvature $\kappa(s)$ and its arc length s satisfy the relation

$$\frac{s^2}{\tan^2\alpha} + \frac{1}{\kappa^2(s)} = R^2.$$

12. Let $\gamma(s) : [0, L] \to \mathbb{R}^2 = \mathbb{C}$ be a closed plane curve, and let p be a point in the plane not belonging to the curve. Prove that the integral

$$\frac{1}{2\pi i}\int_0^L \frac{\vec{t}(s)}{\gamma(s) - p}\, ds$$

is an integer. This is called the *index* or *winding number* of the curve $\gamma(s)$ around the point p.

13 (Jacobi, 1842). Let $\gamma \subset \mathbb{R}^3$ be a closed curve with non-vanishing curvature, and suppose that the curve of principal normal vectors $\vec{h}$ is a simple closed curve in the sphere S^2. Prove that $\vec{h}$ divides the sphere into two parts of equal area.

14. Classify all surfaces of revolution with constant Gaussian curvature using the results from Examples 3 and 11. Altogether there are 9 geometric types. Which of these can be compact manifolds without boundary?

15. Let $F : U \to \mathbb{R}^3$ be a parametrized surface such that the tangent vectors $\partial/\partial u$ and $\partial/\partial v$ are orthogonal at each point $(u, v) \in U$. Set

$$E = \left\langle \frac{\partial}{\partial u}, \frac{\partial}{\partial u} \right\rangle, \quad G = \left\langle \frac{\partial}{\partial v}, \frac{\partial}{\partial v} \right\rangle$$

and choose the orthonormal frame

$$e_1 = \frac{1}{\sqrt{E}} \frac{\partial}{\partial u}, \quad e_2 = \frac{1}{\sqrt{G}} \frac{\partial}{\partial v}.$$

a) In the dual frame $\sigma_1 = \sqrt{E}\, du$, $\sigma_2 = \sqrt{G}\, dv$, the connection form ω_{12} is given by

$$\omega_{12} = \frac{1}{\sqrt{EG}} \left[-\frac{\partial \sqrt{E}}{\partial v} \sigma_1 + \frac{\partial \sqrt{G}}{\partial u} \sigma_2 \right].$$

b) The Gaussian curvature is equal to

$$K = -\frac{1}{\sqrt{EG}} \left[\frac{\partial}{\partial v} \left(\frac{\partial \sqrt{E}/\partial v}{\sqrt{G}} \right) + \frac{\partial}{\partial u} \left(\frac{\partial \sqrt{G}/\partial u}{\sqrt{E}} \right) \right].$$

c) As an application, determine the Gaussian curvature of a surface of revolution whose generating curve is not necessarily parametrized by arc length.

16. Let $M^2 \subset \mathbb{R}^3$ be a surface with non-vanishing Gaussian curvature. Prove:

a) There exist two orthogonal vector fields $\mathcal{V}$ and $\mathcal{W}$ with lengths $||\mathcal{V}|| = 1$ and $||\mathcal{W}|| = |H|$ whose flows are volume-preserving, $\mathcal{L}_{\mathcal{V}}(dM^2) = \mathcal{L}_{\mathcal{W}}(dM^2) = 0$.

b) The integral curves of the vector field $\mathcal{W}/|H|$ are geodesic lines.

c) Let $B : M^2 \to \mathbb{R}$ be the function measuring the geodesic curvature of the integral curve of the vector field $\mathcal{V}$ at a point. Then $\mathcal{W}(H) = H^2 \cdot B$.

d) The following commutator formula holds: $[\mathcal{W}/|H|, \mathcal{V}] = B \cdot \mathcal{V}$.

17. Discuss the geometric properties of the helicoid and the catenoid, and study the map transforming the one into the other.

18. *Enneper's minimal surface* is parametrized by $(z = a + ib)$

$$F(a, b) = \big(\mathrm{Re}(3z - z^3),\, \mathrm{Re}\, i(3z + z^3),\, \mathrm{Re}(-3z^2)\big)\,.$$

Compute its Gaussian curvature. Prove, moreover, that its curvature lines (i.e., the curves on the surface obtained by fixing one of the two parameters a or b) all lie in a plane through the y- and z-axis and have polynomial arc length. Determine the equation for the tangent plane to a point $x \in F$ and transform it into

$$\langle p - q, x\rangle = \langle p, p\rangle - \langle q, q\rangle\,.$$

This is the equation of a plane with respect to which the two points p and q are mirror images of one another. The curves $p(a)$, $q(b)$ are parabolas in the perpendicular planes $x = 0$, $y = 0$, and the summit of each of them coincides with the focus of the other (so-called *focal parabolas*). Illustrate these properties using Maple or Mathematica.

19. The minimal surface studied by Bour (1862) arises from the Weierstrass representation by inserting $f(z) = cz^m$, $g(z) = z$. Prove that it can be mapped to a surface of revolution in a length-preserving way.

20. Prove the following properties of the modular surface of a holomorphic function $f : U \to \mathbb{C}$:

a) The modular surface is hyperbolically curved in a sufficiently small neighborhood of a finite pole of f.

b) Let z_0 be a finite zero of f. If this zero is multiple, the modular surface is elliptically curved in a sufficiently small neighborhood of z_0. If the zero is simple, every sufficiently small neighborhood of z_0 decomposes into $2k - 2$ sector-like pieces, in which the modular surface is alternatingly curved elliptically and hyperbolically. Here k is the exponent of the non-vanishing term succeeding the summand $a_1(z - z_0)$ in the series expansion of f.

In the next two exercises it will be helpful to use the position functions p_i from the proof of the Minkowski-Steiner theorem.

21. Let M^2 be a surface with normal vector e_3. Prove that the 1-form $\eta^1 = \langle x, de_3\rangle$ defined on the surface is closed, and conclude that, for a curve $\gamma \subset M^2$, the relation

$$\int_\gamma \langle x, de_3\rangle = 0$$

holds if one of the following two conditions is satisfied:

a) The curve γ is the boundary of a two-dimensional submanifold $G^2 \subset M^2$.

b) M is simply connected.

22. Let M^2 be a compact surface without boundary. Prove that there exists a point $x_0 \in M^2$ with the following properties:

a) The normal vector e_3 is parallel to the position vector x_0 at the point x_0.

b) The Gaussian curvature is positive at x_0, $G(x_0) > 0$.

Hint: The function $f : M^2 \to \mathbb{R}$, $f(x) = \langle x, x \rangle$, has to have a critical point on M^2.

23. Let the group $G = \mathrm{SL}(2, \mathbb{R})$ act on the hyperbolic plane $\mathcal{H}^2$ by

$$g \cdot z \;=\; \begin{bmatrix} a & b \\ c & d \end{bmatrix} \cdot z \;=\; \frac{az+b}{cz+d}.$$

Verify that the image point $g \cdot z$ actually does belong to $\mathcal{H}^2$, and that superposition of two of these transformations corresponds to matrix multiplication in G. Show, moreover, that each $g \in G$ leaves the metric invariant; hence G is an 3-dimensional group of isometries from $\mathcal{H}^2$ onto itself.

24. Let $M^3 = \mathbb{R}^3$ and denote by $D_Y X$ the directional derivative of the vector-valued function $X : \mathbb{R}^3 \to \mathbb{R}^3$ in the direction of the vector X. Prove that

$$\nabla_X Y \;:=\; D_X Y + \frac{1}{2} X \times Y$$

defines a covariant derivative having all the properties (1)–(4) from Theorem 42, but violating property (5).

25. Consider on the set $\{(x, y) \in \mathbb{R}^2 : -\pi/2 < y < \pi/2\}$ the pseudo-Riemannian metric

$$\mathrm{I} \;=\; \frac{dx^2 - dy^2}{\cos^2(y)}.$$

a) Prove that $E = \dfrac{x'^2 - y'^2}{\cos^2(y)}$ and $P = \dfrac{x'}{\cos^2(y)}$ are first integrals of the geodesic flow, satisfying, in addition, the inequality $P^2 - E \geq 0$.

b) Discuss the geodesic lines on M^2. To do so, assume that y is a function of x and integrate the resulting ordinary differential equation.

c) On M^2 there exist points that cannot be joined by a geodesic line.

26. Prove that every three-dimensional Einstein space is a space of constant curvature.

27. Let M^m be a non-flat Einstein space (for example, S^m). Show that $M^m \times M^m$ with the product metric is an Einstein space, but not a space of constant curvature.

28. It is well-known that a symmetric bilinear form $h(x,y)$ on a vector space is completely determined by its quadratic form $q(x) := h(x,x)$ (via the polarization formula: $2h(x,y) = q(x+y) - q(x) - q(y)$). Prove, in a similar way, that the Riemannian curvature tensor $\mathcal{R}(\mathcal{U},\mathcal{V},\mathcal{W}_1,\mathcal{W}_2)$ is completely determined by the quadratic form $K(\mathcal{U},\mathcal{V}) := \mathcal{R}(\mathcal{U},\mathcal{V},\mathcal{V},\mathcal{U})$ corresponding to the sectional curvature.

29. Prove that a four-dimensional Riemannian manifold is an Einstein space if and only if for every 2-plane $E^2 \subset TM^4$ and its orthogonal complement $(E^2)^\perp$ the corresponding sectional curvatures coincide, $K\big(E^2\big) = K\big((E^2)^\perp\big)$.

30. Because of its symmetry properties, the curvature tensor of a pseudo-Riemannian manifold can be interpreted as a transformation $\mathcal{R} : \bigwedge^2(M^m) \to \bigwedge^2(M^m)$ on 2-forms,

$$\mathcal{R}\big(\sigma_i \wedge \sigma_j\big) \;=\; \frac{1}{2}\sum_{\alpha,\beta=1}^{n} R^j_{\alpha\beta i}\cdot\sigma_\alpha\wedge\sigma_\beta\,.$$

In dimension $m = 4$, this gives rise to an endomorphism $\mathcal{R} : \bigwedge^2(M^4) \to \bigwedge^2(M^4)$. On the other hand, the Hodge operator $* : \bigwedge^2(M^4) \to \bigwedge^2(M^4)$ acts on $\bigwedge^2(M^4)$, and its square depends only on the index k of the metric (see Theorem 5, Chapter 1): $** = (-1)^k$. In the cases $k = 0$ (positive definite metric) and $k = 2$ (neutral metric), the Hodge operator decomposes the real bundle $\bigwedge^2(M^4)$ into the corresponding eigenspaces $\bigwedge^2_\pm(M^4)$ (see Exercise 8, Chapter 1). Prove that the block representation of the curvature tensor,

$$\mathcal{R} \;=\; \begin{pmatrix} \mathcal{R}_{++} & \mathcal{R}_{-+} \\ \mathcal{R}_{+-} & \mathcal{R}_{--} \end{pmatrix},$$

with respect to this decomposition of $\bigwedge^2(M^4)$ has the following properties:

a) $\mathcal{R}$ is symmetric, i.e., $\mathcal{R}^*_{++} = \mathcal{R}_{++}$, $\mathcal{R}^*_{--} = \mathcal{R}_{--}$ and $\mathcal{R}^*_{+-} = \mathcal{R}_{-+}$.

b) The traces of $\mathcal{R}_{++}$ and $\mathcal{R}_{--}$ coincide, $\mathrm{tr}\,(\mathcal{R}_{++}) = \mathrm{tr}\,(\mathcal{R}_{--}) = -\tau/12$.

c) M^4 is an Einstein space if and only if $\mathcal{R}_{+-}$ vanishes.

Literature: Th. Friedrich, Self-duality of Riemannian manifolds and connections, in: *Riemannian geometry and instantons*, Teubner-Verlag, Leipzig, 1981, 56-104.

The *Einstein equation* in the general theory of relativity combines the geometric curvature quantities of a four-dimensional pseudo-Riemannian manifold of signature $(1,3)$ with its physical properties encoded in the energy-momentum tensor T,

$$\mathrm{Ric} - \frac{1}{2} g \cdot \tau \;=\; \kappa T.$$

Here Ric, g, τ are the Ricci tensor, the metric, and the scalar curvature; κ is a constant depending on the chosen system of units. Already for the vacuum, $T = 0$, there are non-trivial, physically very interesting solutions of this equation, among others the Schwarzschild metric to be discussed now. Obviously, a vacuum solution of the Einstein equation has to be an Einstein space with vanishing Ricci tensor in the sense of the definition given before.

31 (Schwarzschild metric). On a spherically symmetric and static space-time manifold M^4 (this is a pseudo-Riemannian manifold of signature $(1,3)$ with isometry group $\mathrm{SO}(3,\mathbb{R})$ and a distinguished time direction), it is always possible to introduce coordinates from $\mathbb{R} \times \mathbb{R}^+ \times S^2$ with respect to which the metric can be written as

$$g \;=\; e^{2a(r)}\, dt^2 - \Big[e^{2b(r)}\, dr^2 \,+\, r^2 \big(d\theta^2 \,+\, \sin^2\theta\, d\varphi^2 \big) \Big]\,.$$

Here, the functions $a(r)$ and $b(r)$ asymptotically tend to zero for $r \to \infty$ (the metric is "asymptotically flat"). We introduce the following basis of 1-forms:

$$\sigma_0 \;=\; e^a\, dt, \quad \sigma_1 \;=\; e^b\, dr, \quad \sigma_2 \;=\; r\, d\theta, \quad \sigma_3 \;=\; r \sin\theta\, d\varphi\,.$$

a) Check that the metric satisfies $g \;=\; \sigma_0^2 - \sigma_1^2 - \sigma_2^2 - \sigma_3^2\,.$

b) Compute the forms $d\sigma_i$ and show, using the first structural equation, that the connection forms are given by

$$\omega_{01} \;=\; -a' e^{-b} \sigma_0, \quad \omega_{02} \;=\; \omega_{03} \;=\; 0,$$

$$\omega_{12} \;=\; \frac{e^{-b}}{r}\,\sigma_2, \quad \omega_{13} \;=\; \frac{e^{-b}}{r}\,\sigma_3, \quad \omega_{23} \;=\; \frac{\cot\theta}{r}\,\sigma_3.$$

c) We introduce the notation $\Omega_{ij} := \frac{1}{2}\sum_{\alpha,\beta} R^j_{\alpha\beta i} \sigma_\alpha \wedge \sigma_\beta$. ($\Omega$ is the so-called *curvature form*). Show, using the second structural equation, that

$$\Omega_{10} \;=\; e^{-2b}(a'b' - a'' - a'^2)\sigma_0 \wedge \sigma_1, \quad \Omega_{20} \;=\; -a'\tfrac{e^{-2b}}{r}\sigma_0 \wedge \sigma_2,$$

$$\Omega_{30} \;=\; -a'\tfrac{e^{-b}}{r}\sigma_0 \wedge \sigma_3, \quad \Omega_{21} \;=\; b'\tfrac{e^{-b}}{r}\sigma_1 \wedge \sigma_2,$$

$$\Omega_{31} \;=\; b'\tfrac{e^{-b}}{r}\sigma_1 \wedge \sigma_3, \quad \Omega_{32} \;=\; \tfrac{1-e^{-b}}{r^2}\sigma_2 \wedge \sigma_3,$$

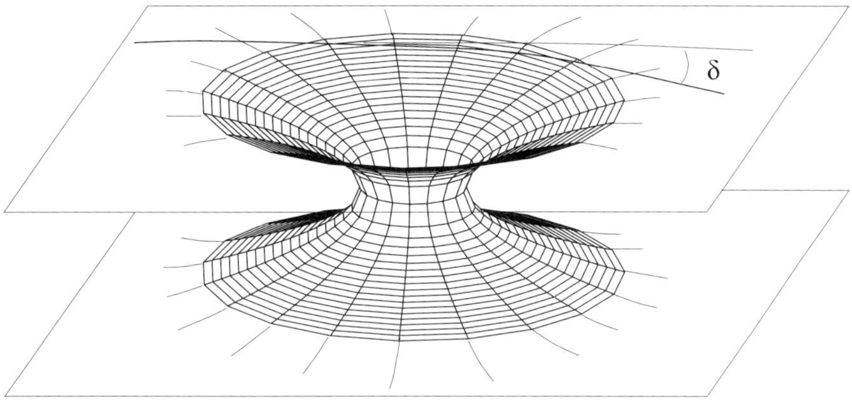

and compute from these formulas the components R^i_{jkl} of the curvature tensor in this basis.

d) Compute the components of the Einstein tensor $G := \mathrm{Ric} - \frac{1}{2} g \cdot \tau$:

$$G_{00} = \tfrac{1}{r^2} - e^{-2b}\left(\tfrac{1}{r^2} - \tfrac{2b'}{r}\right), \quad G_{11} = \tfrac{1}{r^2} - e^{-2b}\left(\tfrac{1}{r^2} + \tfrac{2a'}{r}\right),$$
$$G_{22} = G_{33} = -e^{-2b}\left(a'^2 - a'b' + a'' + \tfrac{a'-b'}{r}\right).$$

e) Solve the vacuum equation by means of this Ansatz. The result is (M is a constant of integration)

$$g = \left[1 - \frac{2M}{r}\right] dt^2 - \left[1 - \frac{2M}{r}\right]^{-1} dr^2 - r^2\left(d\theta^2 + \sin^2\theta\, d\varphi^2\right).$$

32. Restrict the Schwarzschild metric g to the two-dimensional submanifold defined by $\theta = \pi/2$ and $t = \mathrm{const}$. Prove that this yields the metric of a paraboloid of revolution (a "lying" parabola!) with the equation $z^2 = 8M(r - 2M)$ (see the picture).

33. Light moves along geodesic lines whose tangent vectors have length zero. Making use of the fact that the equations defining a geodesic are precisely the Euler-Lagrange equations of the length function $\mathcal{L} = \sum_{i,j} g_{ij}\dot{x}^i\dot{x}^j$, prove the following assertions:

a) A particle moving at $t = 0$ in the equatorial plane $\theta = \pi/2$ stays there forever.

b) The quantities $L := r^2\dot{\varphi}$ and $E := \dot{t}(1 - 2M/r)$ are first integrals. Moreover, the second Kepler law holds: The orbit ray covers equal areas in equal times.

c) Set $r = r(\varphi)$ and derive the equation describing the motion of light rays. Result: It is reasonable to set $u := 1/r$: $u'' + u = 3Mu^2$.

d) Solve this differential equation approximately up to second order in φ. Result:

$$u_0 = \frac{1}{r_0} \sin\varphi + \frac{3M}{2r_0^2}\left[1 + \frac{1}{3}\cos 2\varphi\right] .$$

Interpret the constant of integration r_0 as the scattering length. Which asymptotic value arises for φ if r tends to ∞ and φ is supposed to stay small? Twice this value, denoted by δ in the picture, is the relativistic deviation of light in the gravitational field of a very large mass, which can be described by the Schwarzschild metric.

Chapter 6

Lie Groups and Homogeneous Spaces

6.1. Lie Groups and Lie Algebras

In the preceding chapter Noether's theorem showed that symmetry considerations simplify the study of geometric problems, and sometimes it is only by symmetry considerations that a solution is possible at all. In fact, beginning in the 1870s, the conviction grew that the basic principle organizing geometry ought to be the study of its symmetry groups. In his inaugural lecture at the University of Erlangen, which later became known as the "Erlanger Programm", Felix Klein said, in 1872,

> "Es ist eine Mannigfaltigkeit und in derselben eine Transformationsgruppe gegeben; man soll die der Mannigfaltigkeit angehörigen Gebilde hinsichtlich solcher Eigenschaften untersuchen, die durch die Transformationen der Gruppe nicht geändert werden[1]."

One has to distinguish whether the groups under consideration are discrete (for example permutation groups) or continuous (for example, one-parameter groups of isometries). The latter were systematically introduced by the Norwegian mathematician Sophus Lie (1842-1899), which is why they bear his name today.

[1]"Let a manifold and a transformation group in it be given; the objects belonging to the manifold ought to be studied with respect to those properties which are not changed by the transformations of the group."—quoted from F. Klein, *Das Erlanger Programm*, Ostwalds Klassiker der exakten Wissenschaften, Band 253, Verlag H. Deutsch, Frankfurt a. M., 1995, p. 34.

The fundamental idea of a Lie group is a very simple one. It ought to be a group which is at the same time a manifold, and hence allows a differential calculus. Moreover, the manifold structure has to be compatible with the group structure, i. e., the product is a differentiable map.

Definition 1. A *Lie group* is a group G which, at the same time, is a differentiable manifold such that the map

$$G \times G \longrightarrow G, \quad (g, h) \longmapsto g \cdot h^{-1},$$

is (infinitely often) differentiable.

Remark. Obviously, the last condition is equivalent to requiring that the product map and the inversion

$$(g, h) \longmapsto g \cdot h, \quad g \longmapsto g^{-1},$$

be differentiable maps.

Example 1. Every finite-dimensional vector space V is an abelian Lie group whose group product is exactly the addition of vectors $(v, w) \mapsto v + w$.

Example 2. The unit circle $S^1 = \{z \in \mathbb{C} : |z| = 1\}$ is an abelian Lie group with the usual multiplication of complex numbers as product.

Example 3. Most Lie groups can be realized as matrix groups. The set of all invertible matrices with entries in $\mathbb{K} = \mathbb{R}$ or $\mathbb{C}$ is an open subset of $\mathbb{K}^{n^2}$ and hence a manifold. Endowed with matrix multiplication as product, it forms a Lie group, the *general linear group*

$$\mathrm{GL}(n, \mathbb{K}) \;:=\; \{A \in \mathcal{M}_n(\mathbb{K}) : \; \det A \neq 0\}.$$

More generally, $\mathrm{GL}(V)$ is meant to denote the group of invertible endomorphisms of the vector space V.

Example 4. The vector space $\mathbb{K}^n$ and the general linear group $\mathrm{GL}(n, \mathbb{K})$ combine to form a new Lie group, the *affine group* $\mathrm{Aff}(\mathbb{K}^n) := \mathrm{GL}(n, \mathbb{K}) \times \mathbb{K}^n$ with multiplication

$$(A, v) \cdot (B, w) \;:=\; (AB, Aw + v).$$

This product rule arises in a natural way by defining an action of the affine group on the vector space $\mathbb{K}^n$ through

$$\mathrm{Aff}(\mathbb{K}^n) \times \mathbb{K}^n \longrightarrow \mathbb{K}^n, \quad (A, v)x \;:=\; Ax + v,$$

and then applying the transformations determined by (B, w) and (A, v) on a vector x one after the other.

Each Lie group G acts on itself by means of the left and the right translation with a fixed element $g \in G$,

$$L_g, R_g : \; G \longrightarrow G, \quad L_g(h) \;=\; g \cdot h \;\text{ and }\; R_g(h) \;=\; h \cdot g.$$

The corresponding differentials are the following maps in the tangent bundle of G:

$$(dL_g)_h : T_hG \longrightarrow T_{gh}G, \quad (dR_g)_h : T_hG \longrightarrow T_{hg}G.$$

To avoid double indices, in this chapter we will use the notation $(df)_h$ instead of $f_{*,h}$ for the differential of a map f.

Definition 2. A vector field $\mathcal{X}$ on G is called *left-invariant* (or *right-invariant*, respectively) if it is transformed into itself by dL_g (or dR_g, i. e., $dL_g\mathcal{X} = \mathcal{X}$. At a point $h \in G$, this means

$$(dL_g)_h\mathcal{X}(h) = \mathcal{X}(g \cdot h).$$

Since left translation is obviously a diffeomorphism of G, Theorem 35 in §3.9 can be applied to yield, for the commutator of two left-invariant vector fields, the formula

$$dL_g[\mathcal{X}, \mathcal{Y}] = [dL_g\mathcal{X}, dL_g\mathcal{Y}] = [\mathcal{X}, \mathcal{Y}].$$

This property, together with the fact that vector fields satisfy the Jacobi identity (Theorem 34, §3.9), endows the vector space of left-invariant vector fields with the structure of a *Lie algebra*, the Lie algebra $\mathfrak{g}$ of the Lie group G.

Theorem 1. *The vector space of left-invariant vector fields on a Lie group G is canonically isomorphic to the tangent space at the neutral element,*

$$\mathfrak{g} \cong T_eG.$$

Proof. With each left-invariant vector field $\mathcal{X}$, we associate its evaluation at the neutral element e, $\mathcal{X} \mapsto \mathcal{X}(e) =: X \in T_eG$. Conversely, every element $X \in T_eG$ determines a vector field $\mathcal{X}_X$ on G by setting

$$\mathcal{X}_X(g) := (dL_g)_e(X).$$

This satisfies the relation

$$\mathcal{X}_X(gh) = (dL_{gh})_e(X) = (dL_g)_h(dL_h)_e(X) = (dL_g)_h\mathcal{X}_X(h);$$

hence $\mathcal{X}_X$ is left-invariant. □

Remark. Because of this fact, we will no longer distinguish between the Lie algebra of left-invariant vector fields and the tangent space to the group at the neutral element. Its elements will be denoted by upper case Latin letters $X, Y, \ldots \in \mathfrak{g}$.

Choosing a basis $X_1, \ldots, X_r$ of $\mathfrak{g}$, we can again write their commutators as linear combinations of the basis elements,

$$[X_i, X_j] = \sum_{k=1}^{r} C_{ij}^k X_k.$$

The antisymmetry of the commutator and the Jacobi identity imply that the constants C_{ij}^k have to satisfy the relations

$$C_{ij}^k \;=\; -\,C_{ji}^k, \quad C_{ij}^k C_{km}^l + C_{jm}^k C_{ki}^l + C_{mi}^k C_{kj}^l \;=\; 0\,.$$

Following E. Cartan, the numbers C_{ij}^k are called the *structure constants* of the Lie group G, since Cartan's structural equations are simple to formulate in their terms. To see this, we agree to call a differential form ω on G *left-invariant* if it satisfies the condition

$$L_g^* \omega \;=\; \omega\,.$$

Following the argument in the case of vector fields, it is easy to see that the r-dimensional vector space $\mathfrak{g}^*$ of left-invariant 1-forms is canonically isomorphic to T_e^*G. Now let $\sigma_1, \ldots, \sigma_r$ be the basis of $\mathfrak{g}^*$ dual to $X_1, \ldots, X_r$.

Theorem 2 (Maurer-Cartan Equations)**.** *Let C_{ij}^k be the structure constants of a Lie group G with respect to the basis $X_1, \ldots, X_r$ of its Lie algebra $\mathfrak{g}$. Then the exterior derivatives of the forms in the dual basis $\sigma_1, \ldots, \sigma_r$ of $\mathfrak{g}^*$ are given by*

$$d\sigma_i \;=\; -\sum_{j<k} C_{jk}^i \,\sigma_j \wedge \sigma_k\,.$$

Proof. From Theorem 33, §3.9, we know the general formula for the differential of a 1-form σ_i:

$$d\sigma_i(X_j, X_k) \;=\; X_j\big(\sigma_i(X_k)\big) - X_k\big(\sigma_i(X_j)\big) - \sigma_i\big([X_j, X_k]\big)\,.$$

Since the forms σ_i are dual to the basis vectors X_j, the first two terms on the right-hand side vanish. Now

$$d\sigma_i(X_j, X_k) \;=\; -\,\sigma_i\big([X_j, X_k]\big) \;=\; -\,\sigma_i\big(\textstyle\sum_m C_{jk}^m X_m\big) \;=\; -\,C_{jk}^i\,,$$

which is precisely the assertion. □

We introduce a 1-form $\Theta : T(G) \to \mathfrak{g}$ defined by the formula

$$\Theta(t) \;:=\; dL_{g^{-1}}(t)$$

for every tangent vector $t \in T_g(G)$ at the element $g \in G$. The form Θ shifts the tangent vectors of the Lie group via left translations to the space $T_e(G) = \mathfrak{g}$. It is the so-called *Maurer-Cartan form* of the Lie group. In general, if Θ_1, Θ_2 are two 1-forms with values in the Lie algebra $\mathfrak{g}$, we define its "exterior" product as a 2-form with values in the Lie algebra by the formula

$$[\Theta_1, \Theta_2](t_1, t_2) \;:=\; [\Theta_1(t_1), \Theta_2(t_2)] - [\Theta_1(t_2), \Theta_2(t_1)]\,.$$

In particular,

$$[\Theta, \Theta](t_1, t_2) \;=\; 2 \cdot [\Theta(t_1), \Theta(t_2)],$$

and the Maurer-Cartan equations can be formulated equivalently as

$$d\Theta + \frac{1}{2}\cdot[\Theta,\Theta] \;=\; 0\,.$$

Example 5. The special linear group in dimension two,

$$\mathrm{SL}(2,\mathbb{R}) \;=\; \{A \in \mathcal{M}_2(\mathbb{R}):\; \det A = 1\}\,,$$

is a 3-dimensional Lie group with Lie algebra (see Exercise 5, Chapter 3)

$$\mathfrak{sl}(2,\mathbb{R}) \;=\; \{X \in \mathcal{M}_2(\mathbb{R}):\; \operatorname{tr} A = 0\}\,.$$

The matrices

$$H \;=\; \begin{bmatrix} 1 & 0 \\ 0 & -1 \end{bmatrix}, \quad E \;=\; \begin{bmatrix} 0 & 1 \\ 0 & 0 \end{bmatrix}, \quad F \;=\; \begin{bmatrix} 0 & 0 \\ 1 & 0 \end{bmatrix}$$

form a basis of $\mathfrak{sl}(2,\mathbb{R})$ and satisfy the commutator relations

$$[H,E] \;=\; 2E, \quad [H,F] \;=\; -2F, \quad [E,F] \;=\; H\,.$$

If we denote the forms of the dual basis of $\mathfrak{sl}(2,\mathbb{R})^*$ by σ_H, σ_E, and σ_F, the Maurer-Cartan structural equations read as

$$d\sigma_H \;=\; -\,\sigma_E\wedge\sigma_F, \quad d\sigma_E \;=\; -2\,\sigma_H\wedge\sigma_E, \quad d\sigma_F \;=\; 2\,\sigma_H\wedge\sigma_F\,.$$

Example 6. The Lie group $\mathrm{GL}(n,\mathbb{K})$ is an open subset of $\mathcal{M}_n(\mathbb{K})$. Its Lie algebra is therefore

$$\mathfrak{gl}(n,\mathbb{K}) \;\cong\; T_e\,\mathrm{GL}(n,\mathbb{K}) \;\cong\; \mathcal{M}_n(\mathbb{K})\,.$$

Theorem 3. *Let X be a left-invariant vector field on G and denote by $\Phi_X(t)$ the maximal integral curve of X such that $\Phi_X(0) = e$. Then*

(1) *$\Phi_X(t)$ is a group homomorphism, $\Phi_X(t_1+t_2) = \Phi_X(t_1)\circ\Phi_X(t_2)$;*

(2) *the vector field X is complete, i. e., $\Phi_X(t)$ is defined on all of $\mathbb{R}$.*

Proof. Let $\Phi_X(t)$ be defined on the interval $t_a < t < t_b$. We start by showing the first assertion for two values t_1, t_2 such that they together with their sum t_1+t_2 belong to the domain of definition of $\Phi_X(t)$. To do so, fix t_1, set $g := \Phi_X(t_1)$, and consider the curve

$$\eta(s) \;=\; g\cdot\Phi_X(s) \;\; \text{on the interval} \;\; (t_a,\, t_b)$$

as well as the integral curve obtained by shifting the argument by t_1,

$$\eta_1(s) \;=\; \Phi_X(t_1+s) \;\; \text{on the interval} \;\; (t_a-t_1,\, t_b-t_1)\,.$$

From their derivatives,

$$\frac{d\eta}{ds} \;=\; dL_g\left(\frac{d\Phi_X}{ds}\right) \;=\; dL_g(X(\Phi_X(s))) \;=\; X\big(g\cdot\Phi_X(s)\big) \;=\; X\big(\eta(s)\big)$$

and

$$\frac{d\eta_1}{ds} \;=\; \frac{d\Phi_X(t_1+s)}{ds} \;=\; X\big(\Phi_X(t_1+s)\big) \;=\; X\big(\eta_1(s)\big),$$

we conclude that η and η_1 both are integral curves of X with equal initial conditions,

$$\eta_1(0) \;=\; \Phi_X(t_1) \;=\; g \;=\; \eta(0)\,.$$

Hence η and η_1 have to coincide on $(t_a,\, t_b) \cap (t_a - t_1,\, t_b - t_1)$. One easily checks that t_2 has to belong to this intersection, i. e.,

$$\Phi_X(t_1 + t_2) \;=\; \Phi_X(t_1) \circ \Phi_X(t_2)\,.$$

Now we show that $t_b = \infty$; to prove $t_a = -\infty$ one proceeds analogously. Suppose that t_b is finite. Then,

$$\eta(s) \;=\; \Phi_X\big(\frac{t_b}{2}\big) \cdot \Phi_X\big(s - \frac{t_b}{2}\big)$$

is a curve defined on $[t_a + t_b/2,\, 3\,t_b/2)$ with tangent vector

$$\frac{d\eta}{ds} \;=\; dL_{\Phi_X(t_b/2)} \frac{d\Phi_X}{ds}\big(s - \frac{t_b}{2}\big) \;=\; X\left(\Phi_X\big(\frac{t_b}{2}\big) \cdot \Phi_X\big(s - \frac{t_b}{2}\big)\right) \;=\; X\big(\eta(s)\big)\,.$$

Hence $\eta(s)$ is an integral curve of X defined on $[t_a + t_b/2,\, 3\,t_b/2)$ such that $\eta(s) = \Phi_X(s)$ for $s \in (t_a, t_b) \cap [t_a + t_b/2,\, 3\,t_b/2)$. Thus the maximal integral curve has to be defined at least on $3\,t_b/2$, a contradiction to the maximality of t_b. □

This theorem enables us to define the exponential map for Lie groups.

Definition 3. The *exponential map* $\exp : \mathfrak{g} \to G$ is the evaluation at $t = 1$ of the integral curve Φ_X of $X \in \mathfrak{g}$ satisfying $\Phi_X(0) = e$,

$$\exp(X) \;:=\; \Phi_X(1)\,.$$

Theorem 4. *The exponential* $\exp : \mathfrak{g} \to G$ *is a smooth map with the following properties:*

(1) $\exp(0) = e$;

(2) $\exp(-X) = [\exp(X)]^{-1}$;

(3) $\exp\big((s + t)X\big) = \exp(sX) \cdot \exp(tX)\,.$

In addition, it is a diffeomorphism from an open neighborhood V of zero in $\mathfrak{g}$ onto an open neighborhood of the neutral element in G.

Proof. For the smoothness claim, we have to refer to the fact that the solution of a system of differential equations depends smoothly on the initial conditions. Under scalar multiplication, the flow transforms as

$$\Phi_{\lambda X}(t) \;=\; \Phi_X(\lambda \cdot t)\,,$$

since

$$\frac{d\Phi_X(\lambda t)}{dt} \;=\; \lambda \cdot X\big(\Phi_X(\lambda t)\big)$$

together with $\Phi_X(\lambda \cdot 0) = e$ implies that $\Phi_X(\lambda \cdot t)$ is an integral curve of $\lambda \cdot X$. Restricting the exponential map to a straight line through the origin yields a one-parameter subgroup of G. This proves property (3):

$$\begin{aligned}\exp\big((s+t)X\big) &= \Phi_{(s+t)X}(1) = \Phi_X(s+t) = \Phi_X(s)\cdot\Phi_X(t)\\ &= \Phi_{sX}(1)\cdot\Phi_{tX}(1) = \exp(sX)\cdot\exp(tX)\,.\end{aligned}$$

Property (2) is an immediate consequence of (3), and property (1) is trivial. The final assertion follows from the inverse function theorem if we can show that the differential of the exponential map at the origin,

$$d(\exp)_0 :\ T_0\mathfrak{g} \cong \mathfrak{g} \longrightarrow T_eG \cong \mathfrak{g}\,,$$

is invertible. But this is just the identity map: Consider the curve $\gamma(s) = s \cdot X$ in $\mathfrak{g}$ with tangent vector $\dot\gamma(0) = X$. Then

$$d(\exp)_0(s\cdot X) = \frac{d\exp(sX)}{ds}\Big|_{s=0} = \frac{d\Phi_{sX}(1)}{ds}\Big|_{s=0} = \frac{d\Phi_X(s)}{ds}\Big|_{s=0} = X(e)\,.$$

Since we already know that the evaluation of a left-invariant vector field at the neutral element e is an isomorphism, the proof is completed. □

It turns out that all one-parameter subgroups of a Lie group G have the form $\exp(tX)$ for some $X \in \mathfrak{g}$.

Theorem 5. *Every continuous group homomorphism $\Psi : \mathbb{R} \to G$ has the form*

$$\Psi(t) = \exp(tX)$$

for a certain $X \in \mathfrak{g}$. In particular, every continuous group homomorphism $\mathbb{R} \to G$ is smooth.

Proof. Let $\Psi : \mathbb{R} \to G$ be a continuous group homomorphism, and let $V \subset \mathfrak{g}$ be a neighborhood of the origin so small that $\exp : \mathfrak{g} \to G$ is a diffeomorphism on $2V$. Let $W = \exp(V)$ be the image of V in G. Since Ψ is continuous, there exists an $\varepsilon > 0$ such that $\Psi([-\varepsilon, \varepsilon]) \subset W$. In particular, $\Psi(\varepsilon)$ has to have a pre-image $X \in V$ under the exponential map, $\Psi(\varepsilon) = \exp(X)$. The smooth group homomorphism

$$f :\ \mathbb{R} \longrightarrow G, \quad f(t) = \exp\left(\frac{t}{\varepsilon}X\right),$$

coincides with $\Psi(\varepsilon)$ at the point ε. Consider the set

$$K = \{t \in \mathbb{R} :\ \Psi(t) = f(t)\}\,.$$

This is a closed subgroup of $\mathbb{R}$, since Ψ and f are continuous group homomorphisms. Therefore, K is equal either to $\mathbb{R}$ or to a discrete subgroup of it.

The latter case will be excluded, implying the equality of f and Ψ. Assume for a contradiction that K is generated by a number $a > 0$,

$$K = \{n \cdot a : n \in \mathbb{Z}\}.$$

As ε already belongs to K, a has to be less than ε, and $\Psi(a)$ belongs to W. We show $a/2 \in K$ to obtain the desired contradiction. To do so, start from the Ansatz

$$f(a/2)^2 = f(a) = \Psi(a) = \Psi(a/2)^2 .$$

Then, applying the inverse of the exponential map,

$$2\exp^{-1}(f(a/2)) = 2\exp^{-1}(\Psi(a/2)) ,$$

and noting that $\Psi(a/2)$ belongs to W, we conclude that $2\exp^{-1}(\Psi(a/2)) \in 2V$. On $2V$, the exponential map is a diffeomorphism, and hence

$$f(a/2) = \Psi(a/2) .$$

This means that $a/2 \in K$. □

Definition 4. Let G_1, G_2 be two Lie groups, and let $\psi : G_1 \to G_2$ be a group homomorphism. We define the differential of ψ as a map between the Lie algebras of left-invariant vector fields on G_1 and G_2 as follows: Consider for $X \in \mathfrak{g}_1$ the one-parameter subgroup $\exp(tX) \in G_1$. By assumption, its image curve $\psi(\exp(tX))$ is a one-parameter subgroup of G_2, which by Theorem 5 has to have the form $\exp(tY)$ for some $Y \in \mathfrak{g}_2$. Now we set

$$\psi_* : \mathfrak{g}_1 \longrightarrow \mathfrak{g}_2, \quad \psi_*(X) := Y .$$

This definition immediately implies that the following diagram commutes:

$$\begin{array}{ccc} \mathfrak{g}_1 & \xrightarrow{\psi_*} & \mathfrak{g}_2 \\ {\scriptstyle \exp}\downarrow & & \downarrow{\scriptstyle \exp} \\ G_1 & \xrightarrow{\psi} & G_2 \end{array}$$

Identifying $\mathfrak{g}_1 \cong T_eG_1$ as well as $\mathfrak{g}_2 \cong T_eG_2$, we see that this definition is compatible with the usual one of the differential at the neutral element, $d\psi_e : T_eG_1 \to T_eG_2$. More precisely, the diagram extended by $d\psi_e$ remains commutative:

$$\begin{array}{ccccccc} \mathfrak{g}_1 & & & \xrightarrow{\psi_*} & & & \mathfrak{g}_2 \\ & \searrow^{\cong} & & & & \swarrow^{\cong} & \\ {\scriptstyle \exp}\downarrow & & T_eG_1 & \xrightarrow{d\psi_e} & T_eG_2 & & \downarrow{\scriptstyle \exp} \\ G_1 & & & \xrightarrow{\psi} & & & G_2 \end{array}$$

To see this, take any left-invariant vector field $X \in \mathfrak{g}_1$ and compute

$$\psi_*(X)(e) = \frac{d\exp(t\psi_*(X))}{dt}\bigg|_{t=0} = \frac{d\psi(\exp(tX))}{dt}\bigg|_{t=0} = d\psi_e\big(X(e)\big).$$

A further important property of the differential ψ_* is its compatibility with the commutator.

Theorem 6. *The differential of a homomorphism between Lie groups, $\psi : G_1 \to G_2$, is a homomorphism of Lie algebras, i. e.,*

$$\psi_*[X, Y] = [\psi_* X, \psi_* Y].$$

Proof. The commutativity of the diagram just discussed is equivalent to the following identity, holding for every left-invariant vector field $X \in \mathfrak{g}_1$:

$$\psi_*(X)(e) = d\psi_e\big(X(e)\big).$$

From this, we deduce the analogous identity holding at *each* point $g \in G_1$,

$$\begin{aligned} d\psi_g\big(X(g)\big) &= d\psi_g(dL_g)_e\big(X(e)\big) = d(\psi \circ L_g)_e\big(X(e)\big) = \big(dL_{\psi(g)}\big)_e d\psi_e\big(X(e)\big) \\ &= \big(dL_{\psi(g)}\big)_e \psi_*(X)(e) = \psi_*(X)\big(\psi(g)\big). \end{aligned}$$

This means that the vector fields X and $\psi_* X$ are ψ-related in the sense of the remark following Theorem 35, §3.9. In particular, this remark implies the assertion. □

6.2. Closed Subgroups and Homogeneous Spaces

The exponential map of Lie groups does not satisfy the functional equation familiar from calculus; it turns out that a second order correction has to be made.

Lemma 1. *Let G be a Lie group and let $\exp : \mathfrak{g} \to G$ be the exponential map. Then, for two arbitrary left-invariant vector fields $X, Y \in \mathfrak{g}$, the following identity holds:*

$$\exp(tX) \cdot \exp(tY) = \exp\big(t(X + Y) + O(t^2)\big).$$

Here $O(t^2)$ indicates that $O(t^2)/t^2$ stays bounded for $t \to 0$.

Proof. Choose a neighborhood $V \subset \mathfrak{g}$, $0 \in V$, such that $\exp : V \to W \subset G$ is a chart around the neutral element $e \in G$, and also choose a basis $X_1, \ldots, X_r$ of $\mathfrak{g}$. By means of

$$(x^1, \ldots, x^r) \longmapsto \exp(x^1 X_1 + \ldots + x^r X_r) \in G,$$

we introduce coordinates in W. Let $U \subset W$ be so small that $U \cdot U \subset W$. (By "$\cdot$" we denote the product in the group G.) On $U \times U \subset G \times G$, we then have the coordinates

$$y^i = x^i \circ \mathrm{pr}_1, \quad z^i = x^i \circ \mathrm{pr}_2,$$

and, because $U \cdot U \subset W$, we can restrict the group multiplication to obtain a map $\mu : U \times U \to W$. Let

$$x^i \;=\; f^i(y^1, \ldots, y^r, z^1, \ldots, z^r)$$

be the coordinate representation of μ. Then

$$f^i(0, \ldots, 0, z^1, \ldots, z^r) \;=\; z^i, \quad f^i(y^1, \ldots, y^r, 0, \ldots, 0) \;=\; y^i\,,$$

and hence f^i has the Taylor expansion

$$f^i(y, z) \;=\; f^i(0,0) + \sum \frac{\partial f^i}{\partial y^k}(0,0) y^k + \sum \frac{\partial f^i}{\partial y^l}(0,0) z^l + R\,,$$

where the remainder R consists of second order terms. Inserting $tY = \sum y^i X_i$, $tZ = \sum z^i X_i$ into the Taylor expansion, we now obtain

$$\exp(tY + tZ + R) \;=\; \exp(tY)\exp(tZ)\,,$$

and the term R tends to zero like $O(t^2)$. □

The following example shows that not every continuous subgroup of a Lie group has to be a Lie group itself.

Example 7. Let $T^2 = S^1 \times S^1$ be the two-dimensional torus, parametrized by $(t, s) \in [0,1] \times [0,1]$. For every real number q, the formula $\gamma_q : \mathbb{R} \to T^2$, $\gamma_q(t) = (t, q \cdot t)$ defines a curve in T^2, which is isomorphic to the additive group $\mathbb{R}$ of real numbers. We want to show that this curve is dense in T^2 for irrational q, and hence provides an example of a connected, non-closed one-parameter subgroup of a Lie group. Let $(t_0, s_0) \in [0,1]^2$ be an arbitrarily given point on T^2, which we want to approximate by a sequence of points in γ_q. For each integer $n \in \mathbb{Z}$, we have

$$\gamma(t_0 + n) \;=\; (t_0 + n, qt_0 + qn)\,,$$

and, as an angle coordinate in S^1, $t_0 + n$ is equivalent to t_0. Similarly, $qt_0 + qn + m$ is equivalent to $qt_0 + qn$ for every integer m; hence it corresponds to *the same* point on γ_q and on T^2. If we are able to show now that $q\mathbb{Z} + \mathbb{Z}$ is dense in $\mathbb{R}$, then we could conclude the existence of a sequence of pairs $(n_i, m_i) \in \mathbb{Z} \times \mathbb{Z}$ for which

$$\lim_{i \to \infty} qn_i + m_i \;=\; s_0 - qt_0\,.$$

But then $q(t_0 + n_i) + m_i$ converges to s_0 for $i \to \infty$, and

$$\lim_{i \to \infty} \gamma_q(t_0 + n_i) \;=\; \gamma(t_0, s_0)\,,$$

as claimed. Thus it remains to prove the following lemma.

Lemma 2. *For every irrational number q, the set $\mathbb{Z} + q\mathbb{Z}$ is dense in $\mathbb{R}$.*

Proof. Like the set $\mathbb{Z}+q\mathbb{Z}$, its closure is a subgroup of $\mathbb{R}$. The group $\mathbb{Z}+q\mathbb{Z}$ cannot be cyclic; if it were, it would be generated by an element $a > 0$, and there would exist integers k and l such that

$$1 + q \cdot 0 = 1 = k \cdot a, \quad 0 + q \cdot 1 = q = l \cdot a .$$

But this would imply $q = l/k \in \mathbb{Q}$. Hence we arrive at a contradiction. As the closed subgroups of $\mathbb{R}$ are precisely the cyclic groups and $\mathbb{R}$ itself, only this last possibility remains for $\overline{\mathbb{Z} + q\mathbb{Z}}$. □

In Example 4, §3.1, we encountered a parametrization of a torus of revolution by means of $S^1 \times S^1$. The following pictures show the trace of a curve γ_q in this parametrization for two close rational and irrational values of q, respectively. In Example 8, §7.4, we show that the motion of a spherical pendulum can be parametrized precisely by such a curve on a torus.

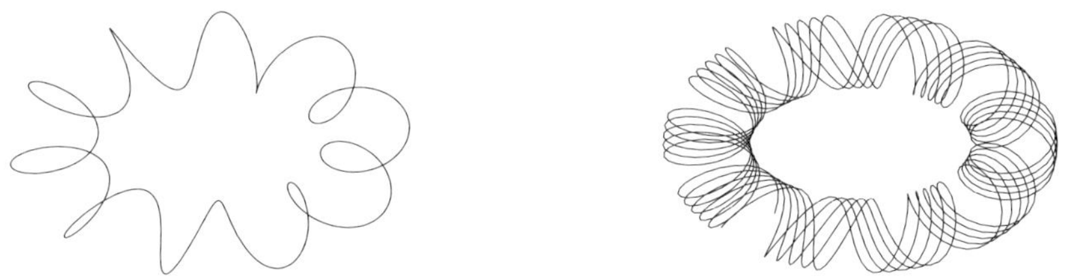

The main objective of this section is to prove that a *closed* subgroup H of a Lie group G is a submanifold of it. Consequently, it is itself a Lie group, and the quotient space G/H carries the structure of a manifold with a smooth G-action. As a preparation, we need a few technical lemmas. Let $|| - ||$ be any norm on the vector space $\mathfrak{g}$.

Lemma 3. *Let H be a closed subgroup of G, and let $X_n \neq 0$ form a sequence converging to zero in $\mathfrak{g}$ such that $\exp(X_n)$ belongs to H and $X_n/||X_n||$ tends to an element $X \in \mathfrak{g}$. Then*

$$\exp(tX) \in H \quad \textit{for all } t \in \mathbb{R} .$$

Proof. For a fixed $t > 0$ we define a sequence of natural numbers m_n by

$$m_n := \max\{k \in \mathbb{N} : \; k \cdot ||X_n|| < t\} .$$

Then the following estimate holds:

$$m_n||X_n|| \; < \; t \; \leq \; (m_n + 1)||X_n|| \; = \; m_n||X_n|| + ||X_n|| .$$

But on the right-hand side, the sequence $||X_n||$ tends to zero, and hence

$$\lim_{n\to\infty} m_n||X_n|| \; = \; t .$$

This implies

$$\lim_{n\to\infty} m_n X_n \;=\; \lim_{n\to\infty} m_n ||X_n|| \cdot \frac{X_n}{||X_n||} \;=\; t \cdot X\,,$$

and, since the exponential map is continuous,

$$\lim_{n\to\infty} \exp(m_n X_n) \;=\; \exp(t \cdot X)\,.$$

Each term in the sequence $\exp(m_n X_n) = [\exp(X_n)]^{m_n}$ is an element of H. As H is closed by assumption, the limit of this sequence, $\exp(t \cdot X)$, has to lie in H, too. The proof for the case $t < 0$ proceeds along the same lines. □

Lemma 4. *For every closed subgroup $H \subset G$, the set*

$$\mathfrak{h} \;:=\; \{X \in \mathfrak{g} : \; \exp(tX) \in H \textit{ for all } t \in \mathbb{R}\}$$

is a linear subspace of $\mathfrak{g}$.

Proof. For any vector $X \in \mathfrak{h}$, its scalar multiples $s \cdot X$ also belong to $\mathfrak{h}$. It thus suffices to show that $\mathfrak{h}$ is closed under addition. To this end, let X, Y be elements of $\mathfrak{h}$, and suppose that $X + Y \neq 0$. In any case, the product $\exp(tX)\exp(tY)$ belongs to the subgroup H, and for sufficiently small t we have, by Lemma 1,

$$\exp(tX)\exp(tY) \;=\; \exp(t(X+Y) + O(t^2))\,.$$

Then $Z(t) := O(t^2)/t$ apparently converges to zero for $t \to 0$, and we can rewrite the preceding equation as

$$\exp(tX)\exp(tY) \;=\; \exp\big(t(X + Y + Z(t))\big) \in H\,.$$

Choose a sequence of positive numbers that converges to zero, $t_n \to 0$, and define $X_n := t_n(X + Y + Z(t_n))$. Each term $\exp(X_n)$ lies in H, and

$$\lim_{n\to\infty} \frac{X_n}{||X_n||} \;=\; \lim_{n\to\infty} \frac{X + Y + Z(t_n)}{||X + Y + Z(t_n)||} \;=\; \frac{X+Y}{||X+Y||}\,.$$

Obviously, we have $X_n \neq 0$ and $X_n \to 0$. Thus, Lemma 3 applies, and we can conclude that for every $t \in \mathbb{R}$

$$\exp\left(t \cdot \frac{X+Y}{||X+Y||}\right)$$

is an element of H. Hence $X + Y$ belongs to $\mathfrak{h}$. □

Suppose that H is, in addition, a closed subgroup of G, and let $\mathfrak{h}$ be defined as in the preceding lemma. We then choose any linear complement $\mathfrak{h}'$ of $\mathfrak{h}$ in $\mathfrak{g}$,

$$\mathfrak{g} \;=\; \mathfrak{h} + \mathfrak{h}'\,.$$

Lemma 5. *There exists a neighborhood $V' \subset \mathfrak{h}'$ of 0 such that, for every $X' \neq 0$ in V', the element $\exp(X')$ does not belong to H.*

Proof. If the assertion were false, there would exist a sequence $X'_n \in \mathfrak{h}'$ converging to zero and satisfying $\exp(X'_n) \in H$. Now consider the compact set

$$K := \{X' \in \mathfrak{h}' : 1 \leq ||X'|| \leq 2\}$$

and choose natural numbers p_n such that $p_n X'_n \in K$. Since K is compact, we may assume that the sequence $p_n X'_n$ converges to some $0 \neq X' \in K$. Again, $\exp(p_n X'_n) = [\exp(X'_n)]^{p_n}$ is an element of H and

$$\lim_{n\to\infty} \frac{p_n X'_n}{||p_n X'_n||} = \frac{X'}{||X'||}.$$

Then Lemma 3 implies that $X'/||X'|| \in \mathfrak{h}$, contradicting $0 \neq X' \in \mathfrak{h}'$. □

Now we can turn to the main theorem of this section.

Theorem 7. *Let G be a Lie group, and let H be a closed subgroup. Then:*

(1) *H is a submanifold of G and thus itself a Lie group.*

(2) *There exists precisely one differential structure on G/H such that*
   (a) *the projection $\pi : G \to G/H$ is smooth,*
   (b) *for every $p \in G/H$ there exist a neighborhood $W_p \subset G/H$ of p and a smooth map $\varphi : W_p \to G$ such that $\pi \circ \varphi = \mathrm{Id}_{W_p}$,*
   (c) *the action of G on G/H defined by $(g, kH) \to gkH$ is smooth.*

Proof. It obviously suffices to show that there exists a neighborhood $W \subset G$ of e for which $H \cap W$ is a submanifold (left translation is a diffeomorphism of G). As before, ddecompose the Lie algebra into $\mathfrak{g} = \mathfrak{h} + \mathfrak{h}'$ and consider the map corresponding to this decomposition,

$$\Phi : \mathfrak{g} = \mathfrak{h} + \mathfrak{h}' \longrightarrow G, \quad \Phi(X + X') = \exp(X) \cdot \exp(X').$$

In $\mathfrak{h}'$ we choose a neighborhood $V' \subset \mathfrak{h}'$ as in Lemma 5, as well as a subset $V \subset \mathfrak{h}$ so small that the exponential map still is a diffeomorphism on $V + V'$. The image W of $V + V'$ under Φ is an open neighborhood of $e \in G$, and

$$H \cap W = \Phi(V + \{0\}),$$

by the definition of $\mathfrak{h}$ and Lemma 5. The set $H \cap W$ is thus parametrized by the chart $\left(V, \Phi\big|_{V+\{0\}}\right)$, and hence a submanifold of G.

Now we turn to the proof of the second assertion. Let $\pi : G \to G/H$ denote the projection. We define a topology on G/H by the condition

$$A \subset G/H \text{ is open } :\Leftrightarrow \pi^{-1}(A) \subset G \text{ is open}.$$

It is called the *quotient topology* on G/H, and it is designed to render the map π continuous. Endowed with this topology, G/H is a Hausdorff space (see Duistermaat/Kolk, Lemma 1.11.3). To verify the properties a manifold has to satisfy, consider the distinguished point $x_0 := e \cdot H \in G/H$ together with the sets V, V' introduced in the first part of the proof. The map

$$\psi : V' \longrightarrow G/H, \quad X' \longmapsto \pi(\exp X'),$$

is continuous and maps V' onto an open neighborhood U of x_0. Moreover, ψ is injective, since $\psi(X') = \psi(Y')$ implies the existence of an element $h \in H$ such that $\exp(X') = \exp(Y') \cdot h$. Hence

$$h \;=\; \exp(X')\exp(-Y') \;=\; \Phi\big(0 + (X' - Y')\big).$$

Thus h also belongs to the set W, which was defined as the image of $V + V'$ under Φ. Since we already proved $H \cap W = \Phi(V + \{0\})$, this implies $X' = Y'$. In summary, the map $\psi : V' \to U$ is continuous and bijective. For an arbitrary point $gH \in G/H$, consider the left translation by $g \in G$ on G/H,

$$\tilde{L}_g : G/H \longrightarrow G/H, \quad kH \longmapsto gkH,$$

and introduce a chart around $gH \in G/H$ by

$$U_{gH} := \tilde{L}_g(U), \quad \psi_{gH} : V' \longrightarrow U_{gH}, \quad \psi_{gH} := \tilde{L}_g \circ \psi.$$

For two points gH and kH, the chart transition can be rewritten as follows,

$$\begin{aligned}\psi_{kH}^{-1} \circ \psi_{gH} &= \psi^{-1} \circ \tilde{L}_{k^{-1}} \circ \tilde{L}_g \circ \psi = \exp^{-1} \circ (\pi^{-1} \circ \tilde{L}_{k^{-1}} \circ \tilde{L}_g \circ \pi) \circ \exp \\ &= \exp^{-1} \circ L_{k^{-1}h} \circ \exp .\end{aligned}$$

Therefore, as a superposition of smooth maps, the chart transition is also smooth. Hence we have proved that G/H is a differentiable manifold, the projection $\pi : G \to G/H$ is smooth, and G acts smoothly from the left on G/H. It remains to show (b). For the distinguished coset $p = x_0 = e \cdot H$, define φ for each x in $U =: W_p$ by

$$\varphi(x) \;=\; \exp(\psi^{-1}(x)) \;=\; \pi^{-1}(x).$$

For an arbitrary point $p = gH$ one again uses the left translation $\tilde{L}_g$. □

Definition 5. The action of a Lie group G on a manifold M is called *transitive* if, for two arbitrarily given points x and y in M, the one can always be written as the image of the other under the action of G, i.e., there exists a $g \in G$ such that $y = g \cdot x$. An equivalent formulation of this requirement is to say that M consists of a single G-orbit, $G \cdot x = M$. A manifold together with a transitive group action is also called a *homogeneous space*.

Obviously, the left translation on the quotient $M = G/H$ is a transitive group action, and thus G/H is a homogeneous space. Theorem 7 can be applied to show that some well-known matrix groups are Lie groups: The following groups apparently are closed subgroups of $\mathrm{GL}(n, \mathbb{K})$.

Example 8. The subgroup of $\mathrm{GL}(n, \mathbb{K})$ consisting of all matrices with determinant 1 is a Lie group, the *special linear group*,

$$\mathrm{SL}(n, \mathbb{K}) \;:=\; \{A \in \mathcal{M}_n(\mathbb{K}) : \; \det A = 1\}\,.$$

Example 9. Let $\mathbb{H} = \left\{ \begin{bmatrix} u & v \\ -\bar{v} & \bar{u} \end{bmatrix} =: h \;:\; u, v \in \mathbb{C} \right\}$ be the vector space of Hamilton's quaternions with standard basis

$$E \;=\; \begin{bmatrix} 1 & 0 \\ 0 & 1 \end{bmatrix}, \quad I \;=\; \begin{bmatrix} i & 0 \\ 0 & -i \end{bmatrix}, \quad J \;=\; \begin{bmatrix} 0 & 1 \\ -1 & 0 \end{bmatrix}, \quad K \;=\; \begin{bmatrix} 0 & i \\ i & 0 \end{bmatrix}$$

and norm $N(h) := u\bar{u} + v\bar{v}$. The group of all quaternions with norm 1 is isomorphic to the Lie group

$$\mathrm{SU}(2) \;:=\; \{A \in \mathrm{GL}(n, \mathbb{C}) : A\bar{A}^t = 1_2 \text{ and } \det(A) = 1\}\,.$$

Example 10. The preceding example can be generalized as follows. The *unitary group* is embedded into the space of complex matrices as

$$\mathrm{U}(n) \;:=\; \{A \in \mathcal{M}_n(\mathbb{C}) : \; A\bar{A}^t = 1_n\}\,.$$

The condition $A\bar{A}^t = 1_n$ immediately implies $|\det A| = 1$, hence $\det A \in S^1$; the *special unitary group* is defined as the group of all unitary matrices A satisfying $\det A = 1$:

$$\mathrm{SU}(n) \;:=\; \{A \in \mathrm{U}(n) : \; \det A = 1\}\,.$$

Example 11. The *orthogonal group* $\mathrm{O}(n, \mathbb{K})$ consists of the matrices $A \in \mathcal{M}_n(\mathbb{K})$ leaving the euclidean standard scalar product of $\mathbb{K}^n$ invariant,

$$\langle Ax, Ay \rangle \;=\; \langle x, y \rangle\;.$$

Realizing the scalar product as $\langle x, y \rangle = x^t y$, we see that this condition is equivalent to $AA^t = 1_n$. Hence we obtain

$$\mathrm{O}(n, \mathbb{K}) \;=\; \{A \in \mathcal{M}_n(\mathbb{K}) : \; AA^t = 1_n\}\,.$$

Obviously, an orthogonal matrix has determinant $+1$ or -1. The subgroup of all orthogonal matrices with determinant $+1$ is called the *special orthogonal group* $\mathrm{SO}(n, \mathbb{K})$,

$$\mathrm{SO}(n, \mathbb{K}) \;=\; \{A \in \mathcal{M}_n(\mathbb{K}) : \; AA^t = 1_n \text{ and } \det A = 1\}\,.$$

6.3. The Adjoint Representation

Definition 6. Let G be a Lie group with Lie algebra $\mathfrak{g}$, and let V be a finite-dimensional vector space.

(1) A *representation of the Lie group G on V* is a smooth group homomorphism $\varrho : G \to \mathrm{GL}(V)$, i.e., a smooth map compatible with the group structure,

$$\varrho(g \cdot h) \;=\; \varrho(g) \cdot \varrho(h) .$$

(2) A *representation of the Lie algebra $\mathfrak{g}$ on V* is a homomorphism of Lie algebras, $\varrho : \mathfrak{g} \to \mathfrak{gl}(V)$, i.e., a linear map compatible with the commutator,

$$\varrho([X,Y]) \;=\; [\varrho(X), \varrho(Y)] \;=\; \varrho(X) \cdot \varrho(Y) - \varrho(Y), \varrho(X) .$$

Sometimes, V is then also called a *G-module* or a *$\mathfrak{g}$-module*, respectively.

Example 12. The *trivial* representation of a Lie group G is the group homomorphism that maps every element $g \in G$ to the neutral element in $\mathrm{GL}(V)$: $\varrho(g) = 1_V$; the *trivial* representation of $\mathfrak{g}$ associates the zero map with every element $\mathcal{X}$, $\varrho(\mathcal{X}) = 0_V$.

Example 13. Matrix groups are defined by means of one of their representations, often called the *defining representation.* In fact, we introduced the groups $\mathrm{GL}(n,\mathbb{R})$, $\mathrm{SL}(n,\mathbb{R})$ and $\mathrm{SO}(n,\mathbb{R})$ in a way endowing them naturally with a representation on $\mathbb{R}^n$. A simple example illustrates that these matrix groups and their Lie algebras have many more representations. The Lie algebra $\mathfrak{sl}(2,\mathbb{R})$, for example, has representations in *all* dimensions: For every natural number n, define $\varrho : \mathfrak{sl}(2,\mathbb{R}) \to \mathfrak{gl}(n+1,\mathbb{R})$ by

$$\varrho(H) \;=\; \mathrm{diag}\big(n,\, n-2,\, \ldots,\, -(n-2), -n\big),$$

$$\varrho(E) \;=\; \begin{bmatrix} 0 & 1 & & & \\ & 0 & 2 & & \\ & & \ddots & \ddots & \\ & & & 0 & n \end{bmatrix}, \quad \varrho(F) \;=\; \begin{bmatrix} 0 & & & & \\ n & 0 & & & \\ & n-1 & 0 & & \\ & & \ddots & \ddots & \\ & & & 1 & 0 \end{bmatrix}.$$

These matrices satisfy the commutator relations of $\mathfrak{sl}(2,\mathbb{R})$,

$$[\varrho(H), \varrho(E)] = 2\varrho(E), \quad [\varrho(H), \varrho(F)] = -2\varrho(F), \quad [\varrho(E), \varrho(F)] = \varrho(H),$$

and hence form an $(n+1)$-dimensional representation of $\mathfrak{sl}(2,\mathbb{R})$. Properties that cannot be expressed by the Lie bracket do not have to be preserved under a representation: For example, we have $E^2 = 0$, but $\varrho(E)^2 \neq 0$. Nevertheless, the property that $\varrho(E)$ is a nilpotent matrix is preserved. There is also a a representation of the Lie group $\mathrm{SL}(2,\mathbb{R})$ corresponding to this representation of the Lie algebra; this will be the subject of Exercise 4.

Apart from left and right translation, there is a third remarkable action of a given Lie group G on itself, the so-called *conjugation action*,

$$\alpha_g : G \longrightarrow G, \quad \alpha_g(h) := ghg^{-1} \;=\; L_g R_{g^{-1}} h .$$

It is smooth and satisfies $\alpha_g(e) = e$, and in contrast to left and right translation, it is far from being transitive. In the case $G = \mathrm{GL}(V)$, it decomposes the invertible matrices precisely into their similarity classes. In addition, the relation $\alpha_g(e) = e$ implies that its differential at e is a map from $\mathfrak{g}$ to $\mathfrak{g}$,

$$d(\alpha_g)_e : \ T_eG \cong \mathfrak{g} \longrightarrow T_eG \cong \mathfrak{g},$$

which is obviously invertible, since $d(L_g)_e$ and $d(R_{g^{-1}})_e$ are invertible. We define the *adjoint representation* of G on $\mathfrak{g}$ by

$$\mathrm{Ad} : \ G \longrightarrow \mathrm{GL}(\mathfrak{g}), \quad \mathrm{Ad}(g) \ = \ d(\alpha_g)_e \in \mathrm{GL}(\mathfrak{g}) \,.$$

Before we verify that this actually is a representation, recall the definition of the *center* of a group. It consists of those elements which commute with all the others:

$$\mathcal{Z}_G \ = \ \{g \in G : \ gh = hg \ \forall h \in G\} \,.$$

Theorem 8. *The map* $\mathrm{Ad} : G \to \mathrm{GL}(\mathfrak{g})$ *is a representation of* G *on the vector space* $\mathfrak{g}$. *The center* $\mathcal{Z}_G$ *of* G *is contained in its kernel,*

$$\mathcal{Z}_G \subset \ker \mathrm{Ad},$$

and equality holds if and only if G *is connected.*

Proof. First we check the homomorphism property:

$$\begin{aligned} \mathrm{Ad}(gh) \ &= \ d(L_{gh}R_{gh}^{-1})_e \ = \ d(L_gL_hR_{h^{-1}}R_{g^{-1}})_e \ = \ d(\alpha_g\alpha_h)_e \\ &= \ d(\alpha_g)_e d(\alpha_h)_e \ = \ \mathrm{Ad}(g)\mathrm{Ad}(h) \,. \end{aligned}$$

Let z belong to the center $\mathcal{Z}_G$. Then $\alpha_z = \mathrm{Id}_G$, and hence $\mathrm{Ad}(z) \ = \ \mathrm{Id}_{\mathrm{GL}(\mathfrak{g})}$, i. e., z is in the kernel of Ad. Now suppose that G is connected and that $\mathrm{Ad}(g) \ = \ \mathrm{Id}_{\mathfrak{g}}$. Since $\alpha_g : G \to G$ is a group homomorphism, the map $t \mapsto \alpha_g(\exp tX)$ is a one-parameter subgroup for every $X \in \mathfrak{g}$, and, by Theorem 5, there exists an element $Y \in \mathfrak{g}$ such that

$$\exp(tY) \ = \ \alpha_g(\exp tX) \,.$$

Differentiating this equation with respect to t, we obtain

$$Y(e) \ = \ \frac{d}{dt}\big(\alpha_g(\exp tX)\big)\big|_{t=0} \ = \ \mathrm{Ad}(g)\big(X(e)\big) \ = \ X(e) \,,$$

which proves $X = Y$. For the one-parameter group defined above this means that

$$\exp(tX) \ = \ \alpha_g(\exp tX)$$

for all $t \in \mathbb{R}$ and $X \in \mathfrak{g}$. The exponential map is a local diffeomorphism $\mathfrak{g} \to G$; hence $\alpha_g = \mathrm{Id}_G$ on an open neighborhood W of e. For a connected

Lie group G this implies $\alpha_g = \mathrm{Id}_G$, since G can be represented as the union of all powers of W (with respect to the group products),

$$G = \bigcup_{i=1}^{\infty} W^i .$$

As $\alpha_g = \mathrm{Id}_G$ is equivalent to $g \in \mathcal{Z}_G$, everything is proved. □

The differential of the adjoint representation of G (in the sense of Definition 4) is a representation of the Lie algebra $\mathfrak{g}$ which can now be expressed by the commutator.

Theorem 9. *The differential* $\mathrm{ad} := \mathrm{Ad}_* : \mathfrak{g} \longrightarrow \mathfrak{gl}(\mathfrak{g})$ *of the adjoint representation is a homomorphism of Lie algebras determined by the formula*

$$\mathrm{ad}(X)(Y) = [X, Y].$$

Proof. By the definition of the differential, we have

$$\mathrm{Ad}(\exp tX) = \exp(t\mathrm{Ad}_*(X)) = 1 + t\mathrm{Ad}_*(X) + \ldots ;$$

hence

$$\mathrm{Ad}_*(X)(Y) = \lim_{t \to 0} \frac{\mathrm{Ad}(\exp tX)(Y) - Y}{t} .$$

The flow corresponding to the vector field $-X$ is $\Phi_t = R_{\exp(-tX)}$, since

$$\left.\frac{d\Phi_t(e)}{dt}\right|_{t=0} = \left.\frac{d\exp(-tX)}{dt}\right|_{t=0} = -X .$$

Applied to a left-invariant vector field Y, however, its differential coincides with $\mathrm{Ad}(\exp tX)(Y)$,

$$\mathrm{Ad}(\exp tX)(Y) = dL_{\exp(tX)} dR_{\exp(-tX)}(Y) = dR_{\exp(-tX)}(Y) = d\Phi_t(Y) .$$

Thus the original identity can be rewritten as

$$\mathrm{Ad}_*(X)(Y) = \lim_{t \to 0} \frac{\Phi_t(Y) - Y}{t} .$$

The right-hand side is precisely the definition of the commutator $[Y, -X] = [X, Y]$. □

This representation will also be called the *adjoint representation* (this time of the Lie algebra $\mathfrak{g}$). In case of doubt, the context has to decide whether the representation of the Lie group or that of its Lie algebra is meant.

Remark. The definition of the differential immediately implies the identity

$$\mathrm{Ad}(\exp X) = \exp(\mathrm{ad} X) .$$

This has to be understood as an identity of operators. Applied to an element Y, it means

$$\begin{aligned}\exp(X)\cdot Y\cdot\exp(-X) &= 1+\mathrm{ad}(X)(Y)+\frac{\mathrm{ad}(X)^2(Y)}{2!}+\frac{\mathrm{ad}(X)^3(Y)}{3!}+\dots\\ &= 1+[X,Y]+\frac{[X,[X,Y]]}{2!}+\frac{[X,[X,[X,Y]]]}{3!}+\dots.\end{aligned}$$

Example 14. Let $\mathfrak{g}$ be the three-dimensional Lie algebra which is abstractly defined by the following commutator relations for a basis e_1, e_2, e_3:

$$[e_1,e_2] \;=\; e_3, \quad [e_2,e_3] \;=\; e_1, \quad [e_3,e_1] \;=\; e_2.$$

The representing matrices of the adjoint representation with respect to this basis can be computed from them. For the operator $\mathrm{ad}(e_1)$, we obtain

$$\mathrm{ad}(e_1)\begin{bmatrix}e_1\\e_2\\e_3\end{bmatrix} = \begin{bmatrix}0\\e_3\\-e_2\end{bmatrix} = \begin{bmatrix}0&0&0\\0&0&1\\0&-1&0\end{bmatrix}\begin{bmatrix}e_1\\e_2\\e_3\end{bmatrix} =: L_1\begin{bmatrix}e_1\\e_2\\e_3\end{bmatrix},$$

and similarly for the other two operators,

$$L_2 := \mathrm{ad}(e_2) = \begin{bmatrix}0&0&-1\\0&0&0\\1&0&0\end{bmatrix}, \quad L_3 := \mathrm{ad}(e_3) = \begin{bmatrix}0&1&0\\-1&0&0\\0&0&0\end{bmatrix}.$$

Let us look, on the other hand, more closely at the orthogonal group $\mathrm{O}(n,\mathbb{R})$. It was defined as the set of matrices satisfying $f(A) = AA^t - 1_n = 0$. The differential of this map at the point X is

$$df(A)_X \;=\; AX^t + XA^t,$$

and hence, according to Theorem 5 in §3.2, the Lie algebra of $\mathrm{O}(n,\mathbb{R})$ is

$$\mathfrak{o}(n,\mathbb{R}) \;=\; T_e\,\mathrm{O}(n,\mathbb{R}) \;=\; \{A\in\mathcal{M}_n(\mathbb{R}):\; A+A^t=0\}.$$

The Lie algebra of the orthogonal group consists precisely of the skew-symmetric matrices, which for $n=3$ is apparently spanned by L_1, L_2 and L_3. This proves that the three-dimensional defining representation of $\mathfrak{o}(3,\mathbb{R})$ is isomorphic to the adjoint representation. In higher dimensions, this fact no longer holds as a simple dimensional consideration shows: A skew-symmetric matrix has exactly as many degrees of freedom as entries above the diagonal. Therefore,

$$\dim\mathfrak{o}(n,\mathbb{R}) \;=\; \frac{n^2}{2}-n \;=\; \frac{n(n-1)}{2},$$

and this is equal to n only for $n=3$.

Exercises

1. The exponential map $\exp : \mathfrak{sl}(2,\mathbb{R}) \to \mathrm{SL}(2,\mathbb{R})$ is not surjective. Hint: What are the values that $\mathrm{tr}\,\exp(A)$ can attain for $A \in \mathfrak{sl}(2,\mathbb{R})$?

2. Hamilton's quaternions $\mathbb{H}$ (Example 9) form not only a vector space, but also a (non-abelian) division algebra, i.e., an associative algebra in which each non-trivial element is invertible. Prove that the standard basis E, I, J, K of Hamilton's quaternions obeys the following algebra relations:

$$I \cdot J = K, \quad I \cdot K = -J, \quad J \cdot K = I\,,$$

and compute the inverse of the quaternion $h = x_0{\cdot}E + x_1{\cdot}I + x_2{\cdot}J + x_3{\cdot}K \neq 0$.

3. Identify the quaternions of trace 0,

$$\mathbb{H}_0 \;=\; \{x_1 \cdot I + x_2 \cdot J + x_3 \cdot K \mid x_1, x_2, x_3 \in \mathbb{R}\},$$

with the 3-dimensional euclidean space $\mathbb{R}^3$. Prove that, for every $U \in \mathrm{SU}(2)$, the map

$$\mathbb{H}_0 \longrightarrow \mathbb{H}_0, \quad x \longmapsto UxU^{-1}\,,$$

defines a special orthogonal transformation of $\mathbb{R}^3$. The resulting map $\varrho : \mathrm{SU}(2) \to \mathrm{SO}(3,\mathbb{R})$ is a representation which is not injective ("faithful").

4. The defining representation of $\mathrm{SL}(2,\mathbb{R})$ on $\mathbb{R}^2$ is the usual matrix action on vectors,

$$g \cdot v \;=\; \begin{bmatrix} a & b \\ c & d \end{bmatrix} \begin{bmatrix} x \\ y \end{bmatrix} \;=\; \begin{bmatrix} ax + by \\ cx + dy \end{bmatrix}.$$

Let V_n be the $(n+1)$-dimensional vector space of homogeneous polynomials of degree n in the variables x and y. Define an action of $g \in \mathrm{SL}(2,\mathbb{R})$ on the polynomial $p \in V_n$ by

$$\varrho(g) \cdot p\left(\begin{bmatrix} x \\ y \end{bmatrix}\right) \;=\; p\left(g^{-1} \cdot \begin{bmatrix} x \\ y \end{bmatrix}\right).$$

Prove that ϱ is a representation of $\mathrm{SL}(2,\mathbb{R})$.

5. Let G be a Lie group, and let H be a discrete, normal subgroup of G. Prove that H is necessarily contained in the center of G.

6. Let (ϱ, V) be a representation of the group G. A subspace W of the representation space V is called *invariant* if, for every $g \in G$, the relation $\varrho(g)W \subset W$ holds. For trivial reasons, the subspaces $W = V$ and $W = \{0\}$ are invariant; if the representation has no further invariant subspaces, it is called *irreducible*. Consider the following two-dimensional representation of the additive group $\mathbb{R}$:

$$\varrho(t) = \begin{bmatrix} 1 & t \\ 0 & 1 \end{bmatrix}.$$

Prove that this representation is not irreducible. Does the invariant subspace have an invariant complement?

7. By Theorem 6, the differential of a representation (ϱ, V) of the Lie group G is a representation (ϱ_*, V) of its Lie algebra $\mathfrak{g}$. The tensor product of two representations (ϱ, V) and (μ, W) of G is defined by

$$(\varrho \otimes \mu)(g)(v \otimes w) := \varrho(g)v \otimes \mu(g)w, \quad g \in G,\, v \in V, w \in W.$$

Prove that this determines a representation of G on $V \otimes W$ with differential

$$(\varrho \otimes \mu)_*(X)(v \otimes w) := \varrho_*(X)v \otimes w + v \otimes \mu_*(X)w, \quad X \in \mathfrak{g}.$$

8. In order to describe the hyperbolic plane as a homogeneous space, it is useful to introduce a new model for it, the open unit disc.

a) Let $D = \{z \in \mathbb{C} \mid |z| < 1\}$ be the open unit disc with metric

$$g = \frac{1}{(1-|z|^2)^2} \begin{bmatrix} 1 & 0 \\ 0 & 1 \end{bmatrix}.$$

Show that the *Cayley transform*,

$$c\colon D \longrightarrow \mathcal{H}, \quad c(z) = -i\frac{z+i}{z-i} =: x + iy,$$

is an isometry between D with the metric above and the upper half-plane $\mathcal{H}^2$ with the fourfold of the hyperbolic metric.

b) Let the Lie group

$$\mathrm{SU}(1,1) := \left\{ \begin{bmatrix} a & b \\ \bar{b} & \bar{a} \end{bmatrix} : |a|^2 - |b|^2 = 1 \right\}$$

act on D via the formula

$$\begin{bmatrix} a & b \\ \bar{b} & \bar{a} \end{bmatrix} \cdot z = \frac{az+b}{\bar{a}z+\bar{b}}.$$

Prove that this action is transitive, and that the isotropy group of zero,

$$G_0 := \{g \in \mathrm{SU}(1,1) : g \cdot 0 = 0\},$$

is isomorphic to $\mathrm{SO}(2,\mathbb{R})$. Hence $D \cong \mathrm{SU}(1,1)/\mathrm{SO}(2,\mathbb{R})$.

Chapter 7

Symplectic Geometry and Mechanics

7.1. Symplectic Manifolds

Riemannian geometry is the geometry of a *symmetric*, bilinear form depending on the point of a manifold. The curvature is a measure of how far two symmetric bilinear forms differ locally. Contrary to this, symplectic geometry is that of an *antisymmetric* bilinear form depending on the point of a manifold—hence the geometry of a 2-form ω. It turns out that all symplectic manifolds are locally equivalent; there cannot be any concept similar to curvature in the sense of Riemannian geometry. Symplectic structures differ, if at all, only globally. Historically, the formulation of mechanics in the sense of *Hamilton* led to symplectic geometry, hence its essential role in modern mathematical physics.

Definition 1. A *symplectic manifold* is a pair (M^{2m}, ω) consisting of a manifold M^{2m} of even dimension together with a closed non-degenerate 2-form ω,

$$d\omega = 0 \quad \text{and} \quad \omega^m := \omega \wedge \ldots \wedge \omega \neq 0,$$

called the *symplectic form* or *symplectic structure.*

By Theorem 16 in §3.4, every symplectic manifold is orientable. The *volume form* is understood to be the $2m$-form

$$dM^{2m} = \frac{(-1)^{m(m-1)/2}}{m!} \cdot \omega^m .$$

Example 1. In $\mathbb{R}^{2m}$ with coordinates $\{q_1, \ldots, q_m, p_1, \ldots, p_m\}$, the formula

$$\omega := \sum_{i=1}^{m} dp_i \wedge dq_i$$

defines a symplectic form with highest power

$$\omega^m = m! \cdot (-1)^{m(m-1)/2} \cdot dp_1 \wedge \cdots \wedge dp_m \wedge dq_1 \wedge \ldots \wedge dq_m .$$

The volume form in the sense of symplectic geometry is the ordinary volume form of $\mathbb{R}^{2m}$. This symplectic structure is called the *canonical symplectic structure.*

Example 2. Define a 1-form θ—the so-called *Liouville form*—in the cotangent bundle T^*X^m of an arbitrary m-dimensional manifold as follows: Let $\mathcal{V} \in T_\eta(T^*X^m)$ be a tangent vector at $\eta \in T^*X^m$ and represent it by a curve $\mathcal{V} : (-\epsilon, \epsilon) \to T^*X^m$ such that

$$\mathcal{V}(0) = \eta, \quad \dot{\mathcal{V}}(0) = \mathcal{V}.$$

Project this curve first by means of the projection $\pi : T^*X^m \to X^m$ to the manifold, and, after that, apply the 1-form η to the tangent vector of the projected curve:

$$\theta(\mathcal{V}) := \eta\Big(\frac{d}{dt}(\pi \circ \mathcal{V}(t))\big|_{t=0}\Big) .$$

The 2-form $\omega := d\theta$ is a symplectic structure on T^*X^m. Any system, $\{q_1, \ldots, q_m\}$, of coordinates in X^m determines—representing a 1-form η as $\eta = \sum p_i \cdot dq_i$—coordinates $\{q_1, \ldots, q_m, p_1, \ldots, p_m\}$ in T^*X^m. By the definition of the Liouville form θ, we have

$$\theta = \sum_{i=1}^{m} p_i \cdot dq_i ,$$

and the 2-form

$$\omega = d\theta = \sum_{i=1}^{m} dp_i \wedge dq_i$$

is non-degenerate. In particular, the (co-)tangent bundle of every manifold is an orientable manifold (see Exercise 9 in Chapter 3).

Further examples of symplectic manifolds arise as the orbits of the coadjoint representation of a Lie group G. Starting from the adjoint representation, $\mathrm{Ad} : G \to GL(\mathfrak{g})$, of the group G and passing to the dual of the linear operator, $\mathrm{Ad}^*(g) := \big(\mathrm{Ad}(g^{-1})\big)^* : \mathfrak{g}^* \to \mathfrak{g}^*$, we obtain a representation

$$\mathrm{Ad}^* : G \longrightarrow GL(\mathfrak{g}^*)$$

of the group G in the dual space $\mathfrak{g}^*$ of the vector space $\mathfrak{g}$. Through each functional $F \in \mathfrak{g}^*$ passes an orbit

$$\mathcal{O}^*(F) \ := \ \{\mathrm{Ad}^*(g)F \ : \ g \in G\}\,,$$

on which the group G acts transitively. The *isotropy group*

$$G_F \ := \ \{g \in G \ : \ \mathrm{Ad}^*(g)F = F\}$$

is a closed subgroup of G, and $\mathcal{O}^*(F)$ is diffeomorphic to the homogeneous space G/G_F. Its Lie algebra can be characterized by a similar condition:

Theorem 1. *The Lie algebra $\mathfrak{g}_F \subset \mathfrak{g}$ of the isotropy group $G_F \subset G$ is equal to*

$$\mathfrak{g}_F \ = \ \{X \in \mathfrak{g} \ : F([X,Y]) = 0 \quad \textit{for all} \quad Y \in \mathfrak{g}\}\,.$$

Proof. Suppose that $F([X,Y]) = 0$ holds for all elements $Y \in \mathfrak{g}$. Then from

$$\begin{aligned}\big(\mathrm{Ad}^*(\exp(t\,X))F\big)(Y) \ &= \ F\big(\mathrm{Ad}(\exp(-\,t\,X))Y\big) = F\big(\exp(-\,t\,\mathrm{ad}(X))Y\big) \\ &= \ F\big(Y - \frac{t}{1!}[X,Y] + \frac{t^2}{2!}[X,[X,Y]] - \ldots\big) = F(Y)\end{aligned}$$

we immediately obtain $\mathrm{Ad}^*(\exp(t\cdot X))F = F$. The one-parameter group $\exp(t\cdot X)$ is a subgroup of G_F, and hence its tangent X belongs to the Lie algebra $\mathfrak{g}_F$. This proves one inclusion, the converse is proved analogously. □

If a Lie group G acts smoothly on a manifold M^m, we can associate with each element $X \in \mathfrak{g}$ of the Lie algebra the unique vector field $\tilde{X}$ on M^m whose integral curves coincide with the trajectories of the one-parameter transformation group $\exp(t\cdot X)$:

$$\tilde{X}(x) \ := \ \frac{d}{dt}\big(\exp(-tX)\cdot x\big)\big|_{t=0}\,.$$

The vector field $\tilde{X}$ is called the *fundamental vector field* corresponding to the element $X \in \mathfrak{g}$ of the Lie algebra. If G acts transitively on M^m, every tangent vector $\mathcal{V} \in T_x M^m$ at a fixed point $x \in M^m$ can be realized by a fundamental vector field.

This general construction will now be applied to an orbit $\mathcal{O}^*$ of the coadjoint representation. First, realize a given vector $\mathcal{V} \in T_F\mathcal{O}^*$ as the value of a fundamental vector field, $\tilde{X}(F) = \mathcal{V}$. For a further element $Y \in \mathfrak{g}$ of the Lie algebra such that $\tilde{Y}(F) = \mathcal{V}$, the equality $(\widetilde{X-Y})(F) = 0$ immediately implies

$$\frac{d}{dt}\Big(\mathrm{Ad}^*\big(\exp(t\cdot(Y-X))\big)F\Big)\Big|_{t=0} \ = \ 0\,.$$

Theorem 1 implies $X - Y \in \mathfrak{g}_F$, and the resulting map is injective,

$$\mathfrak{g}/\mathfrak{g}_F \ni X \longmapsto \tilde{X}(F) \in T_F\mathcal{O}^* .$$

For dimensional reasons, it is bijective: the tangent space $T_F\mathcal{O}^*$ to an orbit $\mathcal{O}^*$ at $F \in \mathcal{O}^*$ can be identified with the vector space $\mathfrak{g}/\mathfrak{g}_F$. We now define a symplectic structure $\omega_{\mathcal{O}^*}$ on each orbit $\mathcal{O}^* \subset \mathfrak{g}^*$.

Definition 2. Let $\mathcal{V}, \mathcal{W} \in T_F\mathcal{O}^*$ be two tangent vectors to the orbit at $F \in \mathcal{O}^*$, and choose elements $X, Y \in \mathfrak{g}$ with $\tilde{X}(F) = \mathcal{V}, \tilde{Y}(F) = \mathcal{W}$. The value of the *Kirillov form* $\omega_{\mathcal{O}^*}$ on the vectors $\mathcal{V}, \mathcal{W}$ is determined by the formula

$$\omega_{\mathcal{O}^*}(\mathcal{V}, \mathcal{W}) \ := \ F([X, Y]) .$$

Theorem 2. *The pair $(\mathcal{O}^*, \omega_{\mathcal{O}^*})$ is a symplectic manifold, and the 2-form $\omega_{\mathcal{O}^*}$ is G-invariant.*

Proof. Note first that the 2-form $\omega_{\mathcal{O}^*}$ is uniquely defined. If the elements $X, X_1 \in \mathfrak{g}$ realize the vector $\mathcal{V}$ at F, then the difference $X - X_1$ belongs to the Lie algebra $\mathfrak{g}_F$, and Theorem 1 implies

$$F([X, Y]) \ = \ F([X - X_1, Y]) + F([X_1, Y]) \ = \ F([X_1, Y]) .$$

Moreover, $\omega_{\mathcal{O}^*}$ is a non-degenerate 2-form. If, in fact, $\omega_{\mathcal{O}^*}(\mathcal{V}, \mathcal{W}) = 0$ for every tangent vector $\mathcal{W} \in T_F\mathcal{O}^*$, then we obtain $F([X, Y]) = 0$ for all elements $Y \in \mathfrak{g}$. By Theorem 1, X lies in the Lie algebra $\mathfrak{g}_F$, and hence $\mathcal{V} = \tilde{X}(F) = 0$. It remains to show that $\omega_{\mathcal{O}^*}$ is a closed form. For two elements $X, Y \in \mathfrak{g}$, the function $\omega_{\mathcal{O}^*}(\tilde{X}, \tilde{Y}) : \mathcal{O}^* \to \mathbb{R}$ is determined by the formula

$$\omega_{\mathcal{O}^*}(\tilde{X}, \tilde{Y})(F) \ = \ F([X, Y]) .$$

Differentiate this relation in the direction of a third fundamental vector field:

$$\begin{aligned} \tilde{Z}\big(\omega_{\mathcal{O}^*}(\tilde{X}, \tilde{Y})\big)(F) &= \frac{d}{dt}\big[\mathrm{Ad}^*(\exp(-t \cdot Z)F)[X, Y]\big]\big|_{t=0} \\ &= \frac{d}{dt} F\big(\mathrm{Ad}(\exp(tZ))([X, Y])\big)\big|_{t=0} \ = \ F([Z, [X, Y]]) . \end{aligned}$$

Then the expression for the exterior derivative $d\omega_{\mathcal{O}^*}$ of the 2-form vanishes identically:

$$\begin{aligned} d\omega_{\mathcal{O}^*}(\tilde{X}, \tilde{Y}, \tilde{Z}) &= \tilde{X}(\omega_{\mathcal{O}^*}(\tilde{Y}, \tilde{Z})) - \tilde{Y}(\omega_{\mathcal{O}^*}(\tilde{X}, \tilde{Z})) + \tilde{Z}(\omega_{\mathcal{O}^*}(\tilde{X}, \tilde{Y})) \\ &\quad - \omega_{\mathcal{O}^*}([\tilde{X}, \tilde{Y}], \tilde{Z}) + \omega_{\mathcal{O}^*}([\tilde{X}, \tilde{Z}], \tilde{Y}) - \omega_{\mathcal{O}^*}([\tilde{Y}, \tilde{Z}], \tilde{X}) , \end{aligned}$$

since it reduces to the Jacobi identity of the Lie algebra $\mathfrak{g}$. □

Corollary 1. *Each orbit $\mathcal{O}^* \subset \mathfrak{g}^*$ of the coadjoint representation of a Lie group is a manifold of even dimension.*

Example 3. The affine group of $\mathbb{R}$ has the matrix representation

$$G = \left\{ \begin{bmatrix} a & b \\ 0 & 1 \end{bmatrix} : a > 0 \, , \, b \in \mathbb{R} \right\}$$

with Lie algebra

$$\mathfrak{g} = \left\{ \begin{bmatrix} x & y \\ 0 & 0 \end{bmatrix} : x, y \in \mathbb{R} \right\} .$$

The computation

$$\mathrm{Ad} \begin{bmatrix} a & b \\ 0 & 1 \end{bmatrix} \begin{bmatrix} x & y \\ 0 & 0 \end{bmatrix} = \begin{bmatrix} a & b \\ 0 & 1 \end{bmatrix} \cdot \begin{bmatrix} x & y \\ 0 & 0 \end{bmatrix} \cdot \begin{bmatrix} 1/a & -b/a \\ 0 & 1 \end{bmatrix} = \begin{bmatrix} x & ay - bx \\ 0 & 0 \end{bmatrix}$$

implies that $\mathfrak{g}$ has one-dimensional orbits. To determine the orbits of the coadjoint representation, write any element of $\mathfrak{g}^*$ as a pair (α, β) of real numbers, whose evaluation on the element $(x, y) \in \mathfrak{g}$ is

$$\langle (\alpha, \beta), \, (x, y) \rangle = \alpha x + \beta y \, .$$

By definition, the group element $g = \begin{bmatrix} a & b \\ 0 & 1 \end{bmatrix}$ acts as follows:

$$\begin{aligned} \left\langle \mathrm{Ad}^*(g^{-1})(\alpha, \beta), \, (x, y) \right\rangle &= \langle (\alpha, \beta), \, \mathrm{Ad}(g)(x, y) \rangle = \langle (\alpha, \beta), \, (x, ay - bx) \rangle \\ &= \alpha x + \beta(ay - bx) = \langle (\alpha - \beta b, \beta a), (x, y) \rangle \, . \end{aligned}$$

Summarizing, we have $\mathrm{Ad}^*(g^{-1})(\alpha, \beta) = (\alpha - \beta b, \beta a)$, and hence for $\beta \neq 0$ the coadjoint orbit through (α, β) is two-dimensional. This example shows that the adjoint and the coadjoint representation of a group G are, in general, not equivalent.

After having discussed examples of symplectic manifolds, we now want to introduce the symplectic gradient, which is the analogue of the gradient of a function on a Riemannian manifold. In this situation, we will make use of the fact that the non-degenerate 2-form ω also provides a linear bijection between the tangent bundle TM^{2m} and the cotangent bundle T^*M^{2m}.

Definition 3. Let $H : M^{2m} \to \mathbb{R}$ be a smooth function on a symplectic manifold. The *symplectic gradient* s-grad(H) is the vector field on M^{2m} defined by

$$\omega(\mathcal{V}, \text{s-grad}(H)) := dH(\mathcal{V}) \, .$$

Example 4. Let $\{q_1, \ldots, q_m, p_1, \ldots, p_m\}$ be coordinates on M^{2m} such that the symplectic form ω can be written as $\omega = \sum_i dp_i \wedge dq_i$. Then

$$\text{s-grad}(H) = \sum_{i=1}^{m} \left(\frac{\partial H}{\partial p_i} \cdot \frac{\partial}{\partial q_i} - \frac{\partial H}{\partial q_i} \cdot \frac{\partial}{\partial p_i} \right) .$$

This formula immediately follows from the equation defining the symplectic gradient. A curve $\gamma(t)$ in the symplectic manifold M^{2m}, represented in the fixed coordinates $\gamma(t) = \{q_1(t), \ldots, q_m(t), p_1(t), \ldots, p_m(t)\}$, is thus an integral curve of the vector field s-grad(H) if and only if the so-called *Hamilton equations* hold:

$$\dot{q}_i = \frac{\partial H}{\partial p_i} \quad \text{and} \quad \dot{p}_i = -\frac{\partial H}{\partial q_i} .$$

Theorem 3 (Liouville's Theorem). *Let H be a function on a symplectic manifold (M^{2m}, ω), and suppose that* s-grad(H) *is a complete vector field with flow $\Phi_t : M^{2m} \to M^{2m}$. Then:*

(1) *The Lie derivative of ω vanishes,*

$$\mathcal{L}_{\text{s-grad}(H)}(\omega) = 0 .$$

(2) *The flow Φ_t preserves the symplectic volume,*

$$\int_A dM^{2m} = \int_{\Phi_t(A)} dM^{2m} .$$

Proof. From $d\omega = 0$ and Theorem 32 in §3.9, we conclude that

$$\mathcal{L}_{\text{s-grad}(H)}(\omega) = d(\text{s-grad}(H) \lrcorner\, \omega) = -d\,d\,H = 0 .$$

Hence the diffeomorphisms $\Phi_t : M^{2m} \to M^{2m}$ do not alter the symplectic structure, i. e., $\Phi_t^*(\omega) = \omega$, and so both assertions are proved. □

The existence of this invariant measure has consequences for the dynamics of symplectic gradient fields.

Theorem 4 (Poincaré's Return Theorem). *Let (M^{2m}, ω) be a symplectic manifold of finite volume, and let $\Phi_t : M^{2m} \to M^{2m}$ be the flow of the symplectic gradient of a smooth function H. For any set $A \subset M^{2m}$ of positive measure, the set*

$$B = \{x \in A : \ \Phi_n(x) \notin A \quad \textit{for all} \quad n = 1, 2, \ldots\}$$

has measure zero.

Proof. Note first that the intersections $\Phi_{-n}(B) \cap B$ are empty. Any point $x \in \Phi_{-n}(B) \cap B$ would be a point $x \in B$ such that $\Phi_n(x) \in B \subset A$, contradicting the definition of the set B. This immediately also implies that the intersections $\Phi_{-n}(B) \cap \Phi_{-m}(B)$ are empty for $n \neq m$, and, from the invariance of the measure, we obtain

$$\sum_{n=1}^{\infty} \text{vol}(B) = \sum_{n=1}^{\infty} \text{vol}(\Phi_n(B)) < \text{vol}(M^{2m}) < \infty ,$$

i. e., the measure of the set B has to vanish. □

This result has several famous generalizations, as the only example of which we quote *Birkhoff's ergodicity theorem* (without proof).

Theorem 5 (Birkhoff's Ergodicity Theorem). *Let f be an integrable function on the symplectic manifold (M^{2m}, ω), and let Φ_t be the flow of a symplectic gradient. Then the following limit exists almost everywhere:*

$$\lim_{t\to\infty} \frac{1}{t} \int_0^t f \circ \Phi_t(x) \;=:\; f^*(x)\,.$$

Furthermore, the function f^* is also integrable, and its integral coincides with that of f. Lastly, f^* is invariant under the flow Φ_t.

Definition 4. The *Poisson bracket* of two functions f and g on a symplectic manifold is the function

$$\{f, g\} \;=\; \omega\big(\text{s-grad}(f), \text{s-grad}(g)\big) \;=\; dg\big(\text{s-grad}(f)\big) \;=\; -df\big(\text{s-grad}(g)\big)\,.$$

Example 5. In the $\{q, p\}$-coordinates, we have

$$\{f, g\} \;=\; \sum_{i=1}^m \left(\frac{\partial f}{\partial p_i}\frac{\partial g}{\partial q_i} - \frac{\partial f}{\partial q_i}\frac{\partial g}{\partial p_i}\right).$$

In the next theorem, we summarize the properties of the Poisson bracket.

Theorem 6. *The ring $C^\infty(M^{2m})$ endowed with the Poisson bracket is a Lie-Poisson algebra:*

(1) $\{f, c_1\cdot g_1 + c_2\cdot g_2\} = c_1\cdot\{f, g_1\} + c_2\cdot\{f, g_2\}$ *for constants* $c_1, c_2 \in \mathbb{R}$;

(2) $\{f, g\} = -\{g, f\}$;

(3) $\{f, \{g, h\}\} + \{g, \{h, f\}\} + \{h, \{f, g\}\} = 0$ *(Jacobi identity);*

(4) $\{f, g\cdot h\} = g\cdot\{f, h\} + h\cdot\{f\cdot g\}$;

(5) $\text{s-grad}(\{f, g\}) = [\text{s-grad}(f), \text{s-grad}(g)]$.

Proof. The two first identities result immediately from the definition of the Poisson bracket, and the fourth follows from $d(g\cdot h) = g\cdot dh + h\cdot dg$. We prove (5). We insert the vector fields $\mathcal{V} := \text{s-grad}(f), \mathcal{W} := \text{s-grad}(g)$ together with an additional vector field Y into the equation defining the 3-form $d\omega$,

$$\begin{aligned} 0 \;=\; d\omega(\mathcal{V}, \mathcal{W}, Y) \;=\; & \mathcal{V}\big(\omega(\mathcal{W}, Y)\big) - \mathcal{W}\big(\omega(\mathcal{V}, Y)\big) + Y\big(\omega(\mathcal{V}, \mathcal{W})\big) \\ & - \omega\big([\mathcal{V}, \mathcal{W}], Y\big) + \omega\big([\mathcal{V}, Y], \mathcal{W}\big) - \omega\big([\mathcal{W}, Y], \mathcal{V}\big)\,. \end{aligned}$$

Applying the definition of the symplectic gradient as well as that of the Poisson bracket, we can rewrite this equation as

$$\begin{aligned} 0 = & -\mathcal{V}\big(Y(g)\big) + \mathcal{W}\big(Y(f)\big) + Y\big(\{f, g\}\big) - \omega\big([\mathcal{V}, \mathcal{W}], Y\big) \\ & + [\mathcal{V}, Y](g) - [\mathcal{W}, Y](f) \;=\; -Y\big(\{f, g\}\big) + \omega\big(Y, [\mathcal{V}, \mathcal{W}]\big)\,. \end{aligned}$$

Hence s-grad$(\{f,g\}) = [\mathcal{V},\mathcal{W}] = [$s-grad$(f)$, s-grad$(g)]$. The Jacobi identity is a consequence of formula (5). In fact, we obtain

$$\begin{aligned}&\{f,\{g,h\}\}+\{g,\{h,f\}\}+\{h,\{f,g\}\}\\ &\quad= \text{s-grad}(f)\big(\text{s-grad}(g)(h)\big) - \text{s-grad}(g)\big(\text{s-grad}(f)(h)\big) - \text{s-grad}(\{f,g\})(h)\\ &\quad= [\text{s-grad}(f),\text{s-grad}(g)](h) - [\text{s-grad}(f),\text{s-grad}(g)](h) \;=\; 0\,. \qquad \square\end{aligned}$$

A *Hamiltonian system* consists of a symplectic manifold (M^{2m},ω,H) together with a function H. The integration of the corresponding Hamilton equation relies on determining the integral curves of the vector field s-grad(H). For this, there exists an analogous notion of first integrals as in the Riemannian case.

Definition 5. A function $f : M^{2m} \to \mathbb{R}$ is called a *first integral* of the Hamilton function if it is constant on each integral curve of the vector field s-grad(H).

Theorem 7.

(1) *A function f is a first integral of the Hamilton function H if and only if its Poisson bracket with H vanishes,*

$$\{f,H\} \;=\; 0\,.$$

(2) *The set of all first integrals of a Hamilton function is a Lie-Poisson algebra.*

Proof. We compute the derivative of a function f along an integral curve $\gamma(t)$ of s-grad(H):

$$\frac{d}{dt} f\circ\gamma(t) \;=\; df(\dot\gamma(t)) \;=\; df(\text{s-grad}(H)) \;=\; \{H,f\}\,.$$

This implies the first assertion. The second follows from the Jacobi identity for the Poisson bracket. □

7.2. The Darboux Theorem

The Darboux theorem states that all symplectic manifolds are locally equivalent.

Theorem 8 (Darboux Theorem)**.** *Near each point $x \in M^{2m}$ of a symplectic manifold (M^{2m},ω), there exists a chart $h : U \subset M^{2m} \to \mathbb{R}^{2m}$ in which the symplectic form ω is the pullback of the usual symplectic form,*

$$\omega \;=\; h^*\left(\sum_{i=1}^{m} dp_i \wedge dq_i\right).$$

Coordinates with these properties are called *symplectic (canonical) coordinates.*

Proof. In the cotangent space $T_x^* M^{2m}$ to the manifold at the point $x \in M^{2m}$, we choose a basis $\sigma_1, \ldots, \sigma_m, \mu_1, \ldots, \mu_m$ in which the symplectic form at this point is represented in normal form,

$$\omega(x) = \sum_{i=1}^{m} \sigma_i \wedge \mu_i \,.$$

Consider, moreover, a chart $\Phi : V \to \mathbb{R}^{2m}$ around x such that

$$\Phi(x) = 0 \quad \text{and} \quad \omega(x) = \Phi^* \left(\sum_{i=1}^{m} dp_i \wedge dq_i(0) \right) .$$

Denoting the corresponding symplectic form on V by

$$\omega_1 := \Phi^* \left(\sum_{i=1}^{m} dp_i \wedge dq_i \right) ,$$

we see that there exists a neighborhood $U \subset V$ of x such that for all parameters $t \in [0, 1]$ the form

$$\omega_t := (1-t)\omega + t\omega_1$$

is a symplectic structure on U. In fact, $d\omega_t = 0$, and since $\omega_t(x) = \omega(x)$ for all t, a compactness argument shows that all the forms ω_t $(t \in [0, 1])$ do not degenerate at the same time in a neighborhood of x. Since $d(\omega_1 - \omega_0) = 0$, Poincaré's lemma shows the existence of a 1-form α such that the difference

$$\omega_1 - \omega_0 = d\alpha$$

is the exterior derivative of this 1-form. By subtracting locally, if necessary, a 1-form with constant coefficients from α, we may assume that α vanishes at the point x, $\alpha(x) = 0$. Dualizing α by means of the symplectic forms ω_t, we obtain a family $\mathcal{W}_t$ of vector fields on U parametrized by t,

$$\omega_t(\mathcal{V}, \mathcal{W}_t) = \alpha(\mathcal{V}) \,.$$

Let $\varphi(y, t) \in M^{2m}$ be the solution of the (non-autonomous) differential equation

$$\dot{\varphi}(t) = \mathcal{W}_t(\varphi(t)), \quad \varphi(0) = y \,.$$

All the vector fields $\mathcal{W}_t(x) \equiv 0$ vanish at the point x, and the solution corresponding to the initial condition x is constant, $\varphi(x, t) \equiv x$. Hence there exists a neighborhood $U_1 \subset U$ of x such that, for every initial condition $y \in U_1$, the corresponding solution $\varphi(y, t)$ is at least defined in the interval

$[0,1]$. Let $\varphi_t : U_1 \to M^{2m}$ be the corresponding map. The formula for the Lie derivative of a differential form (Theorem 32, §3.9) implies

$$\begin{aligned}\frac{d}{dt}\varphi_t^*(\omega_t) &= \varphi_t^*\left(\frac{d\omega_t}{dt}\right) + \varphi_t^*\left(\mathcal{L}_{\partial\varphi_t/\partial t}(\omega_t)\right) = \varphi_t^*(\omega_1 - \omega) + \varphi_t^*(\mathcal{L}_{\mathcal{W}_t}(\omega_t)) \\ &= \varphi_t^*\big(\omega_1 - \omega + d(\mathcal{W}_t \lrcorner\, \omega_t) + \mathcal{W}_t \lrcorner\, d\omega_t\big) = \varphi_t^*(\omega_1 - \omega - d\alpha) = 0.\end{aligned}$$

Thus $\varphi_t^*(\omega_1) = \varphi_0^*(\omega) = \omega$, and $\Phi \circ \varphi_1$ is the chart we were looking for,

$$(\Phi \circ \varphi_1)^* \left(\sum_{i=1}^{m} dp_i \wedge dy_i\right) = \omega\,. \qquad \square$$

7.3. First Integrals and the Moment Map

As in the Riemannian case, some first integrals can be derived from symmetry considerations. The isometries (which are not available on symplectic manifolds) giving rise to these first integrals will be replaced by symplectic diffeomorphisms which, however, satisfy a compatibilty condition with respect to the Hamilton function under consideration. We will describe this in detail in the case of an *exact symplectic manifold*, i. e. , a symplectic manifold whose symplectic form is the exterior derivative of a 1-form, $\omega = d\theta$. Suppose that a Lie group G acts from the left on M^{2m} in such a way that each diffeomorphism $l_g : M^{2m} \to M^{2m}$ leaves the form θ invariant,

$$l_g^*(\theta) = \theta\,.$$

These diffeomorphisms l_g are then *symplectic*, i. e., they preserve the symplectic structure ω. Now let $X \in \mathfrak{g}$ be an element of the Lie algebra of the group G, and let $\tilde{X}$ be the fundamental vector field corresponding to X under the G-action on M^{2m}. The evaluation of the 1-form θ on $\tilde{X}$ is a function,

$$\Phi(X) := \theta(\tilde{X})\,.$$

This construction determines a linear map $\Phi : \mathfrak{g} \to C^\infty(M^{2m})$ from the Lie algebra $\mathfrak{g}$ to the space of functions $C^\infty(M^{2m})$ on the symplectic manifold. Its properties are the subject of the following symplectic variant of Noether's theorem.

Theorem 9 (Noether's Theorem)**.**

(1) $\Phi : \mathfrak{g} \to C^\infty(M^{2m})$ *is a homomorphism of Lie algebras,*

$$\Phi([X,Y]) = \{\Phi(X), \Phi(Y)\}\,.$$

(2) s-grad$\circ\Phi$ *corresponds to the transition to fundamental vector fields,*

$$\text{s-grad}(\Phi(X)) = \tilde{X}\,.$$

(3) *If the Hamilton function H is G-invariant, then $\Phi(X)$ is a first integral of H,*

$$\{H, \Phi(X)\} \;=\; 0\,.$$

Proof. Fix an element $X \in \mathfrak{g}$ in the Lie algebra and consider the one-parameter group of diffeomorphisms corresponding to the group elements $\exp(-t \cdot X)$. Its generating vector field is the fundamental vector field $\tilde{X}$. The relation $l_g^*(\theta) = \theta$ implies that the Lie derivative of θ with respect to $\tilde{X}$ vanishes,

$$0 \;=\; \mathcal{L}_{\tilde{X}}(\theta) \;=\; \tilde{X} \lrcorner\, d\theta + d(\tilde{X} \lrcorner\, \theta) \;=\; \tilde{X} \lrcorner\, \omega + d(\Phi(X))\,.$$

Thus, for every vector field $\mathcal{V}$, we have the equation

$$-\,\omega(\tilde{X}, \mathcal{V}) \;=\; \mathcal{V}(\Phi(X)) \;=\; \omega(\mathcal{V}, \text{s-grad}(\Phi(X)))$$

as well as $\tilde{X} = \text{s-grad}(\Phi(X))$. Using this formula, we compute the difference

$$\begin{aligned} \{\Phi(X), \Phi(Y)\} - \Phi([X,Y]) &= \omega(\tilde{X}, \tilde{Y}) - \Phi([X,Y]) \\ &= d\theta(\tilde{X}, \tilde{Y}) - \Phi([X,Y]) \\ &= \tilde{X}(\Phi(Y)) - \tilde{Y}(\Phi(X)) - 2\Phi([X,Y]) \\ &= 2\big(\{\Phi(X), \Phi(Y)\} - \Phi([X,Y])\big)\,, \end{aligned}$$

and this yields

$$\{\Phi(X), \Phi(Y)\} \;=\; \Phi([X,Y])\,.$$

If, finally, the Hamilton function is G-invariant, we obtain

$$\{H, \Phi(X)\} \;=\; -\text{s-grad}(\Phi(X))H \;=\; -\tilde{X}(H) \;=\; 0\,,$$

i.e., $\Phi(X)$ is a first integral. □

The elements of the Lie algebra $\mathfrak{g}$ provide first integrals for every G-invariant Hamilton function. These first integrals can be combined into a single vector-valued first integral by passing to the dual space $\mathfrak{g}^*$.

Definition 6. The *moment map* of a Hamiltonian system with symmetry group G is the map $\Psi : M^{2m} \to \mathfrak{g}^*$ from the symplectic manifold to the dual of the Lie algebra defined by

$$\Psi(m)(X) \;:=\; \Phi(X)(m).$$

Theorem 10.

(1) *The map Ψ is* Ad**-equivariant, i. e., the following diagram commutes:*

$$\begin{array}{ccc} M^{2m} & \xrightarrow{\;l_g\;} & M^{2m} \\ \Psi\downarrow & & \downarrow\Psi \\ \mathfrak{g}^* & \xrightarrow{\;\mathrm{Ad}^*(g)\;} & \mathfrak{g}^* \end{array}$$

(2) Ψ *is a first integral of* H.

Proof. The fundamental vector field $\tilde{X}$ of a G-action has the following invariance property:

$$\begin{aligned}\tilde{X}(l_g(x)) &= \frac{d}{dt}\big[\exp(-tX)\cdot g\cdot x\big]\big|_{t=0} \\ &= d\,l_g\left(\frac{d}{dt}\big[\exp(-t\cdot \mathrm{Ad}(g^{-1})X)\cdot x\big]\big|_{t=0}\right) \\ &= d\,l_g\big(\widetilde{\mathrm{Ad}(g^{-1})X}(x)\big)\,.\end{aligned}$$

The 1-form θ is G-invariant by assumption, and from this we obtain

$$\Psi(l_g x)X \;=\; \theta(\tilde{X}(l_g x)) \;=\; \theta(\widetilde{\mathrm{Ad}(g^{-1})X}(x)) \;=\; \Psi(x)(\mathrm{Ad}(g^{-1})X)\,. \quad \square$$

In Exercise 11, we discuss the case $M^m = T^*\mathbb{R}^3$ with symmetry group $\mathrm{SO}(3,\mathbb{R})$ and its usual representation on $\mathbb{R}^3$. In particular, it is shown that the moment map $\Psi : T^*\mathbb{R}^3 \to \mathfrak{so}(3,\mathbb{R}) \cong \mathbb{R}^3$ coincides with classical angular momentum, hence justifying its name. Closely related to this situation is the following example:

Example 6. Consider the 2-dimensional representation of $G = \mathrm{SL}(2,\mathbb{R})$ on $M^2 = V = \mathbb{R}^2$. Its cotangent bundle is $T^*M = V \times V^* \cong V \times V$, since the representation V is self-dual. A group element $g \in \mathrm{SL}(2,\mathbb{R})$ acts on an element $(p,q) \in V \times V$ of the cotangent bundle by

$$g\cdot(p,q) \;=\; (gp, gq).$$

We call two elements (p,q) and (p',q') equivalent if they lie in the same G-orbit, $(p,q)\sim(p',q')$. Let $p=(p_1,p_2)$ and $q=(q_1,q_2)$ be the components of the vectors p and q, respectively. One easily computes that the moment map is given by

$$\Psi:\; V\times V \longrightarrow \mathfrak{sl}(2,\mathbb{R}),\quad (q,p)\longmapsto \begin{bmatrix} \frac{1}{2}(q_1p_2+q_2p_1) & -q_1p_1 \\ q_2p_2 & -\frac{1}{2}(q_1p_2+q_2p_1)\end{bmatrix}.$$

In particular, this map is equivariant with respect to the adjoint action of $\mathrm{SL}(2,\mathbb{R})$ on $\mathfrak{sl}(2,\mathbb{R})$. The moment map is best studied by examining its action on $\mathrm{SL}(2,\mathbb{R})$-orbits. For this, observe that the quantity

$$\det(q,p) \;:=\; \det\begin{bmatrix} q_1 & p_1 \\ q_2 & p_2\end{bmatrix} \;=\; q_1p_2 - q_2p_1$$

is $\mathrm{SL}(2,\mathbb{R})$-invariant. It thus allows to parametrize the orbit space; we omit the easy proof here[1].

[1] For details on this $\mathrm{SL}(2,\mathbb{R})$-action, we refer to §I.4. of the book by Hanspeter Kraft, *Geometrische Methoden in der Invariantentheorie*, Vieweg, 1985.

Theorem 11.

(1) *If* $\det(q,p) =: \lambda \neq 0$, *then*

$$(q,p) \sim \left(\begin{pmatrix}1\\0\end{pmatrix}, \begin{pmatrix}0\\\lambda\end{pmatrix}\right).$$

(2) $\det(q,p) = 0$ *if and only if* p *and* q *are linearly dependent. In this case, there exist infinitely many* G*-orbits, for which one can choose the following representatives:*

$$\left(\begin{pmatrix}0\\0\end{pmatrix}, \begin{pmatrix}0\\0\end{pmatrix}\right), \quad \left(\begin{pmatrix}1\\0\end{pmatrix}, \begin{pmatrix}0\\0\end{pmatrix}\right), \quad \left(\begin{pmatrix}\mu\\0\end{pmatrix}, \begin{pmatrix}1\\0\end{pmatrix}\right) \quad \text{with } \mu \in \mathbb{R}.$$

Thus, the moment map acts on G-orbits as follows:

$$\left(\begin{pmatrix}1\\0\end{pmatrix}, \begin{pmatrix}0\\\lambda\end{pmatrix}\right) \overset{\Psi}{\longmapsto} \begin{bmatrix}\lambda/2 & 0\\ 0 & -\lambda/2\end{bmatrix}, \quad \left(\begin{pmatrix}\mu\\0\end{pmatrix}, \begin{pmatrix}1\\0\end{pmatrix}\right) \overset{\Psi}{\longmapsto} \begin{bmatrix}0 & -\mu\\ 0 & 0\end{bmatrix},$$

$$\left(\begin{pmatrix}0\\0\end{pmatrix}, \begin{pmatrix}0\\0\end{pmatrix}\right), \left(\begin{pmatrix}1\\0\end{pmatrix}, \begin{pmatrix}0\\0\end{pmatrix}\right) \overset{\Psi}{\longmapsto} \begin{bmatrix}0 & 0\\ 0 & 0\end{bmatrix}.$$

In particular, the generic orbits with parameter $\lambda \neq 0$ are mapped to semi-simple elements of the Lie algebra $\mathfrak{sl}(2,\mathbb{R})$. Their orbits are 2-dimensional closed submanifolds of $\mathfrak{sl}(2,\mathbb{R})$.

7.4. Completely Integrable Hamiltonian Systems

In this section we will make use of the following fact concerning the structure of discrete subgroups Γ of the additive group $\mathbb{R}^n$.

Theorem 12. *Let* $\Gamma \subset \mathbb{R}^n$ *be a discrete subgroup. Then there exist linearly independent vectors* $v_1, \ldots, v_k$ *such that*

$$\Gamma = \left\{ \sum_{i=1}^{k} m_i \cdot v_i \; : \; m_i \text{ an integer} \right\},$$

i. e., Γ *is the lattice generated by the vectors* $v_1, \ldots, v_k$.

Proof. If $\Gamma \neq \{0\}$ is not trivial, we choose a vector $\gamma_1 \in \Gamma$ such that $||\gamma_1|| \neq 0$ and consider the ball $D^n(0;||\gamma_1||)$. The intersection $D^n(0;||\gamma_1||) \cap \Gamma$ is a compact and discrete subset of $\mathbb{R}^n$, hence finite. Thus, on the straight line generated by γ_1, there exists a vector $\gamma_1^* \in D^n(0;||\gamma_1||) \cap \Gamma$ realizing the minimum of the distance to $0 \in \mathbb{R}^n$. For this vector we have

$$\mathbb{R} \cdot \gamma_1 \cap \Gamma = \{m \cdot \gamma_1^* \; : \; m \text{ an integer}\},$$

since any vector $x \neq m\gamma_1^*$ belonging to the intersection $(\mathbb{R} \cdot \gamma_1) \cap \Gamma$ would have to lie in one of the segments $(m \cdot \gamma_1^*, (m+1) \cdot \gamma_1^*)$, and then $(m+1)\gamma_1^* - x$ would be a vector on the line through γ_1 with a smaller distance to 0 than γ_1^*. If the group Γ contains only integer multiples of γ_1^*, the proof is completed.

Otherwise, there has to exist a vector $\gamma_2 \in \Gamma \backslash \{m \cdot \gamma_1^* : m \text{ an integer}\}$. We project γ_2 orthogonally to the straight line passing through γ_1 and denote by $\bar{\gamma}_2$ the resulting vector. It lies in one of the half-closed segments $\bar{\gamma}_2 \in [m \cdot \gamma_1^*, (m+1) \cdot \gamma_1^*)$. Let Σ be the cylinder with axis $[m\gamma_1^*, (m+1) \cdot \gamma_1^*)$ whose radius is equal to the distance from γ_2 to the line through γ_1. In this cylinder, there are again only finitely many elements of the group Γ. Let γ_2^* be the vector in $\Gamma \cap \Sigma$ whose distance to the axis of the cylinder is minimal and which is not a multiple of γ_1^*. Then we have

$$\Gamma \cap \{\mathbb{R} \cdot \gamma_1^* \oplus \mathbb{R}\gamma_2^*\} = \left\{ \sum_{i=1}^{2} m_i \, \gamma_i^* \ : \ m_i \text{ an integer} \right\}.$$

In fact, if there were a point $x \neq m_1 e_1 + m_2 e_2$ in $\Gamma \cap \{\mathbb{R}\gamma_1^* \oplus \mathbb{R}\gamma_2^*\}$, then x would belong to the interior of a parallelogram in the $\{\gamma_1^*, \gamma_2^*\}$-plane. Taking the difference with a vertex of this parallelogram, we obtain a vector in Γ lying closer than γ_2^* to the axis of the cylinder. Repeating this construction finitely many times proves the assertion. □

Corollary 2. *Let $\Gamma \subset \mathbb{R}^n$ be a discrete subgroup. Then $\mathbb{R}^n/\Gamma$ is diffeomorphic to the product of a k-dimensional torus T^k with $\mathbb{R}^{n-k}$,*

$$\mathbb{R}^n/\Gamma \approx T^k \times \mathbb{R}^{n-k}.$$

Theorem 13 (Arnold-Liouville Theorem). *Let (M^{2m}, ω, H) be a Hamiltonian system, and let $f_1 = H, f_2, \ldots, f_m$ be m functions with the following properties:*

(1) *all functions f_i are first integrals of H;*

(2) *the functions f_i commute, $\{f_i, f_j\} = 0$;*

(3) *the differentials $df_1, \ldots, df_m$ are linearly independent at each point;*

(4) *the symplectic gradients* s-grad(f_i) *are complete vector fields.*

For a given point $c = (c_1, \ldots, c_m) \in \mathbb{R}^m$, we consider the level manifold

$$M_c = \{x \in M^{2m} \ : \ f_1(x) = c_1, \ldots, f_m(x) = c_m\}.$$

Then:

a) *The connected components of M_c are diffeomorphic to $T^k \times \mathbb{R}^{m-k}$.*

b) *The vector field* s-grad(H) *is tangent to M_c. In particular, each integral curve of this vector field is completely contained in one of the level manifolds.*

c) *If M_c is compact and connected, angle coordinates $\varphi_1, \ldots, \varphi_m$ can be introduced in $M_c \approx T^m$ so that the integral curves of* s-grad(H)

are described by the system of differential equations

$$\dot{\varphi}_i = \nu_i, \quad \nu_i = \textit{constant.}$$

Proof. Consider the flows $\Phi_t^1, \ldots, \Phi_t^m : M^{2m} \to M^{2m}$ of the symplectic gradients s-grad(f_i). Since

$$0 = \text{s-grad}\{f_i, f_j\} = [\text{s-grad}(f_i), \text{s-grad}(f_j)],$$

all these flows commute with one another (Theorem 36, §3.9). This determines an action of the additive group $\mathbb{R}^m$ on the manifold M^{2m}:

$$(t_1, \ldots, t_m) \cdot x = \Phi_{t_1}^1 \circ \ldots \Phi_{t_m}^m(x)\,.$$

The orbits of this $\mathbb{R}^m$-action coincide with the connected components of the level manifolds. In fact, since

$$0 = \{f_i, f_j\} = \text{s-grad}(f_i)(f_j),$$

each function f_j is constant on every orbit. Hence, the orbits of the $\mathbb{R}^m$-action are contained in the level manifolds. On the other hand, both are m-dimensional submanifolds of M^{2m}, since the differentials $df_1, \ldots, df_m$ are linearly independent. The isotropy group $\Gamma(x_0) = \{t \in \mathbb{R}^m : t \cdot x_0 = x_0\}$ of a point $x_0 \in M^{2m}$ for the $\mathbb{R}^m$-action is discrete. Hence each component of a level manifold is diffeomorphic to the product of a torus and euclidean space:

$$\mathbb{R}^m/\Gamma(x_0) = T^k \times \mathbb{R}^{m-k}\,.$$

Assertions a) and b) are proved, so we turn to the remaining one. Suppose that a level manifold M_c is compact and connected. Choose a point $x_0 \in M_c$ and denote by $v_1, \ldots, v_m \in \mathbb{R}^m$ a basis of the isotropy group $\Gamma(x_0)$. Representing the basis vectors $\{v_i\}$ of the vector space $\mathbb{R}^m$ as linear combinations of the vectors in the standard basis $e_1, \ldots, e_m$ of $\mathbb{R}^m$,

$$v_i = \sum_{\alpha=1}^{m} a_{i\alpha} e_\alpha\,,$$

we obtain a quadratic matrix $\mathcal{A} := (a_{ij})$. Let

$$\Gamma(m) = \left\{ \sum_{i=1}^{m} n_i \cdot e_i \;:\; n_i \text{ an integer} \right\}$$

denote the standard integral lattice in $\mathbb{R}^m$. Then

$$\Phi : \mathbb{R}^m/\Gamma(m) \longrightarrow \mathbb{R}^m/\Gamma(x_0) \xrightarrow{\sim} M_c, \quad \Phi\Big(\sum_{i=1}^{m} x^i \cdot e_i\Big) = \sum_{i=1}^{m} x^i \cdot v_i\,,$$

defines a diffeomorphism. The inverse of this map,

$$\Phi^{-1} : M_c \longrightarrow \mathbb{R}^m/\Gamma(m) = S^1 \times \ldots \times S^1,$$

as well as its components, $\Phi^{-1} = (\varphi_1, \ldots, \varphi_m)$, lead to the angle coordinates for the level manifold M_c. In fact, if $y^1, \ldots, y^m$ are the coordinates in $\mathbb{R}^m/\Gamma(x_0) \cong M_c$ determined by

$$(y^1, \ldots, y^m) \longrightarrow y^1 \cdot v_1 + \ldots + y^m \cdot v_m\,,$$

then, by the construction of the $\mathbb{R}^m$-action on M_c,

$$\text{s-grad}(f_i) \;=\; \frac{\partial}{\partial y^i}\,.$$

With respect to the $\{\varphi_1, \ldots, \varphi_m\}$-coordinates, this yields

$$\text{s-grad}(f_i) \;=\; \sum_{\alpha=1}^{m} a_{i\alpha} \frac{\partial}{\partial \varphi_\alpha}\,.$$

In particular, the symplectic gradient s-grad(H) is a vector field with constant coefficients on the torus $M_c = T^m$, and the third assertion results by taking $\nu_j := a_{1j}$. □

We want to discuss more closely how the angle coordinates $(\varphi_1, \ldots, \varphi_m)$ of a compact and connected level manifold can be determined directly from the commuting first integrals. This will lead to an explicit algorithm for the integration of a Hamiltonian system (M^{2m}, ω, H) provided with sufficiently many commuting first integrals. Because of this procedure, these systems are called *completely integrable* (or *integrable by quadrature*). Denote by $\omega_1(c), \ldots, \omega_m(c)$ the frame of 1-forms on M_c dual to the vector fields s-grad$(f_1), \ldots,$ s-grad(f_m). The representation of the vector fields s-grad(f_i) in terms of the vector fields $\partial/\partial\varphi_j$ immediately implies the following formula for the differentials:

$$d\varphi_i \;=\; \sum_{\alpha=1}^{m} a_{i\alpha} \cdot \omega_\alpha(c)\,.$$

Let γ_k be the closed curve in M_c corresponding to the parameter values $\varphi_1 = \ldots = \varphi_{k-1} = \varphi_{k+1} = \ldots = \varphi_m = 0$. Then

$$\delta_{ik} \;=\; \int_{\gamma_k} d\varphi_i \;=\; \sum_{\alpha=1}^{m} a_{i\alpha} \int_{\gamma_k} \omega_\alpha(c)\,.$$

Hence, first the coefficients a_{ij} and then the angle coordinates can be computed directly from the first integrals. We summarize this in the form of an algorithm comprising five steps.

Step 1. Fix $c = (c_1, \ldots, c_m) \in \mathbb{R}^m$ and let M_c be compact and connected. Choose a homology basis $\gamma_1, \ldots, \gamma_m$ for the first homology group $H_1(M_c; \mathbb{Z})$.

Step 2. Compute the symplectic gradients s-grad(f_i) of the first integrals $f_1 = H, f_2, \dots, f_m$.

Step 3. Determine the frame of 1-forms $\omega_1(c), \dots, \omega_m(c)$ dual to the frame of vector fields s-grad$(f_1), \dots,$ s-grad(f_m) on M_c.

Step 4. Compute the line integrals $\int_{\gamma_k} \omega_\alpha(c)$, and invert the resulting $(m \times m)$ matrix. This yields the matrix $\mathcal{A} = (a_{ij}(c))$.

Step 5. Compute the angle coordinates $\varphi_i(c)$ on the level set M_c from the equations

$$d\varphi_i = \sum_{\alpha=1}^{m} a_{i\alpha}(c) \cdot \omega_\alpha(c), \quad 1 \le i \le m.$$

This procedure computes the angle coordinates $\varphi_i(c)$ on *one* level manifold M_c. Note that, according to Step 5, these are only determined up to constants. As we vary the parameters $c = (c_1, \dots, c_m)$, the $\varphi_1, \dots, \varphi_m$ become functions on an open neighborhood of a level manifold $M_c \subset M^{2m}$. Since the symplectic gradients are tangent to M_c, the Poisson bracket with the original functions $f_1, \dots, f_m$ is computed by

$$\{\varphi_i, f_j\} = d\varphi_i(\text{s-grad} f_j) = a_{ij}(f_1, \dots, f_m).$$

Moreover, it is a function exclusively depending on $f_1, \dots, f_m$. Similarly, we prove that the functions $\{\varphi_i, \varphi_j\}$ are also constant on the level sets.

Lemma 1. *The Poisson brackets* $\{\varphi_i, \varphi_j\} = b_{ij}(f_1, \dots, f_m)$ *are functions depending only on* $f_1, \dots, f_m$. *In particular, they are constant on each level set* M_c.

Proof. We compute the derivative of $\{\varphi_i, \varphi_j\}$ with respect to the vector field s-grad(f_k):

$$\begin{aligned} \text{s-grad}(f_k)\{\varphi_i, \varphi_j\} &= \{\{\varphi_i, \varphi_j\}, f_k\} = -\{\{\varphi_j, f_k\}, \varphi_i\} - \{\{f_k, \varphi_i\}, \varphi_j\} \\ &= -\{a_{jk}(f_1, \dots, f_m), \varphi_i\} + \{a_{ik}(f_1, \dots, f_m), \varphi_j\} \\ &= -\sum_{\alpha=1}^{m} \frac{\partial a_{jk}}{\partial y^\alpha}\{f_\alpha, \varphi_i\} + \sum_{\alpha=1}^{m} \frac{\partial a_{ik}}{\partial y^\alpha}\{f_\alpha, \varphi_j\} \\ &= \sum_{\alpha=1}^{m} \left(\frac{\partial a_{jk}}{\partial y^\alpha} a_{i\alpha} - \frac{\partial a_{ik}}{\partial y^\alpha} a_{j\alpha} \right). \end{aligned}$$

Thus all the derivatives s-grad$(f_k)\{\varphi_i, \varphi_j\}$ $(1 \le k \le m)$ are constant on every torus $T^m = M_c$. But then the Poisson bracket $\{\varphi_i, \varphi_j\}$ itself is constant on every level manifold M_c. □

Now we alter the angle coordinates, which up to now were considered only on a single level manifold, by a suitable constant on adjacent level manifolds. The aim of choosing these constants of integration for the angle coordinates is to obtain functions φ_i commuting on M^{2m}.

Lemma 2. *There exist functions $B_1(y^1, \dots, y^m), \dots, B_m(y^1, \dots, y^m)$ such that the angle coordinates*

$$\varphi_i^* := \varphi_i + B_i(f_1, \dots, f_m)$$

commute on the symplectic manifold M^{2m}, $\{\varphi_i^*, \varphi_j^*\} = 0$.

Proof. Using the notation $\{\varphi_i, f_j\} = a_{ij}$ and $\{\varphi_i, \varphi_j\} = b_{ij}$, we apply the Jacobi identity to the triples $(\varphi_i, \varphi_j, f_k)$ and $(\varphi_i, \varphi_j, \varphi_k)$, and take into account the fact that the Poisson brackets $\{f_i, f_j\}$ vanish. Thus we obtain the relations

$$\sum_{\alpha=1}^{m}\left[\frac{\partial a_{ik}}{\partial y^\alpha}\cdot a_{j\alpha} - \frac{\partial a_{jk}}{\partial y^\alpha}\cdot a_{i\alpha}\right] = 0 \,,$$

$$\sum_{\alpha=1}^{m}\left[\frac{\partial b_{ij}}{\partial y^\alpha}\cdot a_{k\alpha} + \frac{\partial b_{jk}}{\partial y^\alpha}\cdot a_{i\alpha} + \frac{\partial b_{ki}}{\partial y^\alpha}\cdot a_{j\alpha}\right] = 0 \,.$$

Inserting also the coefficients a^{ij} of the inverse of the matrix $\mathcal{A} = (a_{ij})$, we consider the 2-form

$$\Omega := \sum_{i,j=1}^{m}\sum_{\alpha,\beta=1}^{m} b_{ij}a^{i\alpha}a^{j\beta}\cdot dy^\alpha \wedge dy^\beta \,.$$

The above relations say that Ω is a closed 2-form. In fact, the first relation means that the 1-forms

$$\sigma_i := \sum_{\alpha=1}^{m} a^{i\alpha}\cdot dy^\alpha$$

are closed, $d\sigma_i = 0$. Hence we can choose coordinates $z^1, \dots, z^m$ such that $\sigma_i = dz^i$, and Ω becomes

$$\Omega = \sum_{i,j=1}^{m} b_{ij}\cdot dz^i \wedge dz^j \,.$$

Since

$$\frac{\partial b_{ij}}{\partial z^k} = \sum_{\alpha=1}^{m}\frac{\partial b_{ij}}{\partial y^\alpha}\cdot\frac{\partial y^\alpha}{\partial z^k} = \sum_{\alpha=1}^{m}\frac{\partial b_{ij}}{\partial y^\alpha}\cdot a_{k\alpha} \,,$$

the second relation then precisely expresses the vanishing of the exterior derivative $d\Omega = 0$. The angle coordinates we set out to find are now taken to have the form

$$\varphi_i^* := \varphi_i + \sum_{\alpha=1}^{m} a_{i\alpha}B_\alpha^*$$

with functions $B_1^*, \ldots, B_m^*$ depending only on $f_1, \ldots, f_m$. Then

$$\{\varphi_i^*, \varphi_j^*\} = b_{ij} + \sum_{\alpha,\beta=1}^{m} \left[\frac{\partial a_{i\alpha}}{\partial y^\beta} a_{j\beta} - \frac{\partial a_{j\alpha}}{\partial y^\beta} a_{i\beta}\right] B_\alpha^*$$

$$+ \sum_{\alpha,\beta=1}^{m} \frac{\partial B_\alpha^*}{\partial y^\beta} (a_{i\alpha}a_{j\beta} - a_{j\alpha}a_{i\beta}) .$$

Taking into account the first of the relations above, the condition $\{\varphi_i^*, \varphi_j^*\} = 0$ turns out to be equivalent to

$$b_{ij} = \sum_{\alpha,\beta=1}^{m} \frac{\partial B_\alpha^*}{\partial y^\beta} \cdot (a_{i\beta}a_{j\alpha} - a_{i\alpha}a_{j\beta}) .$$

This system of differential equations can be reformulated as

$$\frac{\partial B_\alpha^*}{\partial y^\beta} - \frac{\partial B_\beta^*}{\partial y^\alpha} = \sum_{i,j=1}^{m} b_{ij} a^{i\alpha} a^{j\beta},$$

and, by Poincaré's lemma, it has a solution, since the 2-form Ω considered above is closed. □

Thus we can determine the constants of integration occurring in the transition from the differentials

$$d\varphi_i(c) = \sum_{\alpha=1}^{m} a_{i\alpha}(c) \cdot \omega_\alpha(c)$$

to the angle coordinates on the individual level manifolds in such a way that the functions φ_i, defined in a neighborhood of a level manifold, commute on M^{2m}. Now we add so-called action coordinates $J_1, \ldots, J_m$ and thereby bring the symplectic structure near a level manifold into normal form. The Hamilton function $H = f_1$ is itself constant on the level manifolds and only depends on the action variables, $H = H(J_1, \ldots, J_m)$.

Theorem 14 (Action and Angle Coordinates). *In a neighborhood of any compact, connected level manifold $M_c \subset M^{2m}$ of a Hamiltonian system determined by m commuting integrals $f_1, \ldots, f_m$, there exist angle coordinates $\varphi_1, \ldots, \varphi_m$ and first integrals $J_1, \ldots, J_m$ such that*

$$\omega = \sum_{i=1}^{m} d\varphi_i \wedge dJ_i .$$

In particular, this implies $\{\varphi_i, \varphi_j\} = 0 = \{J_i, J_j\}$ and $\{\varphi_i, J_j\} = \delta_{ij}$.

Proof. First, we determine the angle coordinates near the compact, connected level manifold M_c such that

$$\{\varphi_j, \varphi_j\} = 0 \quad \text{and} \quad \{\varphi_i, f_j\} = a_{ij}(f_1, \ldots, f_m) .$$

We look for the functions $J_1, \ldots, J_m$ using the Ansatz $J_i = A_i(f_1, \ldots, f_m)$ and compute the Poisson bracket

$$\{\varphi_i, J_j\} = \sum_{\alpha=1}^{m} \frac{\partial A_j}{\partial y^\alpha} a_{i\alpha} .$$

The condition $\{\varphi_i, J_j\} = \delta_{ij}$ leads to the system of equations

$$\frac{\partial A_i}{\partial y^j} = a^{ij} ,$$

where a^{ij} is the inverse of the matrix a_{ij}. By Poincaré's lemma, the integrability condition is

$$\frac{\partial a^{ij}}{\partial y^k} = \frac{\partial a^{ik}}{\partial y^j} .$$

On the other hand, we obtain from the Jacobi identity

$$0 = \{\varphi_k, \{\varphi_i, f_j\}\} + \{\varphi_i, \{f_j, \varphi_k\}\} + \{f_j, \{\varphi_k, \varphi_i\}\} ,$$

and, taking into account $\{\varphi_i, \varphi_k\} = 0$, this immediately yields

$$\sum_{\alpha=1}^{m} \frac{\partial a_{ij}}{\partial y^\alpha} a_{k\alpha} = \sum_{\alpha=1}^{m} \frac{\partial a_{kj}}{\partial y^\alpha} a_{i\alpha} .$$

A simple computation shows that this relation is equivalent to the integrability condition for the coefficients a^{ij} of the inverse matrix. □

Example 7 (Two-dimensional Hamiltonian System)**.** Consider in $\mathbb{R}^2$ with the symplectic structure $\omega = dp \wedge dq$ a Hamilton function $H(q,p)$ for which the level curves

$$M_c = \{(q,p) \in \mathbb{R}^2 : \quad H(q,p) = c\}$$

are closed. The action variable $J = J(H)$ is a function of H, and, together with the angle coordinate φ, we have

$$dp \wedge dq = d\varphi \wedge dJ = J'(H) \cdot d\varphi \wedge dH .$$

Applying the Hodge operator $*$ of $\mathbb{R}^2$, we see that $d\varphi$ is proportional to the 1-form $*dH$,

$$d\varphi = -\frac{1}{J'(H)} \frac{1}{\|dH\|^2} * dH .$$

The integral of $d\varphi$ over each level curve is an integer, which we take to be equal to -1. This condition is called the classical *Bohr-Sommerfeld condition.* The equation

$$J'(c) = \int_{M_c} \frac{1}{\|dH\|^2} * dH$$

uniquely determines the action variable $J = J(H)$ in terms of the Hamilton function H. Consider the domain bounded by the level curve M_c,

$$\Omega_c = \{(q,p) \in \mathbb{R}^2 : \quad H(q,p) \leq c\}.$$

The vector field $\mathcal{W} := \operatorname{grad}(H)/\|\operatorname{grad}(H)\|^2$ satisfies $\mathcal{W}(H) \equiv 1$; hence its flow Φ_t maps the set Ω_c onto Ω_{c+t}. We compute the change of the area of the domain:

$$\begin{aligned}\frac{d}{dc}\big(\operatorname{vol}(\Omega_c)\big) &= \lim_{t\to 0}\frac{1}{t}\left(\int_{\Omega_c}\Phi_t^*(d\mathbb{R}^2) - \int_{\Omega_c} d\mathbb{R}^2\right) = \int_{\Omega_c}\mathcal{L}_{\mathcal{W}}(d\mathbb{R}^2)\\ &= \int_{\Omega_c} d(\mathcal{W} \lrcorner\, d\mathbb{R}^2) = \int_{M_c}\mathcal{W} \lrcorner\, d\mathbb{R}^2 = \int_{M_c}\frac{1}{\|dH\|^2} * dH = J'(c)\,.\end{aligned}$$

Thus the action variable $J = J(H)$ can be interpreted as follows: $J(c)$ *is the volume* $\operatorname{vol}(\Omega_c)$ *of the domain bounded by the level curve* $M_c = \{(q,p) \in \mathbb{R}^2 : H(q,p) = c\}$.

Example 8 (Spherical Pendulum)**.** Consider spherical coordinates on the sphere $S^2 \backslash \{N, S\}$ with the north and south pole deleted,

$$h(\varphi, \psi) = (\cos\varphi\cos\psi, \sin\varphi\cos\psi, \sin\psi)\,.$$

The Riemannian metric of the sphere is then described by the matrix (see Example 14 in §3.2)

$$g = \begin{bmatrix} 1 & 0 \\ 0 & \cos^2\psi \end{bmatrix}.$$

Denote the coordinates in the cotangent bundle by $(\varphi, \psi, p_\varphi, p_\psi)$ and consider the Hamilton function

$$H = \frac{p_\psi^2}{2} + \frac{p_\varphi^2}{2\cos^2\psi} + (1 + \sin\psi)\,.$$

H describes the motion of a pendulum of length one suspended in the center of the sphere. The meaning of the angles φ and ψ can be seen in the picture to follow. For simplicity, the gravitational constant was taken to be one.

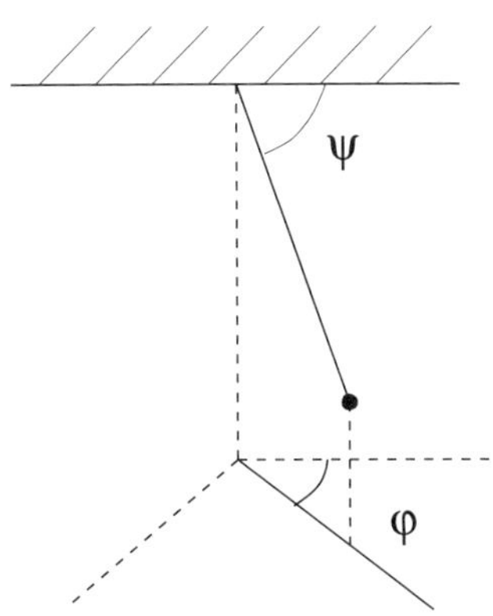

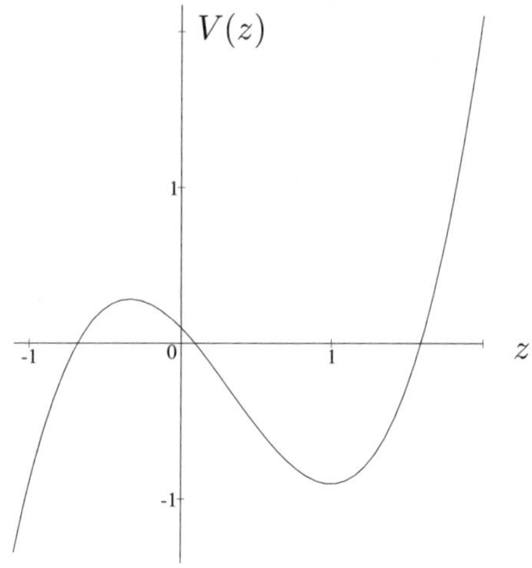

The variable φ does not explicitly occur in the Hamilton function; hence $P := p_\varphi = \dot{\varphi} \cos^2 \psi$ is a first integral, $\{H, P\} = 0$. The Hamiltonian system (T^*S^2, H) is thus completely integrable. The level manifold

$$M^2(c_1, c_2) \;:=\; \{(\varphi, \psi, p_\varphi, p_\psi) \in T^*S^2 \,:\, P = c_1,\ H = c_2\}$$

is empty for negative values of the parameter c_2, and it consists of the south pole remaining at rest in the case $c_2 = 0$. Hence we suppose that the parameter is positive, $c_2 > 0$. The equations describing the manifold $M^2(c_1, c_2)$ are

$$c_1 \;=\; p_\varphi, \quad c_2 \;=\; \frac{p_\psi^2}{2} + \frac{c_1^2}{2\cos^2\psi} + 1 + \sin\psi\,.$$

The relation $c_1 = 0$ implies that φ has to be constant. In this case, the second equation of motion reduces to that of a planar pendulum, so we will henceforth exclude this case. Depending upon the sign of c_1, the function φ is monotone increasing or decreasing, i. e., the pendulum does not change its direction of motion. Rewriting the second equation and setting $z = \sin\psi$, we obtain

$$\frac{p_\psi^2}{2} \;=\; c_2 - 1 - z - \frac{c_1^2}{2(1-z^2)} \;=:\; U(z)$$

and see that the function $U(z)$ thus defined cannot be negative. The limiting case $p_\psi = 0$ corresponds to the pendulum moving in a fixed plane, hence on a meridian. To see where the function $U(z)$ can be strictly positive, we multiply it by the denominator $1-z^2$ and look for the zeroes of the resulting cubic polynomial,

$$V(z) \;=\; (1-z^2)U(z) \;=\; (c_2 - 1 - z)(1-z^2) - \frac{c_1^2}{2}.$$

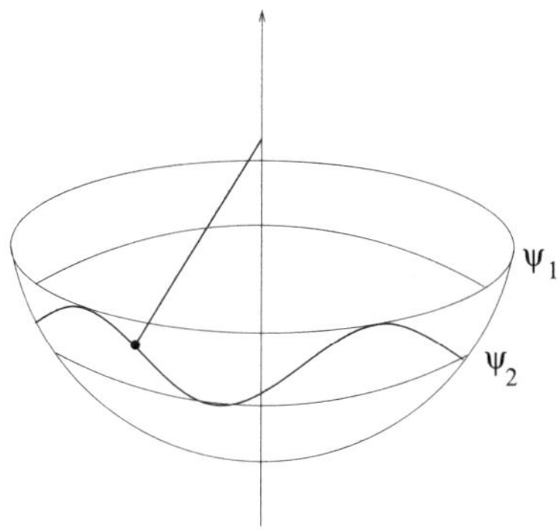

At the boundaries of the interval, $V(\pm 1) = -c_1^2/2 < 0$ is negative, and at $+\infty$ the function V diverges to $+\infty$. Hence one of the three zeroes has to lie above 1, and, since it cannot correspond to any angle ψ, this zero has no physical relevance. In the generic case, the other zeroes belong to the interval $(-1, 1)$; between them V, and hence also U, is positive. We conclude that $V(z)$ has the qualitative behavior of the graph on the previous page.

The mass point can only move between the two meridians corresponding to the zeroes in the interval $(-1, 1)$. The boundary values ψ_1, ψ_2 defined this way are actually reached at the end of every up or down swing. Summarizing, the level manifold can be parametrized by the two parameters $\varphi^* = \varphi$ and $\psi^* = \psi$ via

$$\left(\varphi^*, c_1, \psi^*, \pm\sqrt{2c_2 - \frac{c_1^2}{\cos^2\psi^*} - 2 - 2\sin\psi^*}\right).$$

It consists of 2 two-dimensional tori, where ψ^* only takes values in $[\psi_1, \psi_2]$. The corresponding coordinate vector fields are

$$\frac{\partial}{\partial\varphi^*} = \frac{\partial}{\partial\varphi}, \quad \frac{\partial}{\partial\psi^*} = \frac{\partial}{\partial\psi} + \frac{\partial p_\psi}{\partial\psi} \cdot \frac{\partial}{\partial p_\psi}.$$

We express the symplectic gradients of the first integrals in the coordinates of the level manifold:

$$\begin{aligned}
\text{s-grad}(H) &= \frac{p_\varphi}{\cos^2\psi}\frac{\partial}{\partial\varphi} + p_\psi\frac{\partial}{\partial\psi} - \frac{d}{d\psi}\left(\frac{p_\varphi^2}{2\cos^2\psi} + 1 + \sin\psi\right)\frac{\partial}{\partial p_\psi} \\
&= \frac{p_\varphi}{\cos^2\psi}\frac{\partial}{\partial\varphi} + p_\psi\frac{\partial}{\partial\psi} + p_\psi\frac{\partial p_\psi}{\partial\psi}\frac{\partial}{\partial p_\psi} \\
&= \frac{p_\varphi}{\cos^2\psi}\frac{\partial}{\partial\varphi^*} + p_\psi\frac{\partial}{\partial\psi^*}, \\
\text{s-grad}(P) &= \frac{\partial}{\partial\varphi} = \frac{\partial}{\partial\varphi^*}.
\end{aligned}$$

The dual forms $\omega_1(c_1, c_2)$ and $\omega_2(c_1, c_2)$ are thus

$$\omega_1(c_1, c_2) = \frac{1}{p_\psi} d\psi^*, \quad \omega_2(c_1, c_2) = d\varphi^* - \frac{c_1}{p_\psi \cos^2\psi} d\psi^* .$$

We compute the periods with respect to the homology cycle γ_1 which is parametrized in $M^2(c_1, c_2)$ by ψ^*. The factor 2 is a consequence of the fact that the boundaries, ψ_1, ψ_2, of the interval correspond to the minimum and the maximum of the motion, whereas a cycle is meant to be a motion between two extremal points of the same kind:

$$b_{11} = \int_{\gamma_1} \omega_1 = 2\int_{\psi_1}^{\psi_2} \frac{d\psi}{p_\psi}, \quad b_{12} = \int_{\gamma_1} \omega_2 = -2c_1 \int_{\psi_1}^{\psi_2} \frac{d\psi}{p_\psi \cos^2\psi} .$$

Let, similarly, γ_2 denote the homology cycle determined by $0 \le \varphi^* \le 2\pi$. Then

$$b_{21} = \int_{\gamma_2} \omega_1 = 0, \quad b_{22} = \int_{\gamma_2} \omega_2 = 2\pi .$$

The matrix occurring in the algorithm for computing the angle coordinates is now easily calculated:

$$\begin{bmatrix} b_{11} & b_{12} \\ b_{21} & b_{22} \end{bmatrix}^{-1} = \begin{bmatrix} a_{11} & a_{12} \\ a_{21} & a_{22} \end{bmatrix} = \begin{bmatrix} \dfrac{1}{\int_{\gamma_1}\omega_1} & \dfrac{-\int_{\gamma_1}\omega_2}{2\pi\int_{\gamma_1}\omega_1} \\ 0 & 1/2\pi \end{bmatrix} .$$

The resulting quotient of the basic frequencies is

$$\frac{\nu_1}{\nu_2} = \frac{a_{11}}{a_{12}} = -\frac{1}{2\pi}\int_{\gamma_1} \omega_2 = \frac{c_1}{\pi}\int_{\psi_1}^{\psi_2} \frac{d\psi}{\cos^2\psi \cdot p_\psi} .$$

This is nothing but the perihelion precession, i. e. the total variation of the angle φ for a complete cycle:

$$\Delta\varphi = \int_{\gamma_1} d\varphi = 2\int_{\psi_1}^{\psi_2} \frac{d\varphi}{d\psi} d\psi = 2\int_{\psi_1}^{\psi_2} \frac{\dot\varphi}{\dot\psi} d\psi = 2c_1 \int_{\psi_1}^{\psi_2} \frac{d\psi}{\cos^2\psi \cdot p_\psi} = 2\pi\frac{\nu_1}{\nu_2} .$$

In general, the motion is quasi-periodic. The trajectory is closed if and only if ν_1/ν_2 is rational; otherwise, it is dense on the torus. The transition to action angle coordinates allows us to determine the physically relevant basic frequencies of the system *without* having to explicitly perform the integration. This accounts for their importance in astronomic perturbation theory.

7.5. Formulations of Mechanics

Newton's equations describe the motion of a mechanical system under the impact of a force. The latter is understood as a vector field depending upon position and velocity, and is central for Newton's formulation of mechanics.

During the 18th century, this view changed in that Lagrange considered the action integral as the fundamental quantity for the description of dynamics. Newton as well as Lagrange formulated mechanics within the tangent bundle of configuration space. In the 19th century, by transition to the cotangent bundle, Hamilton succeeded in formulating the dynamics of mechanical systems within the framework of symplectic geometry. The aim of this section is to explain these fundamental ideas of mechanics and the related mathematical structures.

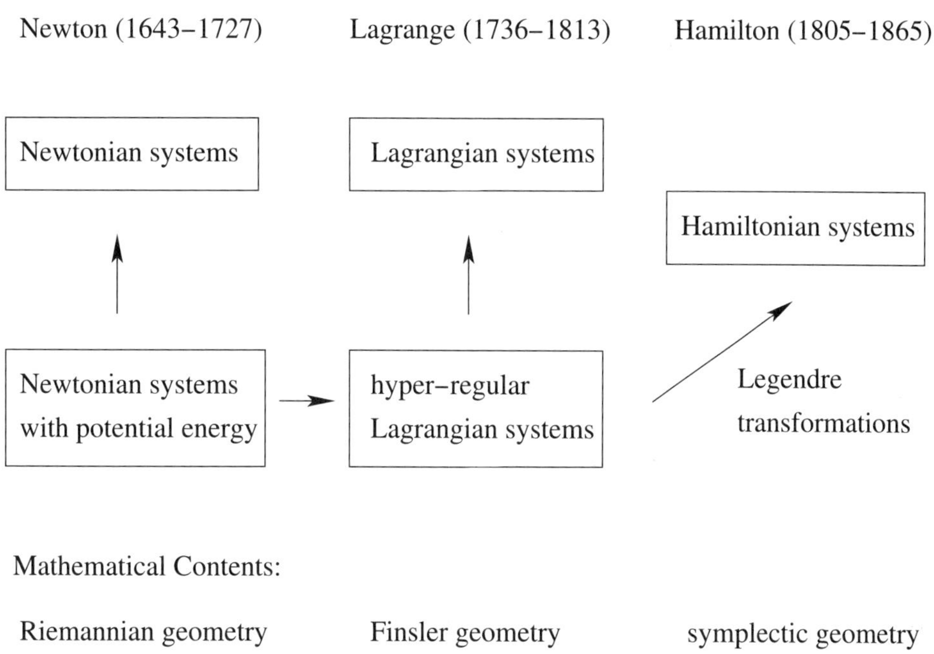

Formulation of Mechanics According to Newton

In Newtonian mechanics, the state of a mechanical system is described by finitely many real parameters. This leads to the notion of *configuration space*, which is a smooth and finite-dimensional manifold M^m. A motion of the mechanical system is a curve $\gamma : (a,b) \to M^m$ in configuration space. Its tangent—the *velocity*—is then a curve $\dot{\gamma} : (a,b) \to TM^m$ in the tangent bundle, and this space is called the *phase space*. According to Newton, the forces acting on the mechanical system are described by vector fields depending only on position and velocity, that is, vector fields X on TM^m. However, not all vector fields are allowed, since a force can act only in space.

The force vector field X has to satisfy the condition

$$d\pi \circ X \;=\; \mathrm{Id}\,.$$

Here $\pi : TM^m \to M^m$ denotes the projection of the tangent bundle, and $d\pi : TTM^m \to TM^m$ is its differential. Summarizing, we arrive at the notion of a Newtonian system.

Definition 7. An autonomous *Newtonian system* is a triple (M^m, g, X) consisting of a manifold M^m, a Riemannian metric g, and a vector field on the space TM^m such that

$$d\pi \circ X \;=\; \mathrm{Id}_{TM^m}\,.$$

The function $T : TM^m \to \mathbb{R}$ defined by

$$T(v) \;:=\; \frac{1}{2}\, g(v, v)$$

is called the *kinetic energy.*

Definition 8. A *motion* of the Newtonian system (M^m, g, X) is understood to be a curve $\gamma : (a, b) \to M^m$ in configuration space whose curve of tangents $\dot\gamma : (a, b) \to TM^m$ is an integral curve of X,

$$\ddot\gamma(t) \;=\; X(\dot\gamma(t))\,.$$

This is an invariant formulation of Newton's equation.

Example 9. Consider $\mathbb{R}$ with the coordinate x and identify $T\mathbb{R} = \mathbb{R}^2$ with $\mathbb{R}^2$. Here the coordinates are denoted by $\{x, \dot x\}$. The vector field

$$X \;=\; \dot x \frac{\partial}{\partial x} + \frac{1}{m}(-k^2 x - \rho \dot x)\frac{\partial}{\partial \dot x}$$

on $T\mathbb{R}$ has the required projection property, and a curve $x(t)$ in $\mathbb{R}$ is a motion in this Newtonian system if $x(t)$ is a solution of the oscillator equation

$$m\, \ddot x(t) \;=\; -k^2 x(t) - \rho \dot x(t)\,.$$

Example 10. Let (M^m, g) be a Riemannian manifold. We define a vector field $S : TM^m \to TTM^m$ on its tangent bundle—the so-called *geodesic spray*—as follows: If $v \in T_x M^m$ is a tangent vector, then there exists precisely one geodesic line $\gamma_{x,v} : (-\varepsilon, \varepsilon) \to M^m$ such that

$$\gamma_{x,v}(0) \;=\; x, \quad \dot\gamma_{x,v}(0) \;=\; v\,.$$

Consider its tangent curve, $\dot\gamma_{x,v} : (-\varepsilon, \varepsilon) \to TM^m$, and set

$$S(v) \;:=\; \frac{d}{dt}\big(\dot\gamma_{x,v}(t)\big)\big|_{t=0}\,.$$

The relation $\pi \circ \dot\gamma_{x,v}(t) = \gamma_{x,v}(t)$ implies $d\pi \circ S = \mathrm{Id}_{TM^m}$. Hence (M^m, g, S) is a Newtonian system, and the motions of this system are the geodesic lines of the Riemannian manifold (M^m, g). In the coordinates $\{x^i\,,\,\dot{x}^i\}$ of the tangent bundle, the geodesic spray is given by the formula

$$S \;=\; \sum_{i=1}^{m} \dot{x}^i\,\frac{\partial}{\partial x^i} \;-\; \sum_{i,j,k=1}^{m} \Gamma^i_{jk}\,\dot{x}^j\dot{x}^k\,\frac{\partial}{\partial \dot{x}^i}\,,$$

a straightforward consequence of the system of differential equations describing geodesic lines—see §5.7.

Many Newtonian systems have a *potential energy.* This is a smooth function $V : M^m \to \mathbb{R}$ defined on configuration space. The gradient $\mathrm{grad}(V)$ is a vector field on M^m locally determined by

$$\mathrm{grad}(V) \;=\; \sum_{i,j=1}^{m} g^{ij}(x)\,\frac{\partial V}{\partial x^i}\,\frac{\partial}{\partial x^j}\,.$$

In the sequel we will need, however, a different vector field, denoted by $\overline{\mathrm{grad}}(V)$. This will be a vector field on phase space. At the point $v \in TM^m$, it is defined by the following equation:

$$\overline{\mathrm{grad}}(V)(v) \;=\; \frac{d}{dt}\big(v + t\cdot \mathrm{grad}(V)(\pi(v))\big)\big|_{t=0}\,.$$

In the manifold M^m, the curve $v + t\cdot\mathrm{grad}(V)(\pi(v))$ projects to the base point $\pi(v) \in M^m$ of the vector $v \in TM^m$. Hence the vector field $\overline{\mathrm{grad}}(V)$ projects to zero under the differential $d\pi$,

$$d\pi \circ \overline{\mathrm{grad}}(V) \;=\; 0\,,$$

and, for any potential energy V, the vector field $X := S - \overline{\mathrm{grad}}(V)$ is an admissible vector field on TM^m in the sense of Newtonian mechanics. In local coordinates, we obtain the formula

$$\overline{\mathrm{grad}}(V) \;=\; \sum_{i,j=1}^{m} g^{ij}(x)\,\frac{\partial V}{\partial x^i}\,\frac{\partial}{\partial \dot{x}^j}\,.$$

Definition 9. A *Newtonian system with potential energy* is a triple (M, g, V) consisting of a Riemannian metric g and a potential energy $V : M^m \to \mathbb{R}$. The corresponding force vector field is

$$X \;=\; S - \overline{\mathrm{grad}}(V)\,.$$

A motion in a Newtonian system with potential energy is defined to be a curve $\gamma : (a, b) \to M^m$ whose tangential curve $\dot{\gamma} : (a, b) \to TM^m$ is an integral curve of X.

Definition 10. The *energy* of a Newtonian system (M^m, g, V) with potential energy is the sum of the kinetic and the potential energy,

$$E \,:\, TM^m \to \mathbb{R}, \quad E \;=\; T + V\circ\pi\,.$$

Theorem 15 (Energy Conservation for Newtonian Systems). *Let (M^m, g, V) be a Newtonian system with potential energy, and let $X = S - \overline{\operatorname{grad}}(V)$ be the force vector field. Then*

$$dE(X) \;=\; 0 \,.$$

In particular, $E\big(\dot\gamma(t)\big)$ is constant for every motion $\gamma(t)$ of the system.

Proof. In local coordinates, the energy E and the force vector field are given by the formulas

$$E \;=\; \frac{1}{2}\sum_{i,j=1}^{m} g_{ij}(x)\dot x^i \dot x^j + V(x)\,,$$

$$X \;=\; \sum_{i=1}^{m} \dot x^i \frac{\partial}{\partial x^i} - \sum_{k=1}^{m}\left[\sum_{i,j=1}^{m}\Gamma_{ij}^k(x)\dot x^i\dot x^j + \sum_{i=1}^{m}\frac{\partial V}{\partial x^i} g^{ik}(x)\right]\frac{\partial}{\partial \dot x^k}\,.$$

Using the expression for the Christoffel symbols Γ_{ij}^k from §5.7,

$$\Gamma_{ij}^k \;=\; \frac{1}{2}\sum_{\alpha=1}^{m} g^{k\alpha}(x)\left(\frac{\partial g_{i\alpha}}{\partial x^j}(x) + \frac{\partial g_{j\alpha}}{\partial x^i}(x) - \frac{\partial g_{ij}}{\partial x^\alpha}(x)\right),$$

we obtain $dE(X) = X(E) = 0$ by an elementary calculation. □

Motions with large energy in a Newtonian system (M^m, g, V) with potential energy are—up to a change of parametrization—geodesic lines with respect to a new Riemannian metric. This construction will lead to the Maupertuis-Jacobi principle. Suppose that the potential energy $V : M^m \to \mathbb{R}$ is bounded from above by E_0,

$$\sup\{V(x) :\; x \in M^m\} \;<\; E_0\,.$$

Then $g^* = (E_0 - V)\cdot g$ is a Riemannian metric on the manifold M^m. Consider a motion $\gamma : (a,b) \to M^m$ of the Newtonian system (M^m, g, V) with energy E_0,

$$E_0 \;=\; \frac{1}{2}g(\dot\gamma(t), \dot\gamma(t)) + V(\gamma(t))\,.$$

Since $E_0 > V(\gamma(t))$, the tangent vector $\dot\gamma(t)$ vanishes nowhere, and the function

$$s(t) \;:=\; \sqrt{2}\int_a^t (E_0 - V(\gamma(\mu)))d\mu$$

becomes a strictly monotone function $s : (a,b) \to (0, b^*)$. We invert this function and thus view $t \in (a,b)$ as a function of the parameter $s \in (0, b^*)$, $t = t(s)$. Let the curve $\gamma^*(s)$ be the initial curve γ in this new parametrization.

Theorem 16 (Maupertuis-Jacobi principle). *Let $\gamma(t)$ be a motion of the Newtonian system (M^m, g, V) with energy E_0. Then $\gamma^*(s)$ is a geodesic line in M^m with respect to the Riemannian metric $g^* = (E_0 - V) \cdot g$.*

Proof. The Christoffel symbols Γ_{ij}^k and ${}^*\Gamma_{ij}^k$ of the metrics g and g^* are, in local coordinates, related by the formula

$$ {}^*\Gamma_{ij}^k = \Gamma_{ij}^k + \frac{1}{2}\frac{1}{(E_0 - V)}\left(-\frac{\partial V}{\partial x^i}\delta_{jk} - \frac{\partial V}{\partial x^j}\delta_{ik} + \sum_{\alpha=1}^{m}\frac{\partial V}{\partial x^\alpha}g^{\alpha k}g_{ij}\right). $$

We write the motion $\gamma(t) = \big(x^1(t), \ldots, x^m(t)\big)$ in local coordinates. Then

$$ \frac{dx^k}{ds} = \frac{dx^k}{dt}\cdot\frac{dt}{ds} = \frac{1}{\sqrt{2}}\frac{1}{(E_0 - V)}\frac{dx^k}{dt}, $$

$$ \frac{d^2x^k}{ds^2} = \frac{1}{2(E_0 - V)^2}\frac{d^2x^k}{dt^2} + \frac{1}{2(E_0 - V)^3}\frac{dx^k}{dt}\sum_{\alpha=1}^{m}\frac{\partial V}{\partial x^\alpha}\cdot\frac{dx^\alpha}{dt}, $$

and, using the equation of motion,

$$ \frac{d^2x^k}{dt^2} = -\sum_{i,j=1}^{m}\Gamma_{ij}^k\frac{dx^i}{dt}\frac{dx^j}{dt} - \sum_{\alpha=1}^{m}\frac{\partial V}{\partial x^\alpha}g^{k\alpha} $$

as well as the energy condition

$$ \sum_{i,j=1}^{m} g_{ij}\frac{dx^i}{dt}\frac{dx^j}{dt} = 2(E_0 - V), $$

we obtain the claimed result:

$$ \frac{d^2x^k}{ds^2} + \sum_{i,j=1}^{m}{}^*\Gamma_{ij}^k\frac{dx^i}{ds}\frac{dx^j}{ds} = -\frac{1}{2(E_0 - V)^2}\sum_{\alpha=1}^{m}\frac{\partial V}{\partial x^\alpha}g^{k\alpha} $$

$$ + \frac{1}{4(E_0 - V)^3}\sum_{\alpha=1}^{m}\frac{\partial V}{\partial x^\alpha}g^{k\alpha}\cdot\sum_{i,j=1}^{m}g_{ij}\frac{dx^i}{dt}\frac{dx^j}{dt} = 0. \qquad \square $$

Formulation of Mechanics According to Lagrange

The transition to Lagrangian mechanics proceeds by considering the Lagrange function of a Newtonian system with potential energy.

Definition 11. Let (M^m, g, V) be a Newtonian system with potential energy. The *Lagrange function* (or *Lagrangian*) $L : TM^m \to \mathbb{R}$ is the difference of kinetic and potential energy,

$$ L = T - V \circ \pi. $$

Theorem 17 (d'Alembert-Lagrange). *A curve* $\gamma : (a, b) \to M^m$ *is a motion of the Newtonian system* (M^m, g, V) *with potential energy if and only if the* Euler-Lagrange equations *hold:*

$$\frac{d}{dt}\left(\frac{\partial L}{\partial \dot{x}^i}(\dot{\gamma}(t))\right) = \frac{\partial L}{\partial x^i}(\dot{\gamma}(t)).$$

Proof. We prove this theorem again using local coordinates. The curve $\gamma(t) = \big(x^1(t), \ldots, x^m(t)\big)$ is a motion of the Newtonian system with potential energy if and only if it solves the system of differential equations

$$\ddot{x}^k = -\sum_{i,j=1}^{m} \Gamma_{ij}^k \dot{x}^i \dot{x}^j - \sum_{\alpha=1}^{m} V_\alpha g^{\alpha k} .$$

For brevity, we denoted by V_α the partial derivative of the potential function, $V_\alpha := \partial V / \partial x^\alpha$. The Lagrange function is

$$L(\dot{\gamma}) = \frac{1}{2}\sum_{i,j=1}^{m} g_{ij}\dot{x}^i\dot{x}^j - V(\gamma),$$

and this leads to the difference

$$\frac{d}{dt}\left(\frac{\partial L}{\partial \dot{x}^i}(\dot{\gamma})\right) - \frac{\partial L}{\partial x^i}(\dot{\gamma}) = \sum_{\alpha,\beta=1}^{m}\left(\frac{\partial g_{i\alpha}}{\partial x^\beta} - \frac{1}{2}\frac{\partial g_{\alpha\beta}}{\partial x^i}\right)\dot{x}^\alpha\dot{x}^\beta + \sum_{\alpha=1}^{m} g_{i\alpha}\ddot{x}^\alpha + V_i .$$

Multiplying the Euler-Lagrange equations by g^{ik} and summing over the index i, this system of equations turns out to be equivalent to

$$0 = \ddot{x}^k + \sum_{i=1}^{m} V_i g^{ik} + \sum_{i=1}^{m}\sum_{\alpha,\beta=1}^{m}\left(\frac{\partial g_{i\alpha}}{\partial x^\beta} - \frac{1}{2}\frac{\partial g_{\alpha\beta}}{\partial x^i}\right) g^{ik}\dot{x}^\alpha\dot{x}^\beta .$$

The claim now immediately follows from the formula for the Christoffel symbols Γ_{ij}^k. □

Thus the equations of motion of Newtonian mechanics are, in the case of a potential force, equivalent to the Euler-Lagrange equations. For the latter, it does not matter that the Lagrange function arises as the difference of a kinetic and a potential energy. Hence we define:

Definition 12. An autonomous *Lagrangian system* is a pair (M^m, L) consisting of a manifold M^m and a smooth function $L : TM^m \to \mathbb{R}$. A *Lagrangian motion* is a curve $\gamma : (a, b) \to M^m$ which solves the system of Euler-Lagrange equations

$$\frac{d}{dt}\left(\frac{\partial L}{\partial \dot{x}^i}(\dot{\gamma}(t))\right) = \frac{\partial L}{\partial x^i}(\dot{\gamma}(t)) .$$

Example 11. Let (M^m, g) be a pseudo-Riemannian manifold and A a 1-form on it. The Lagrange function

$$L(v) \;=\; \frac{m}{2} g(v,v) - A(v)$$

generalizes, in a sense to be discussed in Chapter 9, the motion of a charged particle of mass m under the influence of the Lorentz force of the electromagnetic field generated by A.

Motions in a Lagrangian system can be understood from the point of view that they are critical points of the *action integral* $\mathcal{L}$. This measures for a curve $\gamma : [a,b] \to M^m$ the mean value of the Lagrange function on this curve,

$$\mathcal{L}(\gamma) \;:=\; \int_a^b L(\dot\gamma(t))dt\,.$$

Theorem 18 (Principle of Least Action). *A curve $\gamma : [a,b] \to M^m$ is a motion of the Lagrangian system (M^m, L) if and only if the variation of the action integral vanishes for every variation γ_μ of the curve with fixed initail and end points, $\gamma_\mu(a) = \gamma(a), \gamma_\mu(b) = \gamma(b)$:*

$$\frac{d}{d\mu}\Big(\int_a^b L(\dot\gamma_\mu(t))dt\Big)\Big|_{\mu=0} \;=\; 0\,.$$

Proof. We compute the derivative of the action integral with respect to the parameter μ in coordinates $\gamma_\mu(t) = \big(x^1(\mu,t), \ldots, x^m(\mu,t)\big)$ by partial integration:

$$\frac{d}{d\mu}\big(\mathcal{L}(\gamma_\mu)\big)\big|_{\mu=0} \;=\; \int_a^b \sum_{i=1}^m \Big[\frac{\partial L}{\partial x^i}(\dot\gamma(t)) - \frac{d}{dt}\Big(\frac{\partial L}{\partial \dot x^i}(\dot\gamma(t))\Big)\Big]\frac{\partial x^i}{\partial\mu}(0,t)dt\,.$$

The functions $\partial x^i(0,t)/\partial\mu$ are arbitrary functions vanishing at the end points of the interval $[a,b]$, and hence the Euler-Lagrange equations are equivalent to the condition

$$\frac{d}{d\mu}\big(\mathcal{L}(\gamma_\mu)\big)\big|_{\mu=0} \;=\; 0\,.$$

□

For general Lagrangian systems, there exists a notion of energy which, on the one hand, generalizes the energy of a Newtonian system with potential energy, and is, on the other hand, a preserved quantity. This Lagrangian energy is obtained by first introducing the Legendre transformation. Let $L : TM^m \to \mathbb{R}$ be a Lagrange function, and let $v \in T_xM^m$ be a vector at the point $x \in M^m$. Now restrict L to the tangent space T_xM^m and consider the differential $D(L|_{T_xM^m})(v)$ at the point $v \in T_xM^m$. This is a linear map $T_xM^m \to \mathbb{R}$, hence a covector in $T_x^*M^m$.

Definition 13. The Legendre transformation $\mathcal{L} : TM^m \to T^*M^m$ of an arbitrary Lagrangian system is the map

$$\mathcal{L}(v) := D(L|_{T_xM^m})(v).$$

Example 12. If the Lagrange function $L = \frac{1}{2}g - V$ is the difference of a kinetic and a potential energy, the following relation holds for $v, w \in TM^m$:

$$\mathcal{L}(v)(w) = \frac{1}{2} \cdot D(g)(v)(w) = g(v, w).$$

Hence the Legendre transformation $\mathcal{L} : TM^m \to T^*M^m$ is simply the identification of the tangent bundle with the cotangent bundle via the metric.

Definition 14. The *energy* of a Lagrangian system (M^m, L) is the function $E : TM^m \to \mathbb{R}$ on the tangent bundle defined by

$$E(v) = \mathcal{L}(v)(v) - L(v).$$

In the case of a Newtonian system with potential energy, we have

$$E(v) = \mathcal{L}(v)(v) - L(v) = = \frac{1}{2}g(v, v) + V(\pi(v)).$$

This shows that the energy in the sense of Lagrangian mechanics coincides with the Newtonian energy.

Example 13. We compute the Legendre transformation for the Lagrange function from Example 11. Let $(x, y) \in T_xM^m$ with local coordinates $x^1, \ldots, x^m$ and $y^1, \ldots, y^m$. Then,

$$\left(L|_{T_xM^m}\right)(y^1, \ldots, y^m) = \frac{m}{2}\sum_{i,j}^{m} g_{ij}y^iy^j - \sum_{i}^{m} A_iy^i,$$

and, as an element of T_x^*M, its differential is

$$D\left(L|_{T_xM^m}\right)(y^1, \ldots, y^m) = m\sum_{i,j}^{m} g_{ij}y^i\, dy^j - \sum_{i}^{m} A_i\, dy^i.$$

We evaluate this map at (x, v), $v = (v^1, \ldots, v^m)$:

$$\mathcal{L}(y)(v) = D\left(L|_{T_xM^m}\right)(y)(v) = m \cdot g(y, v) - A(v).$$

By definition, this yields for the energy the value

$$E(v) = \mathcal{L}(v)(v) - L(v) = \frac{m}{2}g(v, v),$$

which has the remarkable property of not depending on A. A more careful physical analysis shows that the energy of a charged particle indeed only depends on the electric, not on the magnetic field. This is related to the fact that the magnetic field does no work on the particle (see §9.5).

Remark. Denoting the coordinates on TM^m by $\{x^1, \ldots, x^m, \dot{x}^1, \ldots, \dot{x}^m\}$ and the coordinates on T^*M^m by $\{q_1, \ldots, q_m, p_1, \ldots, p_m\}$, we see that the Legendre transformation $\mathcal{L}$ is given by

$$q_i = x^i \quad \text{and} \quad p_i = \frac{\partial L}{\partial \dot{x}^i},$$

and the expression for the energy E takes the following form:

$$E(x, \dot{x}) = \sum_{i=1}^{m} \frac{\partial L}{\partial \dot{x}^i} \dot{x}^i - L(x, \dot{x}) .$$

Theorem 19 (Conservation of Energy for Lagrangian Systems). *The energy $E(\dot{\gamma}(t))$ of each motion $\gamma(t)$ of a Lagrangian system (M^m, L) is constant.*

Proof. The energy of a curve is

$$E(\dot{\gamma}(t)) = \sum_{i=1}^{m} \frac{\partial L}{\partial \dot{x}^i} \cdot \frac{dx^i}{dt} - L(\dot{\gamma}(t)),$$

and by differentiation we obtain

$$\frac{d}{dt} E(\dot{\gamma}(t)) = \sum_{i,j=1}^{m} \left(\frac{\partial L}{\partial \dot{x}^i \partial x^j} \frac{dx^i}{dt} \frac{dx^j}{dt} + \frac{\partial L}{\partial \dot{x}^i \partial \dot{x}^j} \frac{dx^i}{dt} \frac{d^2 x^j}{dt^2} \right) - \sum_{i=1}^{m} \frac{\partial L}{\partial x^i} \frac{dx^i}{dt} .$$

Using the Euler-Lagrange equation

$$\frac{\partial L}{\partial x^i} = \sum_{i,j=1}^{m} \frac{\partial L}{\partial \dot{x}^i \partial x^j} \frac{dx^j}{dt} + \sum_{i,j=1}^{m} \frac{\partial L}{\partial \dot{x}^i \partial \dot{x}^j} \frac{d^2 x^j}{dt^2},$$

we immediately obtain the assertion. □

Thus the energy is a first integral for any motion in a Lagrangian system. As in §7.3, further first integrals can be derived from symmetries of the Lagrange function. To this end, consider a one-parameter group of diffeomorphisms $\Phi_s : M^m \to M^m$ of the configuration space as well as its generating vector field,

$$\mathcal{V}(x) := \frac{d}{ds} (\Phi_s(x)) \big|_{s=0},$$

on M^m. The differentials $d(\Phi_s) : TM^m \to TM^m$ are diffeomorphisms of the phase space TM^m into itself.

Theorem 20 (Noether's Theorem). *Let the Lagrange function L be invariant under the action of a one-parameter group of diffeomorphisms, $L(d(\Phi_s)(v)) = L(v)$. Then the function $f_{\mathcal{V}} : TM^m \to \mathbb{R}$ defined by*

$$f_{\mathcal{V}}(w) = \lim_{\mu \to 0} \frac{L(w + \mu \cdot \mathcal{V}(\pi(w))) - L(w)}{\mu}$$

is a constant of motion for the Lagrangian system (M^m, L).

Proof. Let the one-parameter group of diffeomorphisms Φ_t be determined in coordinates by

$$\Phi_t(x^1, \ldots, x^m) = (\Phi_t^1(x^1, \ldots, x^m), \ldots, \Phi_t^m(x^1, \ldots, x^m)).$$

The invariance of the Lagrange function implies that

$$\sum_{i=1}^{m} \frac{\partial L}{\partial x^i} \frac{\partial \Phi_s^i}{\partial s}\bigg|_{s=0} + \sum_{i,j=1}^{m} \frac{\partial L}{\partial \dot{x}^j} \cdot \frac{\partial^2 \Phi_t^j}{\partial x^i \partial s}\bigg|_{s=0} \cdot \frac{dx^i}{dt} = 0,$$

and the vector field $\mathcal{V}$ has the components

$$\mathcal{V} = \sum_{i=1}^{m} \frac{\partial \Phi_s^i}{\partial s}\bigg|_{s=0} \frac{\partial}{\partial x^i}.$$

Thus we obtain

$$f_{\mathcal{V}}(\dot{\gamma}(t)) = \sum_{i=1}^{m} \frac{\partial L}{\partial \dot{x}^i} \frac{\partial \Phi_s^i}{\partial s}\bigg|_{s=0},$$

and from the Euler-Lagrange equation we conclude that

$$\frac{d}{dt}\Big(f_{\mathcal{V}}(\dot{\gamma}(t))\Big) = \sum_{i=1}^{m} \frac{\partial L}{\partial x^i} \frac{\partial \Phi_s^i}{\partial s}\bigg|_{s=0} + \sum_{i,j=1}^{m} \frac{\partial L}{\partial \dot{x}^i} \frac{\partial^2 \Phi_s^i}{\partial x^j \partial s}\bigg|_{s=0} \frac{dx^j}{dt} = 0. \qquad \square$$

Example 14. To each transformation group $\Phi_t : M^m \to M^m$ preserving the metric g and the potential function V there corresponds a first integral,

$$f_{\mathcal{V}}(w) = g(w, \mathcal{V}),$$

which is *linear* in every fiber of the tangent bundle TM^m. We made use of this first integral in Theorem 37, Chapter 5, to integrate the geodesic flow on surfaces of revolution (Clairaut's theorem). Hence first integrals of the geodesic flow which are linear in the fibers arise from isometries of the metric.

Example 15. If Φ_t preserves the pseudo-Riemannian metric g as well as the 1-form A from Example 11, then the Lagrangian of a charged particle in an electromagnetic field is invariant, and hence Noether's theorem can be applied to yield the invariant

$$f_{\mathcal{V}}(w) = m \cdot g(w, \mathcal{V}) - A(\mathcal{V}).$$

Formulation of Mechanics According to Hamilton

We will arrive at the formulation of mechanics according to Hamilton by starting from a Lagrangian system (M^m, L) with bijective Legendre transformation $\mathcal{L} : TM^m \to T^*M^m$. These Lagrangian systems are called *hyperregular*. A regular Lagrangian system is one whose Legendre transformation is locally invertible. For simplicity, we confine ourselves here to the case of a globally invertible Legendre transformation. For example, Newtonian systems with potential energy or the system of a charged particle moving in an electromagnetic field (Example 11) are always hyper-regular.

Definition 15. Let (M^m, L) be a hyper-regular Lagrangian system with Legendre transformation $\mathcal{L} : TM^m \to T^*M^m$ and energy $E : TM^m \to \mathbb{R}$. The function

$$H \;:=\; E \circ \mathcal{L}^{-1}$$

defined on the cotangent bundle is called the *Hamilton function* (or *Hamiltonian*) of the system.

Example 16. In the case of a Newtonian system (M^m, g, V) with potential energy, the Legendre transformation is determined by the formulas

$$q_i \;=\; x^i \quad \text{and} \quad p_i \;=\; \frac{\partial L}{\partial \dot{x}^i} \;=\; \sum_{\alpha=1}^{m} g_{i\alpha}\dot{x}^{\alpha}\,.$$

Inverting this transformation leads to

$$x^i \;=\; q_i \quad \text{and} \quad \dot{x}^i \;=\; \sum_{\alpha=1}^{m} g^{i\alpha} p_{\alpha}\,,$$

and the formulas for the energy E, the Lagrange function L, and the Hamilton function H read as follows:

(1) $E \;=\; \frac{1}{2} \sum\limits_{i,j=1}^{m} g_{ij}\dot{x}^i\dot{x}^j + V(x^1, \ldots, x^m)$,

(2) $L \;=\; \frac{1}{2} \sum\limits_{i,j=1}^{m} g_{ij}\dot{x}^i\dot{x}^j - V(x^1, \ldots, x^m)$,

(3) $H \;=\; \frac{1}{2} \sum\limits_{i,j=1}^{m} g^{ij}p_i p_j + V(q_1, \ldots, q_m)$.

Through this change of phase space—i. e., replacing the tangent bundle TM^m by the cotangent bundle T^*M^m—we enter the realm of symplectic geometry, since T^*M^m always carries the symplectic form $\omega = d\theta$.

Theorem 21 (Hamilton's Theorem). *Let (M^m, L) be a hyper-regular Lagrangian system. A curve $\gamma : (a, b) \to M^m$ is a Lagrangian motion if and only if the curve $\mathcal{L}(\dot{\gamma}) : (a, b) \to T^*M^m$ is an integral curve of the symplectic gradient* s-grad(H) *of the Hamilton function.*

Proof. The Legendre transformation $\mathcal{L}$ is defined by

$$q_i \;=\; x^i, \quad p_i \;=\; \frac{\partial L}{\partial \dot{x}^i}.$$

For its inverse map, we introduce the notation

$$x^i \;=\; q_i \quad \text{and} \quad \dot{x}^i \;=\; \Phi^i(q_1, \ldots, q_m, p_1, \ldots, p_m).$$

The Hamilton function is then

$$H = \sum_{\alpha=1}^{m} p_\alpha \cdot \Phi^\alpha - L(q_1, \ldots, q_m, \Phi^1, \ldots, \Phi^m),$$

and we compute its partial derivatives:

$$\frac{\partial H}{\partial p_i} = \Phi^i + \sum_{\alpha=1}^{m} \left(p_\alpha - \frac{\partial L}{\partial \dot{x}^\alpha} \right) \frac{\partial \Phi^\alpha}{\partial p_i} = \Phi^i,$$

$$\frac{\partial H}{\partial q_i} = \sum_{\alpha=1}^{m} \left(p_\alpha - \frac{\partial L}{\partial \dot{x}^\alpha} \right) \frac{\partial \Phi^\alpha}{\partial q_i} - \frac{\partial L}{\partial x^i} = -\frac{\partial L}{\partial x^i}.$$

Thus we obtain a formula for the symplectic gradient:

$$\text{s-grad}(H) = \sum_{i=1}^{m} \left(\Phi^i \frac{\partial}{\partial q_i} + \frac{\partial L}{\partial x^i} \frac{\partial}{\partial p_i} \right).$$

A curve

$$\mathcal{L}(\dot{\gamma}(t)) = \left(x^1(t), \ldots, x^m(t), \frac{\partial L}{\partial \dot{x}^1}(\dot{\gamma}(t)), \ldots, \frac{\partial L}{\partial \dot{x}^m}(\dot{\gamma}(t)) \right)$$

is an integral curve of the vector field s-grad(h) if and only if

$$\dot{x}^i = \Phi^i \quad \text{and} \quad \frac{d}{dt}\left(\frac{\partial L}{\partial \dot{x}^i}(\dot{\gamma}(t)) \right) = \frac{\partial L}{\partial x^i}(\dot{\gamma}(t)).$$

The first equation is trivialy satisfied by setting $\dot{x}^i = \Phi^i$, and this proves the assertion. □

Exercises

1. In $\mathbb{R}^4$ with the symplectic structure $\omega = dx^1 \wedge dx^3 + dx^2 \wedge dx^4$, we choose the following four diffeomorphisms a, b, c, d:

$$\begin{aligned} a(x^1, x^2, x^3, x^4) &= (x^1, x^2 + 1, x^3, x^4), \\ b(x^1, x^2, x^3, x^4) &= (x^1, x^2, x^3, x^4 + 1), \\ c(x^1, x^2, x^3, x^4) &= (x^1 + 1, x^2, x^3, x^4), \\ d(x^1, x^2, x^3, x^4) &= (x^1, x^2 + x^4, x^3 + 1, x^4), \end{aligned}$$

and denote by Γ the group of motions of $\mathbb{R}^4$ generated by them. Prove that ω is a Γ-invariant 2-form. Hence ω induces a symplectic form on the manifold $M^4 = \mathbb{R}^4/\Gamma$. Prove that M^4 is compact. Finally, denote by $[\Gamma, \Gamma]$ the commutator group of Γ. Then $\Gamma/[\Gamma, \Gamma]$ is a free abelian group of rank three (Thurston, 1971).

2. Prove that the symplectic form ω of a compact symplectic manifold M^{2m} can never be an exact differential form. Hence the second de Rham cohomology $H^2_{DR}(M^{2m})$ is non-trivial. In particular, for $m \neq 1$, even-dimensional spheres S^{2m} have no symplectic structure.

3. Consider $M = \mathbb{R}^2 \backslash \{0\}$ with the symplectic form $\omega = dx \wedge dy$ and the vector field

$$\mathcal{V} = \frac{x}{x^2+y^2} \cdot \frac{\partial}{\partial x} + \frac{y}{x^2+y^2} \cdot \frac{\partial}{\partial y}.$$

a) Let $\varphi = \arctan(y/x)$ be the polar angle defined in every sufficiently small neighborhood of $(x,y) \neq (0,0)$. Compute $\mathrm{grad}(\varphi)$ and $\text{s-grad}(\varphi)$.

b) Conclude from this that $\mathcal{V}$ is *no* Hamiltonian vector field on all of M.

4 (Continuation of Example 3)**.** Prove that the Liouville form on the two-dimensional coadjoint orbit through the element $(\alpha, \beta) \in \mathfrak{g}^*$ $(\beta \neq 0)$ is given by the expression

$$\omega = \frac{1}{\beta}\, d\alpha \wedge d\beta.$$

Hint: Show first that the fundamental vector fields corresponding to the Lie algebra elements $(1,0)$ and $(0,1)$ are

$$\beta \frac{\partial}{\partial \beta} \quad \text{and} \quad -\beta \frac{\partial}{\partial \alpha}.$$

5. Let $\mathcal{V}$ be a vector field on a manifold M^m and $\dot{x} = \mathcal{V}(x)$ the associated differential equation. A first integral of $\mathcal{V}$ is a smooth function $h : M^m \to \mathbb{R}$ which is constant along every solution of the differential equation.

a) Prove that $h : M^m \to \mathbb{R}$ is a first integral of $\mathcal{V}$ if and only if $dh(\mathcal{V}) = 0$.

b) Prove that the set $C^\infty_{\mathcal{V}}(M^m;\mathbb{R})$ of all first integrals of $\mathcal{V}$ is a subring of $C^\infty(M^m;\mathbb{R})$.

c) Show that $h(x,y) = y^2 - 2x^2 + x^4$ is a first integral of the vector field

$$\mathcal{V} = 2y \cdot \partial/\partial x + (4x - 4x^3) \cdot \partial/\partial y$$

in $\mathbb{R}^2$. Describe the geometric shape of the integral curves in the (x,y)-plane by means of h.

d) Find a vector field in the plane which has no non-trivial first integral.

6. Let $f = a + b \cdot i : U \to \mathbb{C}$ be a holomorphic function on an open subset U of $\mathbb{R}^2 \cong \mathbb{C}$. Consider the vector field $\mathcal{V} = \mathrm{grad}\,(a)$.

a) Prove that $b : U \to \mathbb{R}$ is a first integral for $\mathcal{V}$.

b) Describe the geometric shape of the integral curves of this vector field for the functions $f(z) = z^k$ and $f(z) = z + 1/z$.

7. Consider on $\mathbb{R}^3$ a vector field B (a magnetic field) and its 2-form

$$B \;=\; B_x \cdot dy \wedge dz - B_y \cdot dx \wedge dz + B_z \cdot dx \wedge dy\,,$$

as well as the symplectic form on the phase space $\mathbb{R}^3 \times \mathbb{R}^3$ with coordinates $(\vec{q}, \vec{v}) = (x, y, z, v_x, v_y, v_z)$,

$$\omega_B \;=\; m(dx \wedge dv_x + dy \wedge dv_y + dz \wedge dv_z) - \frac{e}{c} \cdot B\,.$$

As Hamilton function, we choose the kinetic energy

$$H_0 \;=\; \frac{m}{2}\left(v_x^2 + v_y^2 + v_z^2\right).$$

a) Show that the defining equation for the Hamiltonian vector field associated with H_0 is equivalent to the *Lorentz equation*

$$(*) \qquad m\,\frac{dv}{dt} \;=\; \frac{e}{c} \cdot v \times B\,.$$

b) If $B = dA$ is exact and $\vec{A}$ denotes the associated vector field, show that the map

$$f:\ \mathbb{R}^3 \times \mathbb{R}^3 \longrightarrow \mathbb{R}^3 \times \mathbb{R}^3, \quad (\vec{q}, \vec{v}) \longmapsto (\vec{q}, m\vec{v} + \frac{e}{c}\vec{A}) \;=:\; (\vec{q}, \vec{p})$$

is a canonical transformation, i. e., for the canonical symplectic form ω_0 and the Hamilton function H_B,

$$\omega_0 \;=\; dx \wedge dp_x + dy \wedge dp_y + dz \wedge dp_z, \quad H_B \;=\; \frac{1}{2m}\left|\left|\vec{p} - \frac{e}{c}\vec{A}\right|\right|^2,$$

the following equations hold:

$$f^*\omega_0 \;=\; \omega_B, \quad f^*H_B \;=\; H_0,$$

and equation $(*)$ does not change.

c) If B is constant, any particle moves on a helix. Hint: Interpret torsion and curvature as first integrals. If both these quantities are constant for a curve, then this curve has to be a helix (Exercise 6 in Chapter 5).

8 (Plane Toda Lattice)**.** Consider in $\mathbb{R}^2$ the second order differential equations

$$\ddot{x} \;=\; -e^{x-y}, \quad \ddot{y} \;=\; e^{x-y}\,,$$

which in phase space $(x, y, \dot{x}, \dot{y}) \in \mathbb{R}^4$ are equivalent to $\dot{\xi} = X(\xi)$ with

$$X(x, y, \dot{x}, \dot{y}) \;=\; \dot{x}\frac{\partial}{\partial x} + \dot{y}\frac{\partial}{\partial y} - e^{x-y}\frac{\partial}{\partial \dot{x}} + e^{x-y}\frac{\partial}{\partial \dot{y}}\,.$$

The energy $E = (\dot{x}^2 + \dot{y}^2)/2 + e^{x-y}$ is a first integral.

a) Show that this system has further first integrals, for example $P = \dot{x} + \dot{y}$ or $K = (\dot{x} - 2\dot{y})(\dot{y} - 2\dot{x})/9 - e^{x-y}$. These quantities are related by $E + K = 5P^2/18$.

b) The set

$$M^2(E, P) = \{(x, y, \dot{x}, \dot{y}) \in \mathbb{R}^4 : (\dot{x}^2 + \dot{y}^2)/2 + e^{x-y} = E, \ \dot{x} + \dot{y} = P\}$$

is not empty if and only if $4E - P^2 > 0$. In this case $M^2(E, P)$ is a smooth two-dimensional submanifold of $\mathbb{R}^4$ without boundary . This leads to a decomposition of $\mathbb{R}^4$ into a family of submanifolds, and every integral curve of X lies completely in one of them.

c) Show that each of the submanifolds $M^2(E, P)$ lies in an affine subspace of dimension three and is diffeomorphic to $\mathbb{R}^2$.

d) If $\xi(t) = (x(t), y(t), \dot{x}(t), \dot{y}(t))$ is an integral curve of X in $M^2(E, P)$, then

$$\dot{x} + \dot{y} = P \quad \text{and} \quad \int \frac{\dot{x} - \dot{y}}{\sqrt{4E - P^2 - 4e^{x-y}}} = t.$$

Show that $(A > 0)$

$$\int \frac{dz}{\sqrt{A - Be^z}} = \frac{1}{\sqrt{A}} \ln \left(\frac{\sqrt{A} - \sqrt{A - Be^z}}{\sqrt{A} + \sqrt{A - Be^z}} \right),$$

and use this formula to integrate the equations for the integral curves in $M^2(E, P)$ completely.

9 (Euler Equations). Let $I : \mathbb{R}^3 \to \mathbb{R}^3$ be a symmetric positive definite operator. Consider the differential equation $I(\dot{\omega}) = I(\omega) \times \omega$, where $\times$ denotes the vector product.

a) Prove that this differential equation has two first integrals, the energy $2E = \langle I(\omega), \omega \rangle$ and the momentum $M^2 = \langle I(\omega), I(\omega) \rangle$.

b) Conclude from this that the integral curves of the differential equation are intersections of two ellipsoids with center $0 \in \mathbb{R}^3$ $(I \neq \mathrm{Id}_{R^3})$. In particular, all integral curves are closed. Let I_1 be the smallest and I_3 the largest eigenvalue of I. Then there exists an integral curve only for $2EI_1 < M^2 < 2EI_3$. .

10 (Motion in a Central Force Field). Let a potential force F act on a point of mass one in $\mathbb{R}^3$,

$$F = -\frac{dU(r)}{dr} \cdot \frac{x}{r} = -\mathrm{grad}(U(r))$$

with a function $U(r)$ depending on the radius $r = ||x||$ only. The energy $E = ||\dot{x}||^2/2 + U(r)$ is a first integral.

a) Show that $M = x \times \dot{x}$ is a further first integral (M is the angular momentum), and conclude from this that every trajectory of the point lies in a plane in $\mathbb{R}^3$.

b) Determine for which values of the parameter $M \in \mathbb{R}^3$ the level surface

$$A^3(M) = \{(x, \dot{x}) \in \mathbb{R}^6 : x \times \dot{x} = M\}$$

is a three-dimensional submanifold of phase space $\mathbb{R}^6$.

c) Let $r(t)$ be the distance to the origin of a motion whose angular momentum is M. Then

$$\frac{d^2 r}{dt^2} = -\frac{dU}{dr} + \frac{||M||^2}{r^3},$$

i. e., $r(t)$ describes the motion of a point in $\mathbb{R}^1$ under the effective force $F_2^* = -\text{grad}(V)$ with effective potential $V(r) = U(r) + ||M||^2/2r^2$. The energy of this motion is

$$E^* = \frac{\dot{r}^2}{2} + U(r) + \frac{||M||^2}{2r^2}.$$

Prove that this energy E^* coincides with the energy E (for fixed angular momentum M).

Hence, if $x(t) \in \mathbb{R}^3$ is the trajectory of a point moving under a force F and $r(t)$ is its distance to $0 \in \mathbb{R}^3$, then

$$\sqrt{2E - \frac{||M||^2}{r^2} - 2U(r)} = \dot{r}, \quad \text{i. e.} \quad \int \frac{\dot{r}}{\sqrt{2E - ||M||^2/r^2 - 2U(r)}}\, dt = t.$$

11 (Classical Moment Map)**.** Consider $\mathbb{R}^3$ with the defining representation of $\mathrm{SO}(3, \mathbb{R})$.

a) Prove that this representation is equivalent to the adjoint representation of $\mathrm{SO}(3, \mathbb{R})$ on its Lie algebra $\mathfrak{so}(3, \mathbb{R})$, if $\mathbb{R}^3$ and $\mathfrak{so}(3, \mathbb{R})$ are identified via the map

$$\widehat{\ }: \mathbb{R}^3 \longrightarrow \mathfrak{so}(3, \mathbb{R}), \quad v = \begin{bmatrix} v_1 \\ v_2 \\ v_3 \end{bmatrix} \longmapsto \widehat{v} = \begin{bmatrix} 0 & -v_3 & v_2 \\ v_3 & 0 & -v_1 \\ -v_2 & v_1 & 0 \end{bmatrix}.$$

Moreover, the adjoint and the coadjoint representation of $\mathrm{SO}(3, \mathbb{R})$ are also equivalent.

b) Prove that this identification satisfies the equation

$$\widehat{v}(w) = v \times w, \quad [\widehat{v}, \widehat{w}] = \widehat{[v, w]}, \quad \langle v, w \rangle = -\frac{1}{2}\text{tr}\,(\widehat{v}\widehat{w}).$$

c) By a), the moment map $\Psi : T^*\mathbb{R}^3 \to \mathfrak{so}(3,\mathbb{R})^*$ can be interpreted as a map from $T^*\mathbb{R}^3$ to $\mathbb{R}^3$. Show that it can then be written in the form

$$\Psi(q,p) = q \times p,$$

and hence it coincides with the classical angular momentum.

Chapter 8

Elements of Statistical Mechanics and Thermodynamics

8.1. Statistical States of a Hamiltonian System

The Hamiltonian formulation of mechanics starts from a configuration space X^m and makes use of the phase space T^*X^m with its canonical symplectic structure. A state of the mechanical system under consideration is a point in the phase space T^*X^m, and the motions of states are the integral curves of the symplectic gradient s-grad(H) of a Hamilton function $H : T^*X^m \to \mathbb{R}$. In this formulation of mechanics, the only essential data to be given are a symplectic manifold M^{2m} and a function H. The point of view of statistical mechanics is based on the idea that, for instance because of the size of the mechanical system or as a consequence of the imprecision of measurements, the state of the mechanical system cannot be determined precisely by fixing $2m$ real parameters. Instead, to each open set $U \subset M^{2m}$, we can only ascribe the probability $\mu(U)$ that the state belongs to the set U. This leads to the concept of considering mechanical states no longer as points in the phase space M^{2m}, but as probability measures on M^{2m}.

Definition 1. Let a Hamiltonian system (M^{2m}, ω, H) consisting of a symplectic manifold and a Hamilton function be given. A *statistical state* is a probability measure μ defined on the σ-algebra of all Borel sets on M^{2m}.

Example 1. For each point $x \in M^{2m}$, consider the measure δ_x concentrated at this point,

$$\delta_x(U) = \begin{cases} 0 & \text{if } x \notin U, \\ 1 & \text{if } x \in U. \end{cases}$$

Therefore, classical mechanical states are particular statistical states.

Let $\Phi_t : M^{2m} \to M^{2m}$ be the flow of the symplectic gradient s-grad(H). The motion of the classical state $x \in M^{2m}$ is the trajectory $\Phi_t(x)$. For the probability measure $\delta_{\Phi_t(x)}$ corresponding to the point $\Phi_t(x)$, we have

$$\delta_{\Phi_t(x)}(U) = \left\{ \begin{array}{l} 0\,, \ x \notin \Phi_t^{-1}(U) \\ 1\,, \ x \in \Phi_t^{-1}(U) \end{array} \right\} = \delta_x(\Phi_t^{-1}(U)),$$

and this formula leads to the following definition.

Definition 2. Let a Hamiltonian system (M^{2m}, ω, H) and a statistical state μ be given. The motion of μ under the impact of the Hamiltonian system is the curve μ_t of measures

$$\mu_t(U) := \mu(\Phi_t^{-1}(U)).$$

Definition 3. An *equilibrium state* of a Hamiltonian system (M^{2m}, ω, H) is a state μ which does not change in the course of the motion of s-grad(H), $\mu_t = \mu$.

Definition 4. A statistical state μ of a Hamiltonian system (M^{2m}, ω, H) has a *stationary terminal distribution* if the equation

$$\mu_\infty(U) := \lim_{t \to \infty} \mu_t(U)$$

defines a Borel measure on M^{2m}.

Theorem 1. *A stationary terminal distribution μ_∞ is always an equilibrium state.*

Consider the volume form

$$dM^{2m} = \frac{(-1)^{m(m-1)/2}}{m!}\,\omega^m$$

of the symplectic manifold and the Borel measure induced from it (see the final remark at the end of §3.5). If the measure μ is absolutely continuous with respect to the volume measure,

$$\mu(U) := \int_U \varrho(x) \cdot dM^{2m}(x),$$

then ϱ is called the *density function* of the state μ.

Theorem 2 (Liouville's Equation). *Let $\mu = \varrho \cdot dM^{2m}$ be a state with density function, and let μ_t be the motion of this state in a Hamiltonian system. Then the measures $\mu_t = \varrho_t \cdot dM^{2m}$ are also states with density function, and*

$$\frac{d}{dt}\varrho_t \;=\; -\{H, \varrho\} \circ \Phi_{-t}\,.$$

Proof. The flow Φ_t consists of symplectic transformations and preserves, in particular, the symplectic volume form dM^{2m}, $\Phi_t^*(dM^{2m}) = dM^{2m}$. Transforming the integral, we see that $\varrho_t = \varrho \circ \Phi_{-t}$ is the density function of the state μ_t. This implies

$$\frac{d}{dt}\varrho_t = \frac{d}{dt}\varrho \circ \Phi_{-t} = d\varrho \circ \frac{d}{dt}\Phi_{-t} = -\,\text{s-grad}(H)(\varrho) \circ \Phi_{-t} = -\{H, \varrho\} \circ \Phi_{-t}\,.$$

□

Corollary 1. *A statistical state μ with density function ϱ is an equilibrium state if and only if ϱ is a first integral of the Hamiltonian H, $\{H, \varrho\} = 0$.*

If $x \in M^{2m}$ is a classical state in Hamiltonian mechanics, the value $H(x)$ of the Hamilton function H is the energy E of the state,

$$E(x) \;=\; H(x) \;=\; \int_{M^{2m}} H(x) \cdot d(\delta_x)\,.$$

The definition of the energy of a statistical state generalizes this relation:

Definition 5. The *energy* of a statistical state μ is the integral

$$E(\mu) \;=\; \int_{M^{2m}} H(x) \cdot d\mu(x)\,,$$

if this integral exists.

Theorem 3 (Conservation of Energy in Statistical Mechanics). *If μ_t is the motion of a statistical state in a Hamiltonian system, the energy $E(\mu_t)$ is constant.*

Proof. After transformation of the integral, we immediately obtain

$$E(\mu_t) \;=\; \int_{M^{2m}} H \circ \Phi_t^{-1}(x) \cdot d\mu(x),$$

and differentiating this with respect to the parameter t yields the formula

$$\frac{d}{dt}E(\mu_t) \;=\; \int_{M^{2m}} \{H, H\}(x) \cdot d\mu(x) \;=\; 0\,.$$ □

Now we want to study how the probability $\mu_t(N^{2m})$ that the state $\mu = \varrho \cdot dM^{2m}$ is in the compact subset $N^{2m} \subset M^{2m}$ at time t changes in time. The following lemma serves as a preparation.

Lemma 1. *Let (M^{2m}, ω) be a symplectic manifold. Then, for any two functions $f, g : M^{2m} \to \mathbb{R}$, the following formula holds:*

$$df \wedge dg \wedge \omega^{m-1} = \frac{1}{m}\{f, g\} \cdot \omega^m .$$

Proof. To prove this formula, we choose on M^{2m} local symplectic coordinates $(q_1, \ldots, q_m, p_1, \ldots, p_m)$,

$$\omega = \sum_{\alpha=1}^{m} dp_\alpha \wedge dq_\alpha .$$

Setting $A_i := dp_i \wedge dq_i$, we see that the exterior product satisfies the relations

$$A_i \wedge A_j = A_j \wedge A_i \quad \text{and} \quad A_i \wedge A_i = 0 .$$

Since the forms A_i commute with one another, we can use the binomial formula to compute the exterior power ω^k,

$$\omega^m = \left(\sum_{i=1}^{m} A_i\right)^m = \sum_{\alpha_1+\ldots+\alpha_m=m} \frac{m!}{\alpha_1!\ldots\alpha_m!} A_1^{\alpha_1} \wedge \ldots \wedge A_m^{\alpha_m} .$$

Because $A_i \wedge A_i = 0$, only summands with $\alpha_i \leq 1$ occur in this sum, and hence it reduces to a single term:

$$\omega^m = m! \cdot A_1 \wedge \ldots \wedge A_m = m! \cdot (dp_1 \wedge dq_1) \wedge \ldots \wedge (dp_m \wedge dq_m) .$$

We compute ω^{m-1} in a similar way, and obtain

$$\omega^{m-1} = (m-1)! \sum_{\beta=1}^{m} A_1 \wedge \ldots \widehat{A}_\beta \ldots \wedge A_m .$$

Thus, the following exterior products vanish:

(1) $dp_i \wedge dp_j \wedge \omega^{m-1} = 0 = dq_i \wedge dq_j \wedge \omega^{m-1}$,

(2) $dp_i \wedge dq_j \wedge \omega^{m-1} = 0 \quad \text{if } i \neq j$,

and we obtain

$$\begin{aligned} df \wedge dg \wedge \omega^{m-1} &= \sum_{i=1}^{m} \left(\frac{\partial f}{\partial p_i} \frac{\partial g}{\partial q_i} dp_i \wedge dq_i + \frac{\partial f}{\partial q_i} \frac{\partial g}{\partial p_i} dq_i \wedge dp_i \right) \wedge \omega^{m-1} \\ &= (m-1)! \sum_{i=1}^{m} \left(\frac{\partial f}{\partial p_i} \frac{\partial g}{\partial q_i} - \frac{\partial f}{\partial q_i} \frac{\partial g}{\partial p_i} \right) \cdot A_1 \wedge \ldots \wedge A_m \\ &= \frac{(m-1)!}{m!} \{f, g\} \cdot \omega^m . \end{aligned}$$

□

The relations $m \cdot d(f \cdot dg \wedge \omega^{m-1}) = m \cdot df \wedge dg \wedge \omega^{m-1} = \{f, g\} \cdot \omega^m$ together with Stokes' theorem imply, after integration,

Theorem 4. *Let $(M^{2m}\omega)$ be a compact symplectic manifold with boundary. Then*

$$\int_{M^{2m}} \{f,g\}\cdot\omega^m \;=\; m\cdot\int_{\partial M^{2m}} f\cdot dg\wedge\omega^{m-1}.$$

In particular, the integral

$$\int_{M^{2m}} \{f,g\}\cdot\omega^m \;=\; 0$$

vanishes for any compact symplectic manifold without boundary.

Corollary 2. *If the Poisson bracket $\{f,g\}$ of two functions defined on a compact symplectic manifold M^{2m} without boundary does not change sign, then it vanishes identically, $\{f,g\}\equiv 0$.*

We will apply these formulas to a statistical state with density function ϱ.

Definition 6. The $(2m-1)$-form

$$j(\mu) \;:=\; \frac{(-1)^{m(m-1)/2}}{(m-1)!}\,\varrho\cdot dH\wedge\omega^{m-1}$$

is called the *probability current* of the statistical state $\mu=\varrho\cdot dM^{2m}$ in the symplectic manifold (M^{2m},ω) with Hamilton function H.

Theorem 5. *Let μ_t be the motion of the statistical state $\mu=\varrho\cdot dM^{2m}$ in the Hamiltonian system (M^{2m},ω,H), and let $N^{2m}\subset M^{2m}$ be a compact submanifold. Then*

$$\frac{d}{dt}\Big(\mu_t(N^{2m})\Big)\Big|_{t=0} \;=\; \int_{\partial N^m} j(\mu)\,.$$

Proof. We compute the derivative with respect to time at $t=0$ and use Liouville's equation as well as the preceding integral formulas:

$$\begin{aligned}\frac{d}{dt}\mu_t(N^{2m})\Big|_{t=0} &= \int\limits_{N^{2m}} \frac{d\varrho_t}{dt}\Big|_{t=0}\cdot dM^{2m} = -\int\limits_{N^{2m}}\{H,\varrho\}\cdot dM^{2m}\\ &= \frac{(-1)^{m(m-1)/2}}{(m-1)!}\int\limits_{\partial N^{2m}}\varrho\cdot dH\wedge\omega^{m-1} = \int\limits_{\partial N^{2m}} j(\mu)\,.\end{aligned}$$ □

The probability current $j(\mu)$ is a hypersurface measure on all $(2m-1)$-dimensional submanifolds of the symplectic manifold M^{2m}. It expresses the infinitesimal change of the probability for the state μ to be in the set $N^{2m}\subset M^{2m}$ as an integral over the boundary of N^{2m}.

Example 2. If the submanifold $N^{2m}\subset M^{2m}$ is described by inequalities of the Hamilton function,

$$N^{2m} \;=\; \{x\in M^{2m}:\; C_1\le H(x)\le C_2\}\,,$$

then the differential dH vanishes on the tangent bundle $T(\partial N^{2m})$ of the boundary, and so does the form $j(\mu)$. We hence obtain, for the motion μ_t of every statistical state $\mu = \varrho \cdot dM^{2m}$,

$$\frac{d}{dt}\mu_t\big(\{x \in M^{2m} : C_1 \leq H(x) \leq C_2\}\big) = 0 .$$

Thus the probability for a state μ to be in N^{2m} at time t is constant.

Example 3. In case $m = 1$, the formulas for the probability current of a two-dimensional Hamiltonian system (M^2, ω, H) simplify:

$$j(\mu) = \varrho \cdot dH \quad \text{and} \quad \frac{d}{dt}\mu_t(N^2) = \int_{\partial N^2} \varrho \cdot dH .$$

Now we turn to the notion of information entropy for a statistical state. As a motivation, we first recall the notion of *information* of an event introduced by C. E. Shannon (1948). The heuristics is the following: If, in a series of experiments, an event occurs with probability close to one, then the "information" contents of this event is small. Conversely, if the probability of this event is close to zero, the occurrence of this event contains a large amount of "information". Modeling the probability computation by a triple $(\Omega, \mathfrak{A}, \mu)$ consisting of a set Ω, a σ-algebra $\mathfrak{A}$ of subsets of Ω, and a measure μ defined on $\mathfrak{A}$ such that $\mu(\Omega) = 1$, we arrive at the following definition for the Shannon information.

Definition 7. Let $(\Omega, \mathfrak{A}, \mu)$ be a probability space. The amount of *information* contained in an event $A \in \mathfrak{A}$ is

$$I(A) := -\log(\mu(A)) .$$

For a finite set Ω, one forms the mean of these information amounts and gets in this way the so-called information entropy.

Definition 8. Let (Ω, μ) be a finite probability space. The *information entropy* of this space is

$$S(\Omega, \mu) := \int_\Omega I(\omega) \cdot d\mu(\omega) = -\sum_{i=1}^{n} p_i \cdot \log(p_i) ,$$

where the set Ω consists of the elements $\{\omega_1, \ldots, \omega_n\}$, and $p_i := \mu(\{\omega_i\})$ is the probability for the event ω_i.

The information entropy is a measure for the uncertainty of the probability space. In fact, if $\Omega = \{\omega_1, \ldots, \omega_n\}$ consists of n points and we denote by μ^* the measure corresponding to the equidistribution, $\mu^*(\omega_i) := 1/n$, then the following holds.

Theorem 6. *The information entropy of a finite probability space* (Ω, μ) *does not exceed the information entropy of the equidistribution:*

$$S(\Omega, \mu) \;\leq\; S(\Omega, \mu^*)\,.$$

Equality occurs if and only if $\mu = \mu^*$ *is the equidistribution.*

Proof. The proof is based on the fact that the function $f(x) := x \cdot \log(x)$ is convex on the half-line $(0, \infty)$, since

$$f'(x) \;=\; 1 + \log(x) \quad \text{and} \quad f''(x) \;=\; \frac{1}{x} \;>\; 0\,.$$

The convexity of f then implies the inequality

$$f\left(\frac{1}{n}\sum_{i=1}^{n} p_i\right) \;\leq\; \frac{1}{n}\sum_{i=1}^{n} f(p_i) \;=\; \frac{1}{n}\sum_{i=1}^{n} p_i \cdot \log(p_i)\,.$$

Using the relation $\sum_{i=1}^{n} p_i = 1$, one immediately deduces from it that

$$S(\Omega, \mu) \;=\; -\sum_{i=1}^{n} p_i \cdot \log(p_i) \;\leq\; -\log\left(\frac{1}{n}\right) \;=\; S(\Omega, \mu^*)\,. \qquad \square$$

The formula $S(\Omega, \mu) \;=\; -\sum p_i \cdot \log(p_i)$ leads to the notion of information entropy for a statistical state with density function.

Definition 9. Let (M^{2m}, ω, H) be a Hamiltonian system, and let $\mu = \varrho \cdot dM^{2m}$ be a statistical state with density function. We define its *information entropy* as the integral

$$S(\mu) \;:=\; -\int_{M^{2m}} \varrho \cdot \log(\varrho) \cdot dM^{2m},$$

if this integral exists.

Remark. This notion of entropy is not without problems in statistical physics. The following result concerning the conservation of entropy claims that the information entropy does not change in the course of the motion of a state. This is a good assumption for *reversible* processes (and processes which can be approximated by reversible ones), but cannot be applied to the vast range of *irreversible* processes, in which the entropy increases. For this reason, we will apply the notion of information entropy $S(\mu)$ only to equilibrium states.

Theorem 7 (Conservation of Information Entropy). *The information entropy* $S(\mu_t)$ *of a statistical state with density function is constant under the motion of a Hamiltonian system.*

Proof. We differentiate the function $S(\mu_t)$ with respect to the parameter t and use Liouville's equation:

$$\begin{aligned}\frac{d}{dt}S(\mu_t) &= -\int_{M^{2m}}\left(\frac{d\varrho}{dt}\log(\varrho)+\frac{d\varrho}{dt}\right)\cdot dM^{2m}\\ &= \int_{M^{2m}}\Big(\{H,\varrho\}\log(\varrho)+\{H,\varrho\}\Big)\cdot dM^{2m}.\end{aligned}$$

But since $\{H,\log(\varrho)\} = \{H,\varrho\}/\varrho$, we have

$$\{H,\varrho\cdot\log(\varrho)\} = \{H,\varrho\}\cdot\log(\varrho)+\{H,\log(\varrho)\}\cdot\varrho = \{H,\varrho\}+\{H,\varrho\}\cdot\log(\varrho),$$

and this yields

$$\begin{aligned}\frac{d}{dt}\,S(\mu_t)\big|_{t=0} &= \int_{M^{2m}}\{H,\varrho\cdot\log(\varrho)\}\cdot dM^{2m}\\ &= \frac{(-1)^{m(m-1)/2}}{m!}\int_{M^{2m}}\{H,\varrho\cdot\log(\varrho)\}\cdot\omega^m\\ &= \frac{(-1)^{m(m-1)/2}}{(m-1)!}\int_{M^{2m}}dH\wedge d(\varrho\cdot\log(\varrho))\wedge\omega^{m-1}.\end{aligned}$$

Since M^{2m} has no boundary, Stokes' theorem immediately implies the assertion. □

Remark. In the case of a manifold M^{2m} with boundary we obtain

$$\frac{d}{dt}\,S(\mu_t)\big|_{t=0} = -\frac{(-1)^{m(m-1)/2}}{(m-1)!}\int_{\partial M^{2m}}\varrho\cdot\log(\varrho)\cdot dH\wedge\omega^{m-1}.$$

A Hamiltonian system (M^{2m},ω,H) has several special equilibrium states. These are characterized as realizing the maximum of information entropy in a particular class of statistical states. We will discuss two of these states, the so-called Gibbs state (or the canonical ensemble) and the microcanonical ensemble. We will, in general, make the following three assumptions for the Hamiltonian system (M^{2m},ω,H):

Assumption 1. The Hamilton function H is non-negative,

$$H(x)\ \geq\ 0 \quad \text{for all } x\in M^{2m}.$$

Assumption 2. For every positive number $B>0$, the following integral exists:

$$Z(B) := \int_{M^{2m}} e^{-H(x)/B}\cdot dM^{2m}(x).$$

The function $Z(B)$ is called the *partition function* of the Hamiltonian system.

Assumption 3. For every positive number $B > 0$, the following integral exists:

$$\int_{M^{2m}} H(x) \cdot e^{-H(x)/B} \cdot dM^{2m}(x)\,.$$

Remark. The parameter B will later be identified with absolute temperature (multiplied by the Boltzmann constant k). Its appearance here indicates the exceptional role that temperature plays among all thermodynamic parameters.

The function

$$E(B) \;:=\; \frac{1}{Z(B)} \int_{M^{2m}} H(x) e^{-H(x)/B} \cdot dM^{2m}(x)$$

is called the *inner energy* of the Hamiltonian system (M^{2m}, ω, H); its *inner entropy* is

$$S(B) \;:=\; \log(Z(B)) + \frac{E(B)}{B}\,.$$

Finally, the *free energy* $F(B)$ is defined by

$$F(B) \;:=\; -B \cdot \log(Z(B)) \;=\; E(B) - B \cdot S(B).$$

First we note that the inner energy $E(B)$ is a non-decreasing function on the interval $(0, \infty)$. Its derivative is

$$\frac{dE}{dB} = \frac{1}{Z^2(B) \cdot B} \Bigg[\Big(\int\limits_{M^{2m}} e^{-\frac{H(x)}{B}} dM^{2m}(x) \Big) \Big(\int\limits_{M^{2m}} H^2(x) e^{-\frac{H(x)}{B}} dM^{2m}(x) \Big) - \Big(\int\limits_{M^{2m}} H(x) e^{-\frac{H(x)}{B}} dM^{2m}(x) \Big)^2 \Bigg],$$

and the Cauchy-Schwarz inequality shows that this derivative is positive ($H \neq$ const). Denote by $E_{\min}$ and $E_{\max}$, respectively, the bounds of the range of the inner energy $E : (0, \infty) \to \mathbb{R}$. We calculate the derivative of the partition function in a similar way:

$$\frac{dZ}{dB} \;=\; \frac{1}{B^2} \int_{M^{2m}} H(x) e^{-H(x)/B} \cdot dM^{2m}(x) \;=\; \frac{1}{B^2} Z(B) \cdot E(B)\,.$$

Hence the inner energy as well as the inner entropy can be expressed in terms of the partition function. We summarize these formulas.

Theorem 8 (Simple Equation of State for a Hamiltonian System).

(1) $E(B) = B^2 \frac{d}{dB}\Big(\log(Z(B))\Big)$;

(2) $S(B) = \frac{d}{dB}\Big(B \cdot \log(Z(B))\Big)$;

(3) *in the sense of* 1*-forms on the one-dimensional manifold* $(0,\infty)$,

$$dS = \frac{1}{B} \cdot dE .$$

Proof. The third equation is a straightforward consequence of the first two. □

The Gibbs state (canonical ensemble) is distinguished by the property that it realizes the maximum of the information entropy among all states of fixed energy.

Theorem 9 (Gibbs State, Canonical Ensemble). *Let an energy value* E_0 *between the minimum and the maximum of the energy function* $E(B)$ *for the Hamiltonian system* (M^{2m}, ω, H) *be given. Among all statistical states* $\mu = \varrho \cdot dM^{2m}$ *with density function of energy* $E(\mu) = E_0$*, there exists precisely one state,* $\mu_{\text{Gibbs}} = \varrho_{\text{Gibbs}} \cdot dM^{2m}$*, of maximal information entropy* $S(\mu_{\text{Gibbs}})$*. The density function of this state is*

$$\varrho_{\text{Gibbs}}(x) = \frac{1}{Z(B_0)} e^{-H(x)/B_0},$$

where the parameter B_0 *is determined as a solution of the equation* $E(B_0) = E_0$*. The value of the maximal entropy* $S(\mu_{\text{Gibbs}})$ *is*

$$S(\mu_{\text{Gibbs}}) = S(B_0) = \log(Z(B_0)) + \frac{E(B_0)}{B_0} .$$

Proof. Choose B_0 as a solution of the equation $E(B_0) = E_0$, and let

$$\mu_{\text{Gibbs}} = \frac{1}{Z(B_0)} e^{-H(x)/B_0} \cdot dM^{2m}$$

be the corresponding Gibbs state. Its inner energy is

$$E(\mu_{\text{Gibbs}}) = \frac{1}{Z(B_0)} \int_{M^{2m}} H(x) e^{-H(x)/B_0} \cdot dM^{2m} = E(B_0) = E_0,$$

and hence μ_{Gibbs} is a statistical state of energy E_0. For every other state $\mu = \varrho \cdot dM^{2m}$ with energy

$$E(\mu) = \int_{M^{2m}} H(x)\varrho(x) \cdot dM^{2m} = E_0,$$

we consider the function

$$f(t) = S\big((1-t)\mu_{\text{Gibbs}} + t\mu\big) ,$$

the information entropy of the statistical state $(1-t)\mu_{\text{Gibbs}} + t\mu$. Differentiating $f(t)$ with respect to the parameter t, we obtain

$$\frac{d^2 f(t)}{dt^2} = -\int_{M^{2m}} \frac{(\varrho - \varrho_{\text{Gibbs}})^2}{(1-t)\varrho_{\text{Gibbs}} + t\varrho} \cdot dM^{2m} \leq 0 .$$

Moreover,

$$\left.\frac{df}{dt}\right|_{t=0} \;=\; -\int_{M^{2m}} \varrho\cdot\log(\varrho_{\mathrm{Gibbs}})\cdot dM^{2m} - S(\mu_{\mathrm{Gibbs}}),$$

and calculating the integral using $E(\mu)=E_0$ leads to

$$\begin{aligned}\int_{M^{2m}} \varrho\cdot\log(\varrho_{\mathrm{Gibbs}})\cdot dM^{2m} \;&=\; \int_{M^{2m}} \varrho\Big(-\frac{H}{B_0}-\log(Z(B_0))\Big)\cdot dM\\ &=\; -\log(Z(B_0)) - \frac{1}{B_0}E_0 \;=\; -S(B_0)\,.\end{aligned}$$

We conclude that the derivative of f vanishes at $t=0$. Hence the derivative is non-positive, and $f(t)$ is decreasing in $[0,1]$. Since $f(0)\geq f(1)$, we obtain

$$S(\mu)\;\leq\; S(\mu_{\mathrm{Gibbs}})\,.$$

If $S(\mu)=S(\mu_{\mathrm{Gibbs}})$, the first derivative of $f(t)$ vanishes identically. Hence the second derivative of this function is also zero, and from the formula stated before we conclude that $\varrho=\varrho_{\mathrm{Gibbs}}$. The assertions of the theorem are proved. □

Example 4 (Maxwell and Boltzmann Distribution)**.** In classical statistics, the Hamilton function of a particle with potential energy depending only on the coordinates q can be written as

$$H(q,p) \;=\; \frac{p^2}{2m}+U(q) \;=\; \frac{p_1^2+p_2^2+p_3^2}{2m}+U(q)\,.$$

Thus, the partition function is equal to

$$Z(B)=\int\limits_{\mathbb{R}^3}\exp\left[-\frac{p_1^2+p_2^2+p_3^2}{2mB}\right]dp_1\wedge dp_2\wedge dp_3\cdot\int\limits_{\mathbb{R}^3}\exp\left[\frac{-U(q)}{B}\right]dq_1\wedge dq_2\wedge dq_3,$$

which, using the integral $\int_{\mathbb{R}} e^{-\alpha x^2}dx=\sqrt{\pi/\alpha}$, can also be written as

$$Z(B)\;=\;(2\pi mB)^{3/2}\cdot Z_q(B)\,;$$

$Z_q(B)$ denotes the value of the integral in which the unknown potential energy occurs. In the same way, the Gibbs state factors into the probability measure μ_p of momentum values and that of coordinate values μ_q,

$$\mu_{\mathrm{Gibbs}} = \mu_p\cdot\mu_q \;=\; \frac{1}{(2\pi mB)^{3/2}}e^{-p^2/2mB}dp\cdot\frac{1}{Z_q}e^{-U/B}dq\,.$$

Going from the momenta to the velocities, we arrive at the so-called *Maxwell distribution*,

$$\mu_p \;=\; \Big(\frac{m}{2\pi B}\Big)^{3/2}\exp\left[-\frac{m(v_1^2+v_2^2+v_3^2)}{2B}\right]dv_1dv_2dv_3,$$

which again is the product of probability distributions for every single velocity component. The *Boltzmann ensemble* is defined to be the probability measure μ_q and is proportional to the particle counting measure ν. For example, for the potential $U(x, y, z) = mgz$ of a homogeneous gravitational field which is parallel to the z-axis, this yields the *barometric formula*

$$\nu = \nu_0 \cdot e^{-mgz/B} dx\, dy\, dz\,.$$

The second natural equilibrium distribution of a Hamiltonian system starts from a subset $A \subset M^{2m}$ of the symplectic manifold which is invariant under the flow Φ_t of the Hamilton function H and has positive finite volume,

$$\Phi_t(A) \subset A, \quad 0 < \mathrm{vol}(A) < \infty\,.$$

Consider the density function

$$\varrho_A(x) = \begin{cases} 1/\mathrm{vol}(A) & \text{if } x \in A, \\ 0 & \text{if } x \in A, \end{cases}$$

as well as the statistical state $\mu_A = \varrho_A \cdot dM^{2m}$. This is an equilibrium state, since A is invariant under the flow Φ_t. Its information entropy is

$$S(\mu_A) = -\int_{M^{2m}} \varrho_A(x) \cdot \log(\varrho_A(x)) \cdot dM^{2m}(x) = \log(\mathrm{vol}(A))\,.$$

Theorem 10 (Microcanonical Ensemble)**.** *Let $A \subset M^{2m}$ be an invariant subset for the flow of the Hamiltonian system (M^{2m}, ω, H) with positive finite measure. If $\mu = \varrho \cdot dM^{2m}$ is a statistical state whose density function has support in A, $\mathrm{supp}(\varrho) \subset A$, then its information entropy is always bounded from above by the information entropy of μ_A,*

$$S(\mu) \leq S(\mu_A)\,.$$

Equality only occurs for $\mu = \mu_A$.

Hence the microcanonical ensemble μ_A realizes the maximum of the information entropy among all statistical states concentrated in the set A.

Proof. The proof proceeds in complete analogy to that of the preceding theorem. Again, we consider the function $f(t) = S((1-t)\mu_A + t\mu)$ and calculate its second derivative:

$$\frac{d^2 f}{dt^2} = -\int_A \frac{(\varrho - \varrho_A)^2}{(1-t)\varrho_A + \varrho} \cdot dM^{2m} \leq 0\,.$$

The derivative of f at $t = 0$ vanishes:

$$\left.\frac{df}{dt}\right|_{t=0} = -\int_A (\varrho - \varrho_A) \cdot \log(\varrho_A) \cdot dM^{2m} = -\log(\varrho_A) \int_A (\varrho - \varrho_A) \cdot dM^{2m} = 0\,,$$

since μ is a probability measure with support in A. □

8.2. Thermodynamical Systems in Equilibrium

First we generalize the simple equations of state of a Hamiltonian system by keeping a fixed symplectic manifold (M^{2m}, ω) but considering a family of Hamilton functions $H(x, y)$ depending on an additional parameter $y \in N^r$. Let the parameter space also be a smooth manifold, and suppose that the Hamilton function $H(x, y_0)$ satisfies the three assumptions of the preceding section for any fixed value of the parameter $y_0 \in N^r$. Then the *partition function* $Z(B, y)$, the *inner energy* $E(B, y)$, the *inner entropy* $S(B, y)$, and the *free energy* $F(B, y)$ become functions on the manifold $(0, \infty) \times N^k$,

$$Z, E, S, F: \ (0, \infty) \times N^k \longrightarrow \mathbb{R},$$

and, of course, the equation of state $B \cdot d_B S = d_B E$ holds for each fixed value of the parameter $y_0 \in N^r$. The general equation of state of a Hamiltonian system depending on parameters is a relation between the complete differentials considered as 1-forms on $(0, \infty) \times N^r$.

Theorem 11 (General Equation of State for Parameter-Dependent Hamiltonian Systems). *In the sense of an equality for* 1*-forms defined on the manifold* $(0, \infty) \times N^r$, *one has*

$$dE \;=\; B \cdot dS + \frac{E}{B} \cdot dB - B \cdot d(\log(Z)) .$$

Proof. By the definition of inner energy, we have

$$B \cdot S \;=\; B \cdot \log(Z) + E .$$

Differentiating this equation,

$$B \cdot dS + S \cdot dB \;=\; \log(Z) \cdot dB + B \cdot d(\log(Z)) + dE ,$$

and again inserting S, yields

$$B \cdot dS + \frac{E}{B} \cdot dB - B \cdot d(\log(Z)) \;=\; dE . \qquad \square$$

Remark. If we allow the until now independent variable B to be a function $B : N^r \to (0, \infty)$ on the parameter space N^r, we may consider all the functions Z, E, S and F as functions on N^r by means of the map

$$N^r \longrightarrow (0, \infty) \times N^r, \quad y \longrightarrow (B(y), y) .$$

The exterior derivative commutes with the pullback of forms and thus implies on the manifold N^r the equation

$$dE \;=\; B \cdot dS + \frac{E}{B} \cdot dB - B \cdot d(\log(Z)) .$$

The macroscopic description of thermodynamic systems starts from a few assumptions. First, all thermodynamic systems should be in mechanical equilibrium. Furthermore, it is supposed that no particle exchange takes place during the change of states. Lastly, the macroscopic state of the thermodynamic system is to be completely described by a parameter $y \in N^k$, which is an element of a finite-dimensional manifold. If this holds, the following phenomenologically observed laws are historically known as the *fundamental theorems of thermodynamics.* In a modern account of theoretical thermodynamics, they are taken as axioms:

Zeroth Fundamental Theorem of Thermodynamics. *There exists a positive function $T : N^r \to \mathbb{R}$, called the temperature.*

First Fundamental Theorem of Thermodynamics. *There exist two (not necessarily exact) 1-forms dA and dQ on the manifold N^r such that their sum*

$$dA + dQ \;=\; dE$$

is an exact form. The function E is called the energy of the thermodynamic system, dA is called the work form, and dQ is called the heat form.

Second Fundamental Theorem of Thermodynamics. *The 1-form*

$$\frac{dQ}{T} \;=\; dS$$

is an exact form, and the resulting function S is called the inner entropy of the thermodynamic system.

At this level of abstraction, a thermodynamic system in its macroscopic description is a 4-tuple (N^r, T, dA, dQ) consisting of a parameter manifold N^r, a function $T : N^r \to \mathbb{R}$, as well as two 1-forms dA and dQ such that the 1-forms

$$dA + dQ \quad \text{and} \quad \frac{dQ}{T}$$

are exact. Of course, the following relation holds:

$$dE \;=\; dA + T \cdot dS\,,$$

and this is *Gibbs' fundamental equation* of thermodynamics.

Now, with each parameter-dependent Hamiltonian system a thermodynamic system can be associated in a canonical way. In fact, let a symplectic manifold (M^{2m}, ω) and an additional manifold N^r be given. Fix two non-negative functions,

$$B : N^r \to (0, \infty) \quad \text{and} \quad H : M^{2m} \times N^r \to [0, \infty)\,,$$

and consider H as a family of Hamilton functions on the symplectic manifold. Using the Gibbs state,

$$\mu_{\text{Gibbs}}^{B(y)} \;=\; e^{-H(x,y)/B(y)} \cdot dM^{2m}(x)\,,$$

we define the function $T : N^r \to \mathbb{R}^1$ and the 1-forms dA and dQ on N^r as follows:

(1) The temperature of the thermodynamic system is the function B divided by the Boltzmann constant k $(= 1.380 \cdot 10^{-23} W/K)$:

$$T(y) \;:=\; \frac{B(y)}{k}\,.$$

(2) The energy of the thermodynamic system is the inner energy of the Gibbs state $\mu_{\text{Gibbs}}^{B(y)}$:

$$E(y) \;=\; E(B(y)) \;=\; \frac{1}{Z(B(y))} \int\limits_{M^{2m}} H(x,y) e^{-H(x,y)/B(y)} \cdot dM^{2m}(x)\,.$$

(3) The entropy of the thermodynamic system is the inner entropy of the Gibbs state $\mu_{\text{Gibbs}}^{B(y)}$:

$$S(y) \;=\; k \cdot S\Big(\mu_{\text{Gibbs}}^{B(y)}\Big) \;=\; k \cdot \log(Z(B(y))) + k \cdot \frac{E(B(y))}{B(y)}\,.$$

(4) The heat form dQ is the product of temperature with the differential of entropy,

$$dQ \;:=\; T \cdot dS\,.$$

(5) The work form dA is

$$dA \;=\; \frac{E}{B} \cdot dB - B \cdot d(\log(Z))\,.$$

The First Fundamental Theorem of Thermodynamics,

$$dA + dQ \;=\; dE\,,$$

is, in the model just described, an automatic consequence of the general equation of state for parameter-dependent Hamiltonian systems. Hence, by means of the Gibbs states, a thermodynamic system is associated with every such Hamiltonian system. Let $\mathcal{V}$ be a vector field defined on the manifold N^r. Then

$$dA(\mathcal{V}) \;=\; \frac{E}{B} \cdot \mathcal{V}(B) - B \cdot \frac{\mathcal{V}(Z)}{Z}\,,$$

and, since

$$\mathcal{V}(Z) = \int\limits_{M^{2m}} \mathcal{V}(-H/B)e^{-H/B}dM^{2m} = -\frac{1}{B}\int\limits_{M^{2m}} \mathcal{V}(H)e^{-H/B}dM^{2m} + \frac{\mathcal{V}(B)}{B^2}E\cdot Z,$$

we obtain an explicit formula for the work form,

$$dA(\mathcal{V}) \;=\; \frac{1}{Z}\cdot\int_{M^{2m}} \mathcal{V}(H)\cdot e^{-H/B}\cdot dM^{2m}\,.$$

The *pressure* $p_{\mathcal{V}}$ of the thermodynamic system generated by the vector field $\mathcal{V}$ on the parameter space N^r is the function

$$p_{\mathcal{V}}(y) \;:=\; -\;\frac{1}{Z(B(y))}\int_{M^{2m}} \mathcal{V}(H(x,y))\cdot e^{-H(x,y)/B(y)}\cdot dM^{2m}(x)\,.$$

Thus the relation $dA(\mathcal{V}) \;=\; -\,p_{\mathcal{V}}$ holds by definition, and the work form can be represented with respect to a frame $\mathcal{V}_1, \ldots, \mathcal{V}_r$ of vector fields of the manifold N^r in the form

$$dA \;=\; -\,\sum_{i=1}^{r} p_{\mathcal{V}_i}\cdot\sigma_i\,,$$

where $\sigma_1, \ldots, \sigma_r$ is the dual frame of 1-forms on N^r.

Example 5. The pressure p as used in thermodynamics is obtained if the parameter manifold N^r is one-dimensional and describes only the volume V, and $\mathcal{V} = \partial/\partial V$ is chosen as the vector field. It is known that then the only contribution to the form dA comes from the mechanical work of volume change,

$$dA \;=\; -p\cdot dV\,.$$

Considering in the continuum limit for a large number of particles their number as a continuous variable, we see that the amounts of matter $n_1, \ldots, n_r$ of a system consisting of r different chemical components can be chosen as parameters for the manifold N^r. In this case, the chemical work is traditionally expressed by the so-called *chemical potentials* $\mu_1, \ldots, \mu_r$ (which the reader should not confuse with statistical states):

$$dA \;=\; \sum_{i=1}^{r} \mu_i\cdot dn_i\,.$$

The chemical potential μ_i of the i-th matter component is the negative of the pressure associated with the vector field $\partial/\partial n_i$. Similar formulas express the electromagnetic work in the presence of magnetic or electric dipol-momentum densities.

Consider, in particular, the case that there exists a vector field $\mathcal{V}$ on the parameter space N^r such that

$$\mathcal{V}(T) \equiv 1 \quad \text{and} \quad \mathcal{V}(H) \equiv 0\,.$$

All the thermodynamic quantities are then computable starting from the partition function

$$Z = \int_{M^{2m}} e^{-H/kT} dM^{2m}$$

or the free energy derived from it,

$$F = -k \cdot T \cdot \log(Z) .$$

Theorem 12. *Suppose that there exists a vector field $\mathcal{V}$ on the parameter space N^r such that $\mathcal{V}(T) \equiv 1$ and $\mathcal{V}(H) \equiv 0$. Then:*

(1) $S = -\mathcal{V}(F) = -dF(\mathcal{V}) = -\mathcal{V} \lrcorner dF.$

(2) $E = F - T \cdot \mathcal{V}(F) = F + S \cdot T.$

(3) $dQ = - T \cdot \mathcal{L}_{\mathcal{V}}(dF).$

(4) $dA = dF - (\mathcal{V} \lrcorner dF) \cdot dT = dF + S \cdot dT.$

(5) *The pressure generated by an arbitrary vector field $\mathcal{W}$ is*

$$p_{\mathcal{W}} = - dF(\mathcal{W}) + dF(\mathcal{V}) \cdot dT(\mathcal{W}) .$$

In particular, the vector field $\mathcal{V}$ generates no pressure, $p_{\mathcal{V}} \equiv 0$.

Proof. Differentiating the partition function Z with respect to the vector field $\mathcal{V}$, we obtain

$$\mathcal{V}(Z) = \frac{1}{k \cdot T^2} \int_{M^{2m}} He^{-H/kT} \cdot dM^{2m} = \frac{E \cdot Z}{k \cdot T^2} .$$

This implies the following formula for the derivative of the free energy,

$$\mathcal{V}(F) = -k \cdot \log(Z) - k \cdot T \cdot \frac{E}{k \cdot T^2} = -k\Big(\log(Z) + \frac{E}{k \cdot T}\Big) = - S,$$

and the first of the assertions is proved. The relation

$$S = -\mathcal{V}(F) = -(\mathcal{V} \lrcorner dF)$$

implies

$$dS = -d(\mathcal{V} \lrcorner dF) = -\mathcal{L}_{\mathcal{V}}(dF).$$

The expression for the heat form is calculated from its definition,

$$dQ = - T \cdot \mathcal{L}_{\mathcal{V}}(dF) .$$

The energy E is determined by the free energy F as well as the entropy,

$$E = F + T \cdot S = F - T(\mathcal{V} \lrcorner dF) .$$

Lastly, we calculate the work form:

$$\begin{aligned} dA &= dE - T \cdot dS = d(F - T(\mathcal{V} \lrcorner dF)) + T\mathcal{L}_{\mathcal{V}}(dF) \\ &= dF - (\mathcal{V} \lrcorner dF)dT . \end{aligned}$$

□

Example 6 (Ideal Gas). We view an *ideal gas* as n points of equal mass M in a volume region $V \subset \mathbb{R}^3$ which do not interact with one another. The configuration space of the mechanical system consists of the product of n copies of the set V factored by the action of the permutation group consisting of $n!$ elements. If we want to avoid this group identification, we can simply consider the phase space $M := T^*V \times \ldots \times T^*V$ equipped with a measure which already takes into account this identification:

$$dM \;:=\; \frac{1}{n!} \sum_{i=1}^{3n} dq_i \wedge dp_i \,.$$

The Hamilton function of the mechanical system is

$$\frac{1}{2M} \cdot \chi(q_1, q_2, q_3) \ldots \chi(q_{3n-2}, q_{3n-1}, q_{3n})(p_1^2 + \ldots + p_{3n}^2)\,.$$

Here χ denotes the characteristic function of the set $V \subset \mathbb{R}^3$. Calculating the partition function of the thermodynamic system depending only on the two parameters T and V leads to the formula

$$\begin{aligned} Z(T,V) \;&=\; \frac{1}{n!} \int_M e^{-H/kT} dq \wedge dp \;=\; \frac{V^n}{n!} \Big(\int_{-\infty}^{\infty} e^{-p^2/2MkT} dp \Big)^{3n} \\ &=\; \frac{V^n}{n!} (2\pi kTM)^{3n/2} \,. \end{aligned}$$

We apply *Stirling's formula* $\log(n!) \simeq n \cdot \big(\log(n) - 1\big)$ and calculate all the thermodynamic quantities for the ideal gas using this approximation. The free energy is

$$F \;=\; -\, k \cdot T \cdot \log(Z) \;=\; -\, nkT \cdot \big(\log(V/n) + 1 + \frac{3}{2} \log(2\pi kTM)\big)\,.$$

We obtain the entropy and the energy from the free energy according to Theorem 12, since the Hamilton function does not depend on the temperature:

$$S \;=\; -\,\frac{\partial F}{\partial T} \;=\; nk \cdot \big(\log(V/n) + \frac{3}{2} \log(T) + \mathrm{const}\big)\,,$$

$$E \;=\; F - T \cdot \frac{\partial F}{\partial T} \;=\; \frac{3n}{2} \cdot k \cdot T\,.$$

The pressure p of the ideal gas corresponds to that of the vector field described by the volume, and is determined by

$$p \;=\; -\,\frac{\partial F}{\partial V} \;=\; k \cdot T \cdot \frac{n}{V}\,.$$

Here n/V is the particle density of the ideal gas.

Example 7 (Solid Body). A *solid body* consists of n points oscillating around a point. If we take this point to be the origin, the Hamilton function is

$$H = \sum_{i=1}^{n} \Big(\frac{p_{3i-2}^2 + p_{3i-1}^2 + p_{3i}^2}{2M_i} + \kappa_i^2 \cdot \frac{q_{3i-2}^2 + q_{3i-1}^2 + q_{3i}^2}{2} \Big) .$$

Introducing the basic *frequencies*

$$\nu_i := \frac{\kappa_i}{2\pi\sqrt{M_i}},$$

we obtain as the value of the partition function

$$Z = \frac{1}{n!}\big(k \cdot T\big)^{3n} \cdot \frac{1}{\nu_1^3 \cdot \ldots \cdot \nu_n^3} .$$

The free energy is

$$F = -k \cdot T \cdot \log(Z) = -kT \cdot \big(n \cdot \log(n/e) + 3n \cdot \log(kT) + \text{const}\big) .$$

Starting from this, one can calculate the other thermodynamic quantities, for example the energy:

$$E = F - T \cdot \frac{\partial F}{\partial T} = -T^2 \cdot \frac{\partial}{\partial T}\big(\frac{F}{T}\big) = 3n \cdot k \cdot T .$$

This formula says that every degree of freedom of an atom carries the mean inner energy $k \cdot T$, a result known as the *Dulong-Petit rule.* However, at temperatures close to absolute zero, this rule no longer holds, since the model for a solid body as being a combination of harmonic oscillators is not adequate anymore.

The inner and the free energy are examples of so-called *thermodynamic potentials.* Classical thermodynamics usually adopts the point of view that the choice of a particular potential already determines the quantities by which the system is to be described. In this sense, the inner energy E is understood to be a function depending upon entropy and volume, the free energy F is a function of temperature and volume, etc. Consider a Hamilton function not depending on temperature. As in the two preceding examples, the free and the inner energy are related to the vector field $\mathcal{V}$ dual to the temperature via the relation (Theorem 12)

$$E = F + T \cdot S = F - T \cdot \frac{\partial F}{\partial \mathcal{V}} .$$

Hence the inner energy E appears to be the Legendre transformation with respect to temperature of the free energy chosen as Lagrange function, $L := -F$ (see §7.5). By means of Legendre transformations with respect to other parameters, one obtains the other thermodynamic potentials, enthalpy and free enthalpy, which we will not discuss in detail here. In the case that only the mechanical work of the volume change contributes to the work form, the First Fundamental Theorem reads as follows:

$$dE = T \cdot dS - p \cdot dV .$$

Hence

$$T = \frac{\partial E}{\partial S} \quad \text{and} \quad p = -\frac{\partial E}{\partial V} .$$

If the energy is even twice continuously differentiable, the interchangeability of second partial derivatives is equivalent to a so-called *thermodynamic* or *Maxwell relation*

$$\frac{\partial^2 E}{\partial V \partial S} = \frac{\partial T}{\partial V} = \frac{\partial p}{\partial S} .$$

Beginning with the free energy,

$$dF = dA - S \cdot dT = -p \cdot dV - s \cdot dT ,$$

we similarly obtain the equations

$$S = -\frac{\partial F}{\partial T}, \quad p = -\frac{\partial F}{\partial V} ,$$

as well as the thermodynamic relation

$$\frac{\partial S}{\partial V} = \frac{\partial p}{\partial T} .$$

Eventually, we want to discuss the *Carnot cycle* briefly, which will lead us to the notion of universal thermal efficiency. Historically this was essential for the understanding of the First and Second Fundamental Theorems of thermodynamics.

We use an arbitrary homogeneous substance and suppose that its state is determined exclusively by the two mechanical variables, the pressure p and the volume V, from which the temperature T can be computed by means of a general equation of state. A Carnot cycle is understood to be a closed path γ in the (V, p)-plane consisting of two isotherms and two adiabatics. More precisely:

(1) From A to B, the substance is expanded up to the volume V_B through contact with a heat reservoir of temperature T_1 while keeping the temperature fixed ("isothermally"). During this process it absorbs the quantity of heat Q_1.

(2) The substance is taken out of the bath T_1 at B and expanded to the point C without absorbing or emitting heat ("adiabatically"). The temperature drops to the value T_2.

(3) The material is put into a heat reservoir of temperature T_2 and compressed isothermally from C to D, emitting the quantity of heat Q_2.

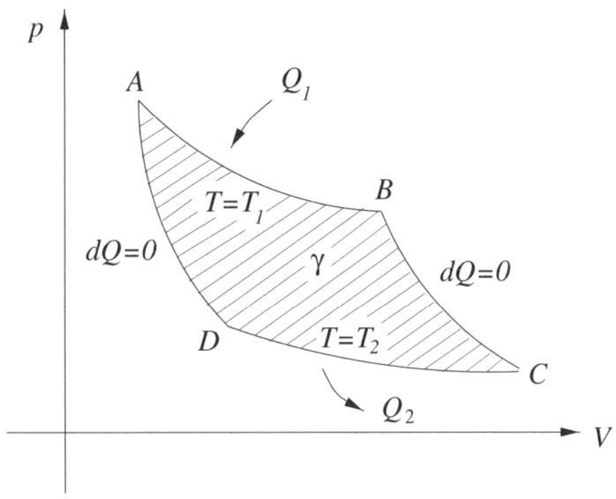

(4) The material is taken out of the second heat reservoir and adiabatically compressed from D to A to reach the temperature T_1 again.

It has to be stressed that we are supposing that the idealized processes discussed throughout this chapter are reversible. In the way explained here, the Carnot cycle describes a power engine (the temperature difference of the heat reservoirs is used to do work); reversing the process, it provides a thermal engine (work is done to further cool down the colder of the two heat reservoirs). By assumption, work is done exclusively in its mechanical form; the total work A is computed as

$$A := \oint_\gamma dA = -\oint_\gamma p\,dV\,,$$

and, since $-d(p\,dV) = dV \wedge dp$, this is equal to the area of the region bounded by the curve γ in the (V,p)-diagram by Stokes' theorem. We integrate the First Fundamental Theorem

$$dE = dA + dQ$$

over the closed path γ. The line integral of the exact form dE along γ vanishes, since the inner energy again reaches the initial value at A:

$$0 = \oint_\gamma dE = \oint_\gamma dA + \oint_\gamma dQ\,.$$

The thermal efficiency η is defined as the ratio of the work done and the added heat, $\eta := A/Q_1$. But the line integral along dQ is precisely $Q_2 - Q_1$; hence we can write the thermal efficiency also as

$$\eta = 1 - \frac{Q_2}{Q_1}\,.$$

As the Carnot cycle was supposed to be reversible, the entropy does not change, and the Second Fundamental Theorem implies the identity

$$0 = \oint_\gamma \frac{dQ}{T} = \frac{Q_2}{T_2} - \frac{Q_1}{T_1} .$$

The thermal efficiency only depends on the temperatures of both heat reservoirs,

$$\eta = \frac{T_1 - T_2}{T_2} ,$$

and not on the particular kind of substance or the particular construction of the power engine.

Exercises

1 (Gay-Lussac Experiment, 1807)**.** Consider a fixed amount of an ideal gas which at time 1 has pressure p_1 and volume V_1. The gas is heated while keeping the volume fixed until it reaches at time 2 the pressure p_2. Then the gas is allowed to expand adiabatically up to the volume V_3 ("overflow"). At the point 3, the gas reaches again the initial pressure p_1. Prove that the entropy change between the states 2 and 3 is equal to

$$\Delta S = S_3 - S_2 = n \cdot k \cdot \log\left(\frac{V_3}{V_2}\right) ,$$

independently of the way in which the physical process is realized in practice.

2. Derive from the Maxwell distribution the probability distribution for the speed $||v|| = (v_1^2 + v_2^2 + v_3^2)^{1/2}$ using spherical coordinates.

3. Consider two containers filled with the same ideal gas having equal temperature T_0 and particle number N, but different volumes V_1 and V_2. The containers are then connected.

a) Calculate the equilibrium temperature of the connected containers from the condition that the entropy remains constant during the connection process.

b) Explain why the work done is equal to the difference of the energies before and after the connection, and calculate the latter.

4. Determine the quantity of heat Q absorbed by an ideal gas in the course of the Carnot cycle. Hint: One of the intermediate results is

$$Q = (T_2 - T_1)(S_2 - S_1)\,.$$

5 (Van der Waals Equation)**.** The free energy of a real gas is heuristically defined as the sum of the free energy of the ideal gas F_{id} and an additional term depending on two constants a and b to be determined experimentally:

$$F = F_{\text{id}} - nkT\log\left(\frac{V-nb}{V}\right) - \frac{n^2a}{V^2}\,.$$

a) Explain why this Ansatz tends for large volumes to the free energy of an ideal gas, and why, on the other hand, it prevents the gas from being compressed unboundedly.
b) Derive the equation of state (the so-called *van der Waals equation*)

$$\left(p + \frac{n^2a}{V}\right)(V - nb) = nkT$$

as well as the formulas for entropy and energy. How can the result for the energy be interpreted physically?

6. The Hamilton function of an ideal gas in the ultra-relativistic limit is

$$c \cdot \chi(q_1, q_2, q_3) \cdot \ldots \cdot \chi(q_{3n-2}, q_{3n-1}, q_{3n}) \cdot \sqrt{p_1^2 + \ldots + p_{3n}^2}$$

(c ist the velocity of light). Calculate the free energy F, the entropy S, and the energy E.

7 (Planck's Radiation Law)**.** An ideal photon gas in thermal equilibrium emits *black-body radiation*, which is described by Planck's law. We describe each photon by a quantum mechanical oscillator and assume as a known fact that the energy levels of an oscillator with basic frequency ω are

$$E_n = \hbar\omega(n + 1/2), \quad n \in \mathbb{N}\,.$$

For photons, it may be supposed that these are distributed within each energy level (up to irrelevant factors of proportionality) according to the Gibbs distribution $\varrho_n = e^{-E_n/kT}$.

a) Calculate the mean energy,

$$\bar{E} := \sum_{n=1}^{\infty} \varrho_n \cdot E_n \Big/ \sum_{n=1}^{\infty} \varrho_n\,,$$

and conclude that the number $\bar{n} = [e^{\hbar\omega/kT} - 1]^{-1}$ can be interpreted as the mean occupation number.

b) The frequency ω is related to the momentum of the photons by $E = c||p|| = \hbar\omega$. Quantum statistics postulates that in the Maxwell distribution μ_p, the Gibbs factor $e^{-c||p||/kT}$ is to be replaced by the mean occupation number $\bar{n}$. Conclude that the probability measure μ_p^{QM} of $||p||$, up to some factors, is

$$\mu_p^{QM} = \frac{\omega^2}{e^{\hbar\omega/kT} - 1} \cdot d\omega \,;$$

the spectral energy density ε is defined as energy per volume unit, hence it is obtained after again multiplying by $\hbar\omega$:

$$\varepsilon = \frac{\hbar\omega^3}{e^{\hbar\omega/kT} - 1} \,.$$

This is *Planck's radiation law* (Planck, 1900).

c) Discuss the behavior of the curve ε depending on the variable ω and the parameter T; describe, in particular, how the position of the maximum changes with T.

d) Discuss the limiting cases $\hbar\omega \ll kT$ (*Rayleigh-Jeans law*) and $\hbar\omega \gg kT$ (*Wien's law*).

Chapter 9

Elements of Electrodynamics

9.1. The Maxwell Equations

The Maxwell equations describe the impact of an electromagnetic field on a distribution of electrical charges in space as well as the interaction between the *electric field* E and the *magnetic field* B. These vector fields are time-dependent vector fields defined on a domain $\Omega \subset \mathbb{R}^3$,

$$E : \Omega \times \mathbb{R} \longrightarrow \mathbb{R}^3 , \quad B : \Omega \times \mathbb{R} \longrightarrow \mathbb{R}^3 ,$$

and the electric charge is described by a time-dependent *density function*

$$\varrho : \Omega \times \mathbb{R} \longrightarrow \mathbb{R} .$$

The electromagnetic field induces a current of the electric charges, and hence also a time-dependent *current density vector*

$$J : \Omega \times \mathbb{R} \longrightarrow \mathbb{R}^3 .$$

The operations of divergence and curl are supposed to be applied to the spatial coordinates of the time-dependent vector fields.

First Group of the Maxwell Equations:

$$\operatorname{curl}(E) = -\frac{1}{c}\frac{\partial B}{\partial t} , \quad \operatorname{div}(B) = 0 .$$

Second Group of the Maxwell Equations:

$$\operatorname{curl}(B) = \frac{1}{c}\frac{\partial E}{\partial t} + \frac{4\pi}{c} J , \quad \operatorname{div}(E) = 4\pi\varrho .$$

Here c denotes the *constant velocity of light.* Since $\operatorname{div}(\operatorname{curl}(B)) = 0$ (see §2.3, Theorem 5), the *continuity equation*

$$\operatorname{div}(J) + \frac{\partial \varrho}{\partial t} = 0$$

is an immediate consequence of the second group of the Maxwell equations.

Now we want to write the Maxwell equations involving differential forms on $\mathbb{R}^3$. For this, we use the transition from a vector field $\mathcal{V}$ to the associated 1-form $\omega_{\mathcal{V}}$ discussed in Chapters 2 and 3. On a manifold M^m, $\omega_{\mathcal{V}}$ is the 1-form defined by the equation (see §§2.3 and 3.4)

$$*\omega_{\mathcal{V}} = \mathcal{V} \lrcorner\, dM^m,$$

and the gradient and divergence satisfy the identities

$$\omega_{\operatorname{grad}(f)} = df, \quad d(*\omega_{\mathcal{V}}) = \operatorname{div}(\mathcal{V}) \cdot dM^m .$$

In $\mathbb{R}^3$, there is an additional relation involving the curl of a vector field $\mathcal{V}$ (§3.11):

$$\omega_{\operatorname{curl}(\mathcal{V})} = *d\omega_{\mathcal{V}}.$$

Passing now from the time-dependent vector fields E and B to the corresponding time-dependent 1-forms ω_E and ω_B, we can restate the Maxwell equations in the following equivalent form:

First Group of the Maxwell Equations:

$$*d\omega_E = -\frac{1}{c}\frac{\partial}{\partial t}(\omega_B), \quad d(*\omega_B) = 0 .$$

Second Group of the Maxwell Equations:

$$*d\omega_B = \frac{1}{c}\frac{\partial}{\partial t}(\omega_E) + \frac{4\pi}{c}\,\omega_J , \quad d(*\omega_E) = 4\pi\varrho \cdot d\mathbb{R}^3 .$$

In these equations, the exterior derivative exclusively refers to the spatial coordinates, and not to the time coordinate. By applying Poincaré's lemma, we can represent the closed 2-form $*\omega_B$ locally as the differential of a time-dependent 1-form A,

$$*\omega_B = dA .$$

Definition 1. The (locally) defined time-dependent 1-form A is called the *magnetic potential.*

The first Maxwell equation can be rewritten equivalently as

$$d\Big(\omega_E + \frac{1}{c}\frac{\partial}{\partial t}(A)\Big) = 0 .$$

Again, Poincaré's lemma shows the local existence of a time-dependent real-valued function ϕ with

$$\omega_E = -\frac{1}{c}\frac{\partial}{\partial t}(A) - d\phi .$$

Definition 2. The (locally) defined time-dependent function ϕ is called the *electric potential.*

The essence of the electromagnetic field is not the same in classical electrodynamics and in quantum mechanics. In classical field theory, only the measurable field strengths E and B are physically relevant; the potentials A and ϕ are auxiliary mathematical functions which turn out to be useful. In quantum mechanics, however, the electric and the magnetic potential acquire a physical meaning on their own; we illustrate this briefly by discussing the Aharonov-Bohm effect.

Remark (Aharonov-Bohm Effect)**.** An infinitely extended coil S, through which a current flows, induces a magnetic field which is almost homogeneous in its interior and vanishes outside of the coil. Consider a closed curve γ around the coil and denote by Ω the surface bounded by γ. Imagine that an electron is prevented from moving inside the coil by an infinitely high potential well. Then the remaining region $\Omega - S$ of the plane is, from the point of view of the electron, no longer simply connected. The vanishing of the magnetic field B in this region implies that the 1-form A of the magnetic potential is closed in $\Omega - S$,

$$0 = *\omega_B = dA ,$$

but nevertheless the line integral is not path-independent, since, for example, the curve segments γ_1 and γ_2 of γ cannot be deformed into one another by a homotopy (see Theorem 9, §2.5). Instead, by Stokes' theorem, the integral

$$\int_\gamma A = \int_{\partial\Omega} A = \int_\Omega dA = \int_\Omega *\omega_B = \Phi_\Omega^B$$

is equal to the magnetic flux through the entire region Ω. Quantum mechanics describes the states of the electron by solutions of the Schrödinger equation called *wave functions*. Let ψ_1 and ψ_2 be the wave functions of the electron along γ_1 and γ_2, respectively, while the current is switched off (i.e., $B = 0$ in S). The covariance principle for the Schrödinger equation with respect to gauge transformations implies that, if the current is switched on, the wave functions have to become

$$\psi_1^* = \psi_1 \exp\left[\frac{ie}{\hbar c}\int_{\gamma_1} A\right], \quad \psi_2^* = \psi_2 \exp\left[\frac{ie}{\hbar c}\int_{\gamma_2} A\right].$$

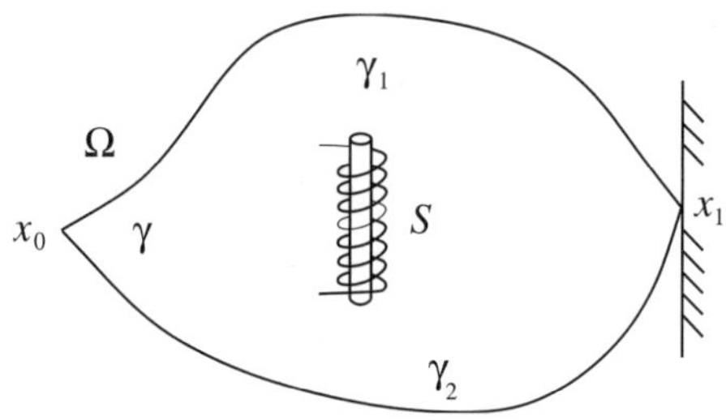

We put a screen at x_1 and form the linear superposition of the two wave functions:

$$\psi_1^* + \psi_2^* = \left(\psi_1 \exp\left[\frac{ie\Phi_\Omega^B}{\hbar c}\right] + \psi_2\right) \cdot \exp\left[\frac{ie}{\hbar c}\int_{\gamma_2} A\right] .$$

As we vary the enclosed magnetic flux Φ_Ω^B, the absolute value $|\psi_1^* + \psi_2^*|$ changes. This corresponds to a shift in the interference picture on the screen, and provides experimental evidence for the intrinsic meaning of the potential A.

A theoretically rigorous explanation of this effect (A does not exist globally on the region Ω!) can only be obtained within the theory of connections on principal S^1-fiber bundles. The integral of A over a closed curve can then be defined correctly and measures the holonomy of the connection[1].

If we put the potentials A and ϕ at the beginning of our study, the electric and magnetic fields are determined by the equations

$$\omega_E = -\frac{1}{c}\frac{\partial}{\partial t}(A) - d\phi, \quad \omega_B = *dA ,$$

and the first group of the Maxwell equations is automatically satisfied. In these formulas, we view the quantities as time-dependent functions or forms in 3-space, and the Hodge operator $*$ and the exterior derivative d refer to that space. We use the adjoint operator δ of the exterior derivative d and note that in $\mathbb{R}^3$ the following formulas hold:

$$\delta = - * d * \quad \text{for 1-forms} \quad \text{and} \quad \delta = *d* \quad \text{for 2-forms} .$$

Then we can write the second group of the Maxwell equations as

$$\delta dA = -\frac{1}{c^2}\frac{\partial^2}{\partial t^2}(A) - \frac{1}{c}\frac{\partial}{\partial t}(d\phi) + \frac{4\pi}{c}\,\omega_J, \quad \frac{1}{c}\frac{\partial}{\partial t}(\delta A) - \Delta\phi = 4\pi\varrho .$$

The Laplacian Δ on functions will be used in this chapter analogously to Chapter 2. Hence we have $\Delta\phi = *d*d\phi = -\delta d\phi$.

---

[1]As a physics reference we recommend: Y. Aharonov and D. Bohm, Phys. Rev. **115** (1959), 485-491, as well as the more detailed discussion of this effect in the textbook by F. Schwabl, *Quantum Mechanics*, Springer, 2nd rev. ed. 1995.

9.2. The Static Electromagnetic Field

If the electric and magnetic fields are time-independent, the Maxwell equations simplify:

$$*d\omega_E = 0\,, \qquad d(*\omega_B) = 0,$$

$$*d\omega_B = \frac{4\pi}{c}\,\omega_J\,, \qquad d(*\omega_E) = 4\pi\varrho\cdot d\mathbb{R}^3\,.$$

Hence the charge density ϱ and the current density vector J do not depend on time, and J is a divergence-free vector field. This system of equations decomposes into a pair of partial differential equations for the electric field and another similar pair of equations for the magnetic field. We begin by solving the equations for the electric field. Using the locally defined electric potential $\omega_E = -d\phi$, this system reduces to the inhomogeneous Laplace equation

$$-\Delta\phi = 4\pi\varrho\,.$$

From now on we suppose that the support of the charge density is compact. Then it is to be expected that the electric field $E(x)$ generated by this charge distribution tends to zero at infinity. Under this condition, the electric field is uniquely determined.

Theorem 1. *Let the support of the charge density ϱ be compact. Then there exists precisely one electric field E such that*

$$\mathrm{div}(E) = 4\pi\varrho, \quad \mathrm{curl}(E) = 0$$

and $\|E(x)\| \to 0$ for $\|x\| \to \infty$. This electric field is given by $E = -\mathrm{grad}(\phi)$ with the electric potential

$$\phi(x) = -\int_{\mathbb{R}^3} \frac{\varrho(y)}{\|x-y\|}\,dy\,,$$

which is also called the Newton potential *for the density ϱ.*

Proof. Consider two electric fields $E_1 = -\mathrm{grad}(\phi_1)$, $E_2 = -\mathrm{grad}(\phi_2)$ generated by one and the same charge density ϱ. The difference $u := \phi_1 - \phi_2$ of the electric potentials is a harmonic function with a gradient vanishing at infinity,

$$\Delta u = 0\,, \quad \|\mathrm{grad}(u)(x)\| \longrightarrow 0 \quad \text{for } \|x\| \longrightarrow \infty\,.$$

Since $\Delta\big(\partial u/\partial x^i\big) = \partial(\Delta u)/\partial x^i = 0$, the partial derivatives $\partial u/\partial x^i$ $(1 \leq i \leq 3)$ are harmonic functions on $\mathbb{R}^3$ also vanishing at infinity. Liouville's theorem for harmonic functions (see Theorem 42, §3.10) immediately implies that all partial derivatives vanish identically. Hence u is a constant function, and the electric fields $E_1 \equiv E_2$ coincide. There exists at most one electric field with the required properties, and it remains to be proved that the electric field $E = -\mathrm{grad}(\phi)$ with potential

$$\phi(x) \;=\; -\int\limits_{\mathbb{R}^3} \frac{\varrho(y)}{\|x-y\|}\, dy$$

is a solution for the problem. To see this, we have to show that $\phi(x)$ is a smooth function in $\mathbb{R}^3$, that it solves the inhomogeneous Laplace equation $-\Delta\phi = 4\pi\varrho$, and that the gradient $\|\mathrm{grad}(\phi)\| \to 0$ at infinity in $\mathbb{R}^3$. We introduce spherical coordinates around a fixed point $x = (x^1, x^2, x^3) \in \mathbb{R}^3$ $(0 \le \varphi \le 2\pi, -\pi/2 < \psi < \pi/2)$:

$$y^1 \;=\; x^1 + r\cos\varphi\cos\psi, \quad y^2 \;=\; x^2 + r\sin\varphi\cos\psi, \quad y^3 \;=\; x^3 + r\sin\psi\,.$$

The volume form of euclidean space is described in these coordinates by the following formulas:

$$d\mathbb{R}^3 \;=\; dy^1 \wedge dy^2 \wedge dy^3 \;=\; r^2\cos\psi \cdot dr \wedge d\varphi \wedge d\psi\,.$$

For every positive number $\varepsilon > 0$, the function

$$f(y) \;:=\; \frac{1}{\|x-y\|^{3-\varepsilon}}$$

is integrable in a neighborhood of its singular point. Integration over the ball $D^3(x,1)$ with center $x \in \mathbb{R}^3$ and radius 1 yields

$$\int\limits_{D^3(x,1)} f(y)\, dy \;=\; \int\limits_0^1 \int\limits_0^{2\pi} \int\limits_{-\pi/2}^{\pi/2} \frac{r^2}{r^{3-\varepsilon}} \cos\psi \cdot dr\, d\varphi\, d\psi \;=\; 4\pi/\varepsilon\,.$$

Because of the compactness of the support of the charge density $\varrho(y)$, the function $\phi(x)$ and its first and second partial derivatives are smooth,

$$\begin{aligned}\frac{\partial\phi}{\partial x^i}(x) \;&=\; \int\limits_{\mathbb{R}^3} \varrho(y)\frac{x^i - y^i}{\|x-y\|^3} dy \;=\; -\int\limits_{\mathbb{R}^3} \varrho(y)\frac{\partial}{\partial y^i}\left(\frac{1}{\|x-y\|}\right) dy \\ &=\; \int\limits_{\mathbb{R}^3} \frac{\partial\varrho}{\partial y^i}(y)\frac{1}{\|x-y\|} dy\,.\end{aligned}$$

Differentiating once again leads to the formula

$$-\Delta\phi(x) \;=\; \lim_{M\to\infty}\lim_{\varepsilon\to 0} \int\limits_{D^3(x,M)\setminus D^3(x,\varepsilon)} \Delta(\varrho)(y) \cdot \frac{1}{\|x-y\|} dy\,.$$

Applying Green's second formula for sufficiently large radius M, we can rewrite the integral over the spherical shell $D^3(x,M)\setminus D^3(x,\varepsilon)$:

$$\int\limits_{D^3(x,M)\setminus D^3(x,\varepsilon)} \Delta(\varrho)(y) \cdot \frac{1}{\|x-y\|} \;=\; \int\limits_{D^3(x,M)\setminus D^3(x,M)} \varrho(y) \cdot \Delta_y \frac{1}{\|x-y\|} \;-$$

$$- \int\limits_{\partial D^3(x,\varepsilon)} \Big[\varrho(y)\Big\langle \mathrm{grad}_y \frac{1}{\|x-y\|}, \mathcal{N}(y)\Big\rangle - \frac{1}{\|x-y\|}\Big\langle \mathrm{grad}_y(\varrho), \mathcal{N}(y)\Big\rangle\Big]\, dy\,.$$

The function $1/\|x-y\|$ is harmonic in the spherical shell; hence the first integral vanishes. Moreover, the normal vector field of the sphere $\partial D^3(x,\varepsilon)$ of radius ε is given by the formula $\mathcal{N}(y) = (y-x)/\varepsilon$, which immediately implies

$$\big\langle \mathrm{grad}_y \frac{1}{\|x-y\|}, \mathcal{N}(y)\big\rangle \;\equiv\; -\,\frac{1}{\varepsilon^2}\,.$$

Inserting these expressions, we obtain

$$\int\limits_{D^3(x,M)\setminus D^3(x,\varepsilon)} \Delta\varrho(y)\frac{1}{\|x-y\|}dy \;=\; \frac{1}{\varepsilon^2}\int\limits_{\partial D^3(x,\varepsilon)} \varrho(y)dy + \frac{1}{\varepsilon^2}\int\limits_{\partial D^3(x,\varepsilon)} \big\langle \mathrm{grad}\,\varrho(y), y-x\big\rangle dy\,.$$

The second integral on the right-hand side is bounded by

$$(1/\varepsilon^2)\cdot\varepsilon\cdot\max\{\|\mathrm{grad}\,\varrho\|\}\cdot\mathrm{vol}(\partial D^3(x,\varepsilon)) = 4\pi\cdot\varepsilon\cdot\max\{\|\mathrm{grad}\,\varrho\|\}\,,$$

and hence it does not contribute in the limit as $\varepsilon \to 0$. Together, this yields

$$-\Delta\phi(x) \;=\; \lim_{\varepsilon\to 0}\; \frac{1}{\varepsilon^2}\int\limits_{\partial D^3(x,\varepsilon)} \varrho(y)dy \;=\; 4\pi\varrho(x)\,,$$

where we used the mean value theorem of integral calculus in the last step. We estimate the length of the gradient of the electric potential using the formula for the partial derivatives $\partial\phi/\partial x^i$ and for points lying far out. To do so we choose a radius $R > 0$ such that the ball with center $0 \in \mathbb{R}^3$ contains the support of the charge density. If now $\|x\| > R$, then

$$\begin{aligned} \|\mathrm{grad}\,\phi(x)\|^2 \;&=\; \sum_{i=1}^{3}\left[\;\int\limits_{D^3(0,R)} \varrho(y)\frac{x^i-y^i}{\|y-x\|^3}dy\right]^2 \\ &\le\; 3\cdot\max\big\{\varrho^2(y) : y\in\mathbb{R}^3\big\}\cdot\frac{4\pi}{3}R^3\cdot\frac{1}{\mathrm{dist}^2(x, D^3(0;R))}\,. \end{aligned}$$

We conclude that the length of the gradient $\|\mathrm{grad}\,\phi(x)\|$ decreases like $1/\|x\|$ for $\|x\| \to \infty$. □

Remark 1. Obviously, the preceding theorem and its proof remain valid for a charge density ϱ which no longer has compact support, but, together with its derivatives, vanishes sufficiently fast at infinity.

Example 1. Consider for a fixed point $x_0 \in \mathbb{R}^3$ a charge density e which is concentrated in a small ball around it,

$$\varrho_\varepsilon(y) \;=\; \begin{cases} e\cdot\big(\mathrm{vol}(D^3(x_0,\varepsilon))\big)^{-1} & \text{if } y\in D^3(x_0,\varepsilon), \\ 0 & \text{otherwise.}\end{cases}$$

We calculate the electric potential:

$$\phi_\varepsilon(x) \;=\; -\int\limits_{\mathbb{R}^3} \frac{\varrho_\varepsilon(y)}{\|x-y\|}dy \;=\; \frac{-e}{\mathrm{vol}(D^3(x_0,\varepsilon))}\int\limits_{D^3(x_0,\varepsilon)} \frac{1}{\|x-y\|}dy\,,$$

and in the limit as $\varepsilon \to 0$ we obtain

$$\lim_{\varepsilon\to 0}\phi_\varepsilon(x) \;=\; \frac{-e}{\|x-x_0\|}\,.$$

In this way, the *Coulomb potential* arises naturally as the field generated by an electric point charge. Let, similarly, n points $x_1, \dots, x_n$ with charges $e_1, \dots, e_n$ be given. Then they generate the potential

$$\phi(x) \;=\; -\sum_{i=1}^{n}\frac{e_i}{\|x-x_i\|}\,.$$

Now we turn to the solution of the two partial differential equations for the static magnetic field $B(x)$ and consider again only the case that the divergence-free current density vector has compact support (see the remark above).

Theorem 2. *Let J be a divergence-free vector field with compact support in $\mathbb{R}^3$. Then there exists precisely one magnetic field B such that*

$$\mathrm{div}(B) \;=\; 0\,, \quad \mathrm{curl}(B) \;=\; \frac{4\pi}{c}J$$

and $\|B(x)\| \to 0$ for $\|x\| \to \infty$. This magnetic field is given by $B = \mathrm{curl}(A)$ with the magnetic potential

$$A(x) \;=\; -\frac{1}{c}\int\limits_{\mathbb{R}^3}\frac{J(y)}{\|x-y\|}\,dy\,.$$

Proof. Let B_1 and B_2 be two magnetic fields with the stated properties. Their difference $\mathcal{V} := B_1 - B_2$ is a divergence- and curl-free vector field,

$$\mathrm{div}(\mathcal{V}) \;=\; 0, \quad \mathrm{curl}(\mathcal{V}) \;=\; 0\,,$$

whose length $\|\mathcal{V}(x)\|$ tends to zero for $\|x\| \to \infty$. By Poincaré's lemma, such a vector field can be represented as the gradient of a harmonic function f,

$$\mathcal{V} \;=\; \mathrm{grad}(f) \quad \text{and} \quad \Delta f \;=\; 0\,.$$

The partial derivatives $\partial f/\partial x^i$ ($1 \le i \le 3$) are thus bounded harmonic functions on the entire space $\mathbb{R}^3$, and by Liouville's theorem they are constant. This immediately implies $\mathcal{V} = B_1 - B_2 \equiv 0$, and this observation shows that there can be at most one magnetic field with the required properties. By the same arguments as in the proof of the preceding theorem, the magnetic potential

$$A(x) = -\frac{1}{c}\int\limits_{\mathbb{R}^3} \frac{J(y)}{\|x-y\|}\,dy$$

is smooth. It remains to be shown that it is divergence-free. The partial derivatives of the components $A = (A^1, A^2, A^3)$ of the magnetic potential are

$$\frac{\partial A^i}{\partial x^i}(x) = \frac{1}{c}\int\limits_{\mathbb{R}^3} \frac{J^i(y)\cdot(x^i-y^i)}{\|x-y\|^3}\,dy,$$

and hence

$$\mathrm{div}(A) = \frac{1}{c}\int\limits_{\mathbb{R}^3} \Big\langle J(y), \frac{x-y}{\|x-y\|^3}\Big\rangle dy = \frac{1}{c}\int\limits_{\mathbb{R}^3} \Big\langle J(y), \mathrm{grad}_y\Big(\frac{1}{\|x-y\|}\Big)\Big\rangle dy\,.$$

Using the formula

$$\mathrm{div}_y\Big(\frac{1}{\|x-y\|}J(y)\Big) = \Big\langle J(y), \mathrm{grad}_y\Big(\frac{1}{\|x-y\|}\Big)\Big\rangle + \frac{1}{\|x-y\|}\mathrm{div}(J)$$

and the fact that J was supposed to be divergence-free, we obtain

$$\mathrm{div}(A) = \frac{1}{c}\int\limits_{\mathbb{R}^3} \mathrm{div}_y\Big(\frac{1}{\|x-y\|}J(y)\Big)dy = 0$$

by Green's formula, since the vector field $J(y)$ has compact support. The curl of the vector field $B = \mathrm{curl}(A)$ can now easily be calculated (see Exercise 6.b, Chapter 2):

$$\mathrm{curl}(B) = \mathrm{curl}\big(\mathrm{curl}(A)\big) = \mathrm{grad}\big(\mathrm{div}(A)\big) - \Delta(A) = -\Delta(A)$$

Here the Laplacian is to be applied component-wise to the vector field A, and the proof of the preceding theorem yields the result

$$\mathrm{curl}(B) = -\Delta(A) = \frac{4\pi}{c}\,J\,. \qquad \square$$

Example 2. Let the current density vector J be constant in a domain $\Omega \subset \mathbb{R}^3$ and zero outside this domain. The magnetic potential and the magnetic field $B = \mathrm{curl}(A)$ are then determined by the formulas

$$A(x) = -\frac{1}{c}\Big(\int\limits_{\Omega} \frac{dy}{\|x-y\|}\Big)\cdot J, \quad B(x) = -\frac{1}{c}\,J \times \int\limits_{\Omega} \frac{(x-y)}{\|x-y\|^3}\,dy\,.$$

The magnetic field induced by an electric charge distribution with constant current density vector is perpendicular to the direction of the current flow. This fact is called the *Biot-Savart law.*

Finally, we interpret both theorems as the Hodge decomposition of a 1-form on $\mathbb{R}^3$ (see Corollary 5, §3.11). Let ω^1 be a 1-form on $\mathbb{R}^3$ with compact support or, more generally, with rapidly decreasing coefficients. Applying Theorem 1 to the function $\varrho = *d(*\omega^1)$, we obtain a closed 1-form ω_E^1 with

$$d\omega_E^1 = 0 \quad \text{and} \quad *d(*\omega_E^1) = *d(*\omega^1).$$

The form ω_E^1 also tends to zero at infinity, and hence, by Theorem 2, the difference $J = \omega^1 - \omega_E^1$ can be represented as

$$*d\omega_B^1 = J = \omega^1 - \omega_E^1$$

with a 1-form ω_B^1 vanishing at infinity. This observation leads to

Theorem 3 (Helmholtz's Theorem, Hodge's Theorem). *Every smooth 1-form ω^1 in $\mathbb{R}^3$ with compact support, or, more generally, with rapidly decreasing coefficients, can uniquely be represented as the sum of a closed and a coclosed form,*

$$\omega^1 = \omega_E^1 + *d\omega_B^1, \quad d\omega_E^1 = 0.$$

ω_E^1 and ω_B^1 both tend to zero at infinity.

Remark 2. Comparing this result with Corollary 5 in §3.11, we stress that, in $\mathbb{R}^3$, the harmonic part does not occur in this decomposition. The reason for this is that there are no harmonic 1-forms in $\mathbb{R}^3$ having compact support or tending to zero sufficiently fast at infinity. If ω^1 were such a 1-form, then, by Poincaré's lemma, we could represent ω^1 as the derivative of a harmonic function, $\omega^1 = df, \Delta f = 0$. The partial derivatives $\partial f/\partial x^i$ would then be bounded harmonic functions, and thus constant. If then the length of ω^1 tended to zero at infinity, then these partial derivatives would vanish and we could conclude that $\omega^1 = 0$.

9.3. Electromagnetic Waves

An *electromagnetic wave* is an electromagnetic field in the vacuum ($\varrho = 0$, $J = 0$). The Maxwell equations in this case read as

$$\operatorname{div}(B) = 0, \quad \operatorname{div}(E) = 0,$$

$$\operatorname{curl}(B) = \frac{1}{c}\frac{\partial E}{\partial t}, \quad \operatorname{curl}(E) = -\frac{1}{c}\frac{\partial B}{\partial t},$$

which implies that the components of the electric and magnetic fields have to be solutions of the *wave equation*

$$\Delta B = \frac{1}{c^2}\frac{\partial^2 B}{\partial t^2}, \quad \Delta E = \frac{1}{c^2}\frac{\partial^2 E}{\partial t^2}.$$

If the wave is known at a certain time $t = 0$ and we want to study its propagation in time, we have to solve the *Cauchy initial value problem for the wave equation*:

$$\Delta u \;=\; \frac{1}{c^2}\frac{\partial^2 u}{\partial t^2}, \quad u(x,0) = u_0(x)\,, \quad \frac{\partial u}{\partial t}(x,0) \;=\; u_1(x)\,.$$

For smooth initial conditions u_0 and u_1, this problem has precisely one solution, which we are going to describe now. To do so we need the *spherical mean* of a function $\varphi : \mathbb{R}^3 \to \mathbb{R}$. This is understood to be the function $I\varphi : \mathbb{R}^3 \times \mathbb{R}_+ \to \mathbb{R}$ which computes the mean value of φ on an arbitrary 2-dimensional sphere $S^2(x;r)$ with center $x \in \mathbb{R}^3$ and radius $r > 0$,

$$(I\varphi)(x,r) \;:=\; \frac{1}{4\pi r^2}\int\limits_{S^2(x,r)} \varphi(y)dy\,.$$

Theorem 4 (Poisson Formula for the Wave Equation)**.** *For any two smooth initial conditions $u_0, u_1 : \mathbb{R}^3 \to \mathbb{R}$, the Cauchy initial value problem for the wave equation*

$$\Delta u \;=\; \frac{1}{c^2}\frac{\partial^2 u}{\partial t^2}, \quad u(x,0) = u_0(x)\,, \quad \frac{\partial u}{\partial t}(x,0) \;=\; u_1(x)$$

has precisely one solution on the space $\mathbb{R}^3 \times \mathbb{R}_+$. Moreover, this solution can be computed explicitely:

$$u(x,t) \;=\; \frac{\partial}{\partial t}\Big(t\cdot(Iu_0)(x, c\cdot t)\Big) + t\cdot(Iu_1)(x,c\cdot t)\,.$$

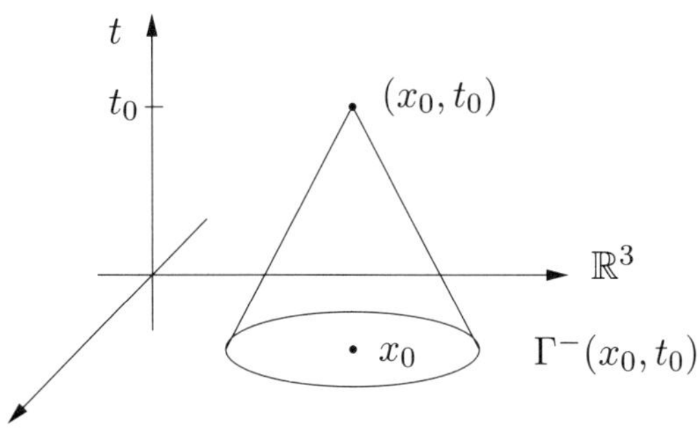

Remark 3. The value $u(x,t)$ of the solution for the wave equation depends only on the behavior of the initial values on the boundary of the base of the backward light cone,

$$\Gamma^-(x_0,t_0) \;:=\; \left\{(x,t) \in \mathbb{R}^3 \times \mathbb{R}_+ : \|x - x_0\| \;\le\; c\cdot|t-t_0|, \quad 0 \le t \le t_0\right\}\,,$$

which is a two-dimensional sphere. Therefore, every wave in $\mathbb{R}^3$ generated by initial conditions with compact support has a forward as well as a backward wave-front (*Huygens' principle*).

Proof of Uniqueness for the solution. We introduce the notation

$$\Box \;:=\; \Delta - \frac{1}{c^2}\frac{\partial^2}{\partial t^2}$$

for the wave operator and

$$Eu(x,t) \;:=\; \sum_{j=1}^{3}\left(\frac{\partial u}{\partial x^j}\right)^2 + \left(\frac{1}{c}\frac{\partial u}{\partial t}\right)^2$$

for the energy. Then the following easy identity holds:

$$2\left(\frac{\partial u}{\partial t}\right)\cdot \Box u \;=\; 2\sum_{j=1}^{3}\frac{\partial}{\partial x^j}\left(\frac{\partial u}{\partial t}\frac{\partial u}{\partial x^j}\right) - \frac{\partial}{\partial t}E(u)\,.$$

Now let u be a solution of the wave equation with initial conditions $u(x,0) = 0$ and $\partial u(x,0)/\partial t = 0$, and denote by $\Gamma_\tau^-(x_0,t_0)$ the truncated cone described by the condition $0 \le t \le \tau$. By integrating the stated identity, we obtain

$$\int\limits_{\Gamma_\tau^-}\frac{\partial}{\partial t}E(u) \;=\; 2\int\limits_{\Gamma_\tau^-}\sum_{j=1}^{3}\frac{\partial}{\partial x^j}\left(\frac{\partial u}{\partial t}\frac{\partial u}{\partial x^j}\right).$$

Gauss' theorem (Theorem 27, §3.8) allows us to transform the volume integrals into surface integrals:

$$\int\limits_{\partial\Gamma_\tau^-} E(u)\cdot\left\langle \mathcal{N}, \frac{\partial}{\partial t}\right\rangle \;=\; 2\int\limits_{\partial\Gamma_\tau^-}\sum_{j=1}^{3}\frac{\partial u}{\partial t}\frac{\partial u}{\partial x^j}\cdot\left\langle \mathcal{N}, \frac{\partial}{\partial x^j}\right\rangle.$$

The boundary $\partial\Gamma_\tau^-$ consists of three parts. On the first part ($t = 0$), the energy $E(u)$ and the derivative $\partial u/\partial t$ vanish; this term does not contribute to either side of the equation. On the second boundary piece ($t = \tau$), the normal vector $\mathcal{N}$ is parallel to $\partial/\partial t$. Hence the integral on the right-hand side of the equation vanishes, whereas the integral on the left-hand side is non-negative. On the third part of the boundary, the lateral surface of the truncated cone, we calculate the scalar products

$$\left\langle \mathcal{N}, \frac{\partial}{\partial x^j}\right\rangle \;=\; \frac{x^j - x_0^j}{\|x - x_0\|}\frac{1}{\sqrt{1+c^2}}\,,\qquad \left\langle \mathcal{N}, \frac{\partial}{\partial t}\right\rangle \;=\; \frac{c}{\sqrt{1+c^2}}\,.$$

Thus we can transform the difference of the left-hand and the right-hand integrand into an expression which is non-negative, too:

$$E\cdot\left\langle \mathcal{N}, \frac{\partial}{\partial t}\right\rangle - 2\sum_{j=1}^{3}\frac{\partial u}{\partial t}\frac{\partial u}{\partial x^j}\left\langle \mathcal{N}, \frac{\partial}{\partial x^j}\right\rangle = \frac{c}{\sqrt{1+c^2}}\sum_{j=1}^{3}\left[\frac{\partial u}{c\partial t}\frac{x^j - x_0^j}{\|x - x_0\|} - \frac{\partial u}{\partial x^j}\right]^2.$$

All in all, the integral formula implies that the energy $E(u)$ vanishes on the upper boundary ($t = \tau$) of the truncated cone Γ_τ^-. The height τ of the truncated cone can be varied, and thus $E(u)$ vanishes identically in the interior of the cone $\Gamma^-(x_0, t_0)$. There the function $u(x,t)$ is constant, and, looking at the initial conditions, identically zero. $\square$

Proof of the existence of a solution. We calculate the derivatives of the spherical mean with respect to time:

$$(Iu_1)(x, ct) = \frac{1}{4\pi c^2 t^2} \int\limits_{S^2(x,ct)} u_1(y)dy = \frac{1}{4\pi} \int\limits_{S^2} u_1(x + ct \cdot z)dz\,,$$

where $S^2(x, ct)$ is the sphere with center $x \in \mathbb{R}^3$ and radius ct. Denote by S^2 the unit sphere. Then Green's formula implies

$$\begin{aligned}\frac{\partial(Iu_1)}{\partial t}(x, ct) &= \frac{c}{4\pi} \int\limits_{S^2} \sum_{j=1}^{3} z^j \cdot \frac{\partial u_1}{\partial x^j}(x + ct \cdot z)dz \\ &= \frac{1}{4\pi ct^2} \int\limits_{S^2(x,ct)} \Big\langle \mathcal{N}(\alpha), \frac{\partial}{\partial x^j} \Big\rangle \cdot \frac{\partial u_1}{\partial x^j}(\alpha)d\alpha = \frac{1}{4\pi ct^2} \int\limits_{D^3(x,ct)} \Delta u_1(y)dy\,.\end{aligned}$$

Introducing spherical coordinates $(r, \alpha) \in (0, \infty) \times S^2$ in the three-dimensional ball $D^3(x, ct)$ and applying the formula $d\mathbb{R}^3 = r^2 \cdot dr \wedge d\alpha$, we obtain

$$\frac{\partial(Iu_1)}{\partial t}(x, ct) = \frac{1}{4\pi ct^2} \int\limits_0^{ct} r^2 \int\limits_{S^2} \Delta u_1(x + r\alpha)d\alpha \cdot dr\,.$$

This leads to

$$\frac{\partial}{\partial t}\big(t \cdot (Iu_1)(x, ct)\big) = (Iu_1)(x, ct) + \frac{1}{4\pi ct} \int\limits_0^{ct} r^2 \int\limits_{S^2} \Delta u_1(x + r\alpha)d\alpha \cdot dr\,,$$

and, after further differentiation,

$$\frac{\partial^2}{\partial t^2}\big(t \cdot (Iu_1)(x, ct)\big) = \frac{c^2 t}{4\pi} \int\limits_{S^2} \Delta u_1(x + ct \cdot \alpha)d\alpha = c^2\Delta(t \cdot (Iu_1)(x, ct))\,.$$

Thus $\omega(x, t) := t \cdot (Iu_1)(x, ct)$ is a solution for the wave equation with initial conditions

$$\omega(x, 0) = 0 = \frac{\partial^2\omega}{\partial t^2}(x, 0) \quad \text{and} \quad \frac{\partial\omega}{\partial t}(x, 0) = (Iu_1)(x, 0) = u_1(x)\,.$$

The derivative $\partial\omega/\partial t$ is also a solution of the wave equation, and it provides the second summand in the asserted formula. $\square$

We have thus solved the wave equation for a scalar function on $\mathbb{R}^3$. The expression for the vector valued functions E and B follows from a tedious but routine computation. The explicit formula may be found in Exercise 3. Now we turn to the inhomogeneous wave equation $\square u = f$ with a smooth function $f : \mathbb{R}^3 \times \mathbb{R}_+ \to \mathbb{R}$, and will solve this for the initial condition $u_0 = 0 = u_1$. Combining the resulting solution with a solution for the homogeneous wave equation with arbitrary initial condition constructed in the preceding theorem, we will arrive at the solution for the inhomogeneous wave equation $\Box u = f$ with arbitrary initial condition u_0, u_1. The *retarded potential* of a smooth function $f : \mathbb{R}^3 \times \mathbb{R}_+ \to \mathbb{R}$ is defined by the integral

$$Rf(x,t) \;:=\; -\frac{1}{4\pi} \int\limits_{\|x-y\|\leq ct} \frac{f(y, t - r/c)}{r}\, dy$$

with the distance function $r := \|x - y\|$.

Theorem 5. *Let $f : \mathbb{R}^3 \times \mathbb{R}_+ \to \mathbb{R}$ be a smooth function. Then the retarded potential Rf is a solution of the Cauchy initial value problem for the inhomogeneous wave equation,*

$$\Box u \;=\; f, \quad u(x,0) \;=\; 0 \;=\; \frac{\partial u}{\partial t}(x,0)\,.$$

Proof. Passing to spherical coordinates with center $x \in \mathbb{R}^3$, we can represent the retarded potential as

$$Rf(x,t) \;=\; -\frac{1}{4\pi} \int\limits_0^{ct} r \int\limits_{S^2} f\big(x + r\cdot\alpha, t - r/c\big)\, d\alpha\, dr\,.$$

Introducing the parameter $\tau := t - \frac{r}{c}$ and the function

$$p(x,t,\tau) \;:=\; \frac{t-\tau}{4\pi} \int\limits_{S^2} f(x + c(t-\tau)\cdot\alpha, \tau)\, d\alpha\,,$$

we obtain from $p(x,t,t) = 0$ the following formulas for the retarded potential and its derivatives:

$$Rf(x,t) \;= -c^2 \int\limits_0^t p(x,t,\tau)\, d\tau\,, \quad \frac{\partial}{\partial t}(Rf)(x,t) \;=\; -c^2 \int\limits_0^t \frac{\partial p}{\partial t}(x,t,\tau)\, d\tau\,,$$

$$\frac{\partial^2}{\partial t^2}(Rf)(x,t) \;=\; -c^2 \int\limits_0^t \frac{\partial^2 p}{\partial t^2}(x,t,\tau)\, d\tau - c^2 \frac{\partial p}{\partial t}(x,t,t)\,.$$

Calculating the derivative of p with respect to t yields

$$\frac{\partial p}{\partial t}(x,t,t) \;=\; \frac{1}{4\pi} \int\limits_{S^2} f(x,t)\, d\alpha \;=\; f(x,t)\;,$$

and, summarizing, we obtain

$$\frac{\partial^2}{\partial t^2}(Rf)(x,t) \;=\; -c^2 \int_0^t \frac{\partial^2 p}{\partial t^2}(x,t,\tau)\,d\tau - c^2 \cdot f(x,t)\,.$$

Applying the wave operator to the retarded potential leads to the formula

$$\Box(Rf)(x,t) \;=\; f(x,t) - c^2 \int_0^t \Box p\,(x,t,\tau)\,d\tau\,.$$

If the parameter τ is fixed, the function

$$p(x,t,\tau_0) \;=\; \frac{t-\tau_0}{4\pi} \int_{S^2} f(x + c(t-\tau_0)\alpha, \tau_0)\,d\alpha$$

is a solution of the homogeneous wave equation. This immediately follows from the preceding theorem, since $p(x,t,\tau_0)$ is defined as a spherical mean. Hence the retarded potential solves the inhomogeneous wave equation. $\Box$

The inhomogeneous wave equation $\Box u = f$ may be interpreted as follows. For simplicity, suppose that f describes the current generated by a charged particle moving along the world line γ. At the point A, the particle produces an electromagnetic field which moves at the speed of light along the surface of the light cone with vertex A. Hence the field in B depends on the motion of the particle in A. In this way, the principle of causality appears as a mathematical consequence of the inhomogeneous wave equation, and its retarded solutions were first described by the physicists Liénard (1898) and Wiechert (1900).

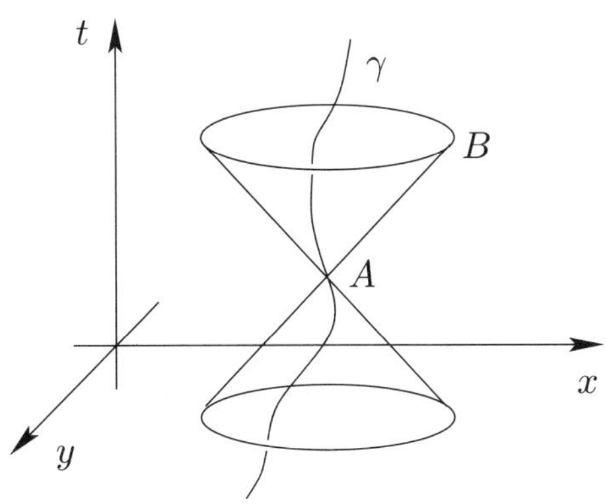

The solution of the two-dimensional wave equation can be constructed starting from that of the three-dimensional wave equation. To this end we transform the spherical mean of a function $\phi(x^1, x^2)$ depending only on the first two spatial variables into a two-dimensional integral by parametrizing the sphere $S^2(x, ct) \subset \mathbb{R}^3$ through

$$y^1 = x^1 + \alpha^1, \quad y^2 = x^2 + \alpha^2, \quad y^3 = x^3 + \sqrt{c^2t^2 - (\alpha^1)^2 - (\alpha^2)^2},$$

where $(\alpha^1)^2 + (\alpha^2)^2 \leq c^2t^2$. A straightforward calculation shows that the volume form of the sphere is described by the formula

$$dS^2 = \frac{ct}{\sqrt{c^2t^2 - (\alpha^1)^2 - (\alpha^2)^2}} \cdot d\alpha^1 \wedge d\alpha^2,$$

from which we obtain the following expression for the spherical mean:

$$I\phi(x^1, x^2, t) = \frac{1}{2\pi ct} \int\limits_{D^2(0,ct)} \frac{\phi(x^1 - \alpha^1, x^2 - \alpha^2)}{\sqrt{c^2t^2 - (\alpha^1)^2 - (\alpha^2)^2}} \, d\alpha^1 \wedge d\alpha^2.$$

We treat the retarded potential of a function only depending on the first two spatial variables and time analogously $f(x^1, x^2, t)$:

$$Rf(x^1, x^2, t) = -\frac{c}{2\pi} \int\limits_{D^2(0,c(t-\tau))} \frac{f(x^1 - \alpha^1, x^2 - \alpha^2, \tau)}{\sqrt{c^2(t-\tau)^2 - (\alpha^1)^2 - (\alpha^2)^2}} \, d\alpha^1 \wedge d\alpha^2 \wedge d\tau.$$

Inserting the resulting expressions into the formulas of the preceding theorem, we obtain a representation of the solution of the inhomogeneous two-dimensional wave equation with prescribed initial conditions $u_0(x^1, x^2)$ and $u_1(x^1, x^2)$ as an integral over the interior of a disk in $\mathbb{R}^2$. From this we conclude that a wave in $\mathbb{R}^2$ generated by initial conditions with compact support has a forward front, but it does not have a backward front. This effect is well-known, for example in the case of water waves. Stated differently, this means that Huygens' principle (see Remark 3) does not hold in two dimensions (in fact, it can be shown to fail in all even dimensions).

Now we turn to the question of which conditions rotating an electromagnetic wave lead to another electromagnetic wave. Let an electromagnetic wave $E(x,t), B(x,t)$ be given. We rotate these vector fields in the 2-plane spanned by the two vectors through the angle $\psi(x,t)$:

$$E^*(x,t) = \cos\psi(x,t) \cdot E(x,t) + \sin\psi(x,t) \cdot B(x,t)\,,$$

$$B^*(x,t) = -\sin\psi(x,t) \cdot E(x,t) + \cos\psi(x,t) \cdot B(x,t)\,.$$

We discuss the conditions to be satisfied so that the pair (E^*, B^*) becomes an electromagnetic wave. Necessarily, ψ has to be a solution of the *eikonal* *equation*

$$\frac{1}{c^2}\left(\frac{\partial\psi}{\partial t}\right)^2 - \|\mathrm{grad}(\psi)\|^2 = 0 .$$

Theorem 6. *(E^*, B^*) is an electromagnetic wave if and only if the phase function ψ satisfies the following conditions:*

(1) $\langle \mathrm{grad}(\psi), B\rangle = 0 = \langle \mathrm{grad}(\psi), E\rangle$;

(2) $\dfrac{1}{c}\dfrac{\partial\psi}{\partial t}\cdot B = -\mathrm{grad}(\psi)\times E$;

(3) $\dfrac{1}{c}\dfrac{\partial\psi}{\partial t}\cdot E = \mathrm{grad}(\psi)\times B$.

Proof. The vector fields E and B are divergence-free, and hence the divergences of E^* and B^* are easily calculated:

$$\mathrm{div}(E^*) = -\sin\psi\cdot\langle \mathrm{grad}(\psi), E\rangle + \cos\psi\cdot\langle \mathrm{grad}(\psi), B\rangle ,$$

$$\mathrm{div}(B^*) = -\cos\psi\cdot\langle \mathrm{grad}(\psi), E\rangle - \sin\psi\cdot\langle \mathrm{grad}(\psi), B\rangle .$$

Thus the vanishing of these divergences $\mathrm{div}(E^*) = \mathrm{div}(B^*) = 0$ is equivalent to the condition $\langle \mathrm{grad}(\psi), E\rangle = 0 = \langle \mathrm{grad}(\psi), B\rangle$. Making use of the formula

$$\mathrm{curl}(f\cdot\mathcal{V}) = \mathrm{grad}(f)\times\mathcal{V} + f\cdot\mathrm{curl}(\mathcal{V}),$$

which generally holds in $\mathbb{R}^3$, we write down the Maxwell equations for the pair (E^*, B^*). This leads to the following pair of equations:

$$\sin\psi\left(-\mathrm{grad}(\psi)\times E - \frac{1}{c}\frac{\partial\psi}{\partial t}\cdot B\right) = \cos\psi\left(\frac{1}{c}\frac{\partial\psi}{\partial t}\cdot E - \mathrm{grad}(\psi)\times B\right),$$
$$\sin\psi\left(-\mathrm{grad}(\psi)\times B + \frac{1}{c}\frac{\partial\psi}{\partial t}\cdot E\right) = \cos\psi\left(\frac{1}{c}\frac{\partial\psi}{\partial t}\cdot B + \mathrm{grad}(\psi)\times E\right),$$

which are equivalent to the second and third conditions of the theorem. □

9.4. The Relativistic Formulation of the Maxwell Equations

Consider Minkowski space $\mathbb{R}^{3,1}$ with coordinates (x, y, z, t) as well as the pseudo-Riemannian metric of index 1 determined by

$$g := (dx^1)^2 + (dx^2)^2 + (dx^3)^2 - c^2dt^2,$$

the so-called *Minkowski metric* (see Example 35, section 3.11). The 1-forms dx^1, dx^2, dx^3 and $c\cdot dt$ form an orthonormal frame. Computing the Hodge operator $\star$ of $\mathbb{R}^{3,1}$ on 2-forms leads to the following result:

$$\begin{aligned} \star(dx^1 \wedge dx^2) &= -dx^3 \wedge c \cdot dt\,, & \star(dx^1 \wedge dx^3) &= dx^2 \wedge c \cdot dt\,, \\ \star(dx^1 \wedge c \cdot dt) &= dx^2 \wedge dx^3\,, & \star(dx^2 \wedge dx^3) &= -dx^1 \wedge c \cdot dt\,, \\ \star(dx^2 \wedge c \cdot dt) &= -dx^1 \wedge dx^3\,, & \star(dx^3 \wedge c \cdot dt) &= dx^1 \wedge dx^2\,. \end{aligned}$$

If ω is a 1-form on $\mathbb{R}^3$, the Hodge operators $*$ of euclidean space $\mathbb{R}^3$ and $\star$ of Minkowski space $\mathbb{R}^{3,1}$, respectively, are related by

$$\star(\omega \wedge c \cdot dt) \;=\; *\omega.$$

In $\mathbb{R}^{3,1}$ we define the 2-form called *field strength form* by

$$F \;:=\; \omega_E \wedge (c \cdot dt) + \star(\omega_B \wedge c \cdot dt)$$

and the 1-form called *density form* by

$$J \;:=\; \frac{1}{c}\,\omega_J - \varrho \cdot dt\,.$$

Theorem 7. *The 2-form F has the following properties:*

(1) $\|F\|^2 \;=\; \|B\|^2 - \|E\|^2$.

(2) $\langle F, \star F\rangle \;=\; 2\langle E, B\rangle$.

(3) *The first group of the Maxwell equations is equivalent to*

$$dF \;=\; 0\,.$$

(4) *The second group of the Maxwell equations is equivalent to*

$$\star\, d \star F \;=\; 4\pi \cdot J\,.$$

Proof. We calculate the exterior derivative dF of F in 4-space:

$$\begin{aligned} dF &= d\omega_E \wedge c \cdot dt + d(*\omega_B) \\ &= d^{\mathbb{R}^3}(\omega_E) \wedge c \cdot dt + d^{\mathbb{R}^3}(*\omega_B) + \frac{\partial(*\omega_B)}{\partial t} \wedge dt \\ &= \left(d^{\mathbb{R}^3}(\omega_E) + \frac{1}{c}\frac{\partial(*\omega_B)}{\partial t} \right) \wedge c \cdot dt + d^{\mathbb{R}^3}(*\omega_B) \end{aligned}$$

and see that $dF = 0$ is equivalent to the pair of equations

$$d^{\mathbb{R}^3}(*\omega_B) \;=\; 0 \quad \text{and} \quad d^{\mathbb{R}^3}(\omega_E) \;=\; -\frac{1}{c}\frac{\partial}{\partial t}(*\omega_B)\,.$$

Similarly, we obtain

$$d \star F = d(*\omega_E - \omega_B \wedge c\,dt) = d^{\mathbb{R}^3}(*\omega_E) + \frac{\partial}{\partial t}(*\omega_E) \wedge dt - d^{\mathbb{R}^3}(\omega_B) \wedge c\,dt,$$

and, after a slight reformulation,

$$\star\, d \star F \;=\; *d^{\mathbb{R}^3}(*\omega_E) \cdot (c \cdot dt) + *d^{\mathbb{R}^3}(\omega_B) - \frac{1}{c}\frac{\partial}{\partial t}(\omega_E)\,.$$

Summarizing, $\star d \star F = 4\pi \cdot J$ is equivalent to

$$d(*\omega_E) \;=\; 4\pi\varrho \quad \text{and} \quad * d(\omega_B) \;=\; -\frac{1}{c}\frac{\partial}{\partial t}(\omega_E) + \frac{4\pi}{c}\omega_J\,. \qquad \square$$

Remark 4. The 2-form F is a closed form in $\mathbb{R}^{3,1}$. Hence by Poincaré's lemma there exists a 1-form $\mathcal{A}$ such that

$$d\mathcal{A} \;=\; -\frac{e}{c}F\,.$$

The 1-form $\mathcal{A}$ comprises the electric as well as the magnetic potential:

$$\begin{aligned} F \;=\; \omega_E \wedge (c\cdot dt) + *\omega_B \;&=\; \left(-\frac{1}{c}\frac{\partial}{\partial t}A - d\phi\right)\wedge (c\cdot dt) + d^{\mathbb{R}^3} A \\ =\; d^{\mathbb{R}^{3,1}}(A) - d\phi\wedge(c\cdot dt) \;&=\; d^{\mathbb{R}^{3,1}}(A - c\cdot\phi\cdot dt), \end{aligned}$$

and thus $\mathcal{A} = -(e/c)A + e\,\phi\,dt$.

The isometry group of the Minkowski metric g is $\mathrm{O}(3,1)$, the *Lorentz group.* It consists of all linear transformations $L : \mathbb{R}^{3,1} \to \mathbb{R}^{3,1}$ which leave g invariant. If we represent g by the matrix

$$A \;=\; \begin{bmatrix} 1 & & & 0 \\ & 1 & & \\ & & 1 & \\ 0 & & & -c^2 \end{bmatrix},$$

then $\mathrm{O}(3,1)$ can be identified with the (4×4) matrices L satisfying

$$L\cdot A\cdot L^t \;=\; A\,.$$

Its Lie algebra $\mathfrak{o}(3,1)$ is spanned by the matrices X with

$$X\cdot A + A\cdot L^t \;=\; 0\,.$$

Of course, $\mathrm{O}(3,1)$ contains the 3-dimensional group of rotations of euclidean space $\mathbb{R}^3$ as a subgroup, embedded as the upper left 3×3 matrices

$$\left\{\left[\begin{array}{c|c} M & 0 \\ \hline 0 & 1 \end{array}\right] : M \in \mathrm{O}(3,\mathbb{R})\right\}$$

and generated by the skew-symmetic matrices L_1, L_2, L_3 described in §6.3, Example 14. The missing part of the Lie algebra $\mathfrak{o}(3,1)$ is spanned by the three elements

$$B_1 := \left[\begin{array}{ccc|c} & & & 1 \\ & 0 & & 0 \\ & & & 0 \\ \hline c^2 & 0 & 0 & 0 \end{array}\right], \quad B_2 := \left[\begin{array}{ccc|c} & & & 0 \\ & 0 & & 1 \\ & & & 0 \\ \hline 0 & c^2 & 0 & 0 \end{array}\right], \quad B_3 := \left[\begin{array}{ccc|c} & & & 0 \\ & 0 & & 0 \\ & & & 1 \\ \hline 0 & 0 & c^2 & 0 \end{array}\right]$$

with commutator relations

$$[B_1, B_2] = c^2 L_3, \quad [B_1, B_3] = -c^2 L_2, \quad [B_2, B_3] = c^2 L_1,$$

$$[L_1, B_2] = -B_3, \quad [L_1, B_3] = B_2, \quad [L_2, B_1] = B_3, \quad [L_2, B_3] = -B_1,$$
$$[L_3, B_1] = -B_2, \quad [L_3, B_2] = B_1, \quad [L_i, B_i] = 0.$$

In the same way as L_1, L_2, L_3 generate ordinary rotations, the elements B_1, B_2, B_3 induce hyperbolic rotations, for example,

$$e^{\theta B_1} = \begin{bmatrix} \cosh(c\,\theta) & 0 & 0 & \sinh(c\,\theta)/c \\ 0 & 1 & 0 & 0 \\ 0 & 0 & 1 & 0 \\ c\sinh(c\,\theta) & 0 & 0 & \cosh(c\,\theta) \end{bmatrix}.$$

The defining condition for $L \in \mathrm{O}(3,1)$ implies that L has determinant 1 or -1. The hyperbolic rotation $e^{\theta B_1}$ and the time reflection $\mathrm{diag}(1,1,1,-1)$ realize these values of the determinant. Hence $\mathrm{O}(3,1)$ has two connectivity components.

Theorem 8.

(1) *The inhomogeneous Maxwell equations $dF = 0$ and $\delta F = 4\pi J$ are invariant under isometries of the Minskowski metric, the Lorentz transformations.*

(2) *The homogeneous Maxwell equations $dF = 0$ and $d \star F = 0$ are invariant under conformal changes of the metric.*

(3) *The quantities $||F||^2$ and $\langle F, \star F\rangle$ are invariant under Lorentz transformations.*

Proof. An element $e^{\theta X} \in \mathrm{O}(3,1)$ acts on differential forms and vectors by its differential $X \in \mathfrak{o}(3,1)$. The 2-form F transforms into its pullback $X^*(F)$. But since pulling back commutes with exterior differentiation, $dF = 0$ is equivalent to $d(X^*F) = 0$, which establishes the Lorentz invariance of the first Maxwell equation. For the second equation, observe that the volume form dM of any pseudo-Riemannian manifold M is invariant under isometries of the metric; hence the adjoint of d, the coderivative δ, also commutes with pullbacks by isometries. Thus, $\delta F = 4\pi J$ is equivalent to $X^*(\delta F) = 4\pi X^*(J)$ and $\delta(X^*F) = 4\pi X^*(J)$. This shows the Lorentz invariance of the second Maxwell equation.

Now consider a conformal change of the metric on a pseudo-Riemannian manifold M, $\tilde{g} = e^{2f}g$ for some function f. The exterior derivative does not depend on the metric; the first Maxwell equation is thus trivially conformally equivalent. An orthonormal basis $e_1, \ldots, e_n$ with dual basis of 1-forms $\sigma_1, \ldots, \sigma_n$ transforms into

$$e^{-f}e_1, \ldots, e^{-f}e_n \quad \text{and} \quad e^f\sigma_1, \ldots, e^f\sigma_n\,.$$

If M has even dimension $n = 2k$, we have the remarkable effect that the Hodge operators $*$ and $\tilde{*}$ with respect to the metrics g and $\tilde{g}$ coincide on differential forms in half the dimension of M. For example,

$$*(\sigma_1\wedge\ldots\wedge\sigma_k) = \varepsilon\cdot\sigma_{k+1}\wedge\ldots\wedge\sigma_{2k}, \quad \tilde{*}(e^f\sigma_1\wedge\ldots\wedge e^f\sigma_k) = \varepsilon\cdot e^f\sigma_{k+1}\wedge\ldots\wedge e^f\sigma_{2k}$$

for some irrelevant sign $\varepsilon = \pm 1$, so that k factors of e^f can be cancelled in the action of $\tilde{*}$, which implies the claim. Its follows for the 2-form F on 4-dimensional Minkowski space that $d * F = 0$ is conformally invariant. □

The conformal group can be identified with $\mathrm{O}(4,2)$, whis has dimension 15. The fact that only the homogeneous Maxwell equations are conformally invariant was the main reason why Lorentz invariance has been established as the fundamental invariance property of electrodynamics. However, the full conformal invariance is useful when dealing with homogeneous situations. The invariance of the scalar quantities $||F||^2$ and $\langle F, \star F\rangle$ is of practical importance, too; it means, for example, that E and B stay orthogonal if they were orthogonal in some frame of reference.

As a 2-form, F is an element of $\Omega^2(\mathbb{R}^{3,1})$. There exists a purely algebraic, Lorentz invariant map from $\Omega^2(\mathbb{R}^{3,1})$ into the space of symmetric $(2,0)$-tensors. Apart from normalizations and an additional multiple of the metric, the image of F under this map is known as the *Maxwell stress tensor*, which we will now discuss.

So let F be any 2-form on Minkowski space, and $e_1, \ldots, e_4$ an orthonormal basis, i.e., satisfying $g(e_i, e_j) = \varepsilon_i \cdot \delta_{ij}$ with $1 = \varepsilon_1 = \varepsilon_2 = \varepsilon_3 = -\varepsilon_4$. We define

$$T^F := \sum_i \varepsilon_i \cdot (e_i \lrcorner F) \odot (e_i \lrcorner F),$$

where $\odot$ denotes the symmetric product of 1-forms (see 5.3). This is an $\mathrm{SO}(3,1)$-invariant, symmetric $(2,0)$-tensor whose trace is a multiple of $||F||^2$. In the example at hand, we may choose

$$e_1 = \frac{\partial}{\partial x^1}, \quad e_2 = \frac{\partial}{\partial x^2}, \quad e_3 = \frac{\partial}{\partial x^3} \quad \text{and} \quad e_4 = \frac{\partial}{c\partial t}.$$

The 2-form F is given by

$$F = \omega_E \wedge c\,dt + B_1\,dx^2 \wedge dx^3 - B_2\,dx^1 \wedge dx^3 + B_3\,dx^1 \wedge dx^2.$$

Hence one computes ($x^4 := c\,dt$)

$$\begin{aligned} T^F &= (E_1 cdt - B_2 dx^3 + B_3 dx^2)^2 + (E_2 cdt + B_1 dx^3 - B_3 dx^1)^2 \\ &\quad + (E_3 cdt + B_1 dx^2 - B_2 dx^1)^2 - (E_1 dx^1 + E_2 dx^2 + E_3 dx^3)^2 \\ &=: \sum_{i,j}^{4} T^F_{ij}\, dx^i \odot dx^j. \end{aligned}$$

In order to obtain a trace-free endomorphism, one typically defines as the *Maxwell stress tensor* the combination

$$T^M \;:=\; -2T^F + \langle F, F\rangle \cdot (dx^1 \odot dx^1 + dx^2 \odot dx^2 + dx^3 \odot dx^3 - c^2 dt \odot dt).$$

Its importance stems from the fact that its components have a physical interpretation. For example, its pure time component

$$T^M_{44} \;=\; -(||E||^2 + ||B||^2)c^2 dt^2$$

is the energy density of the electromagnetic field, and the mixed contributions ($i = 1, \ldots, 3$)

$$T^M_{i4} \;=\; 4(E \times B)_i dx^i \odot cdt$$

are the components of the *Poynting vector* $E \times B$, which has the dimension of an energy flux density. Since T^M contains basically the same information as F, it is plausible that one can rewrite the Maxwell equations in terms of T^M solely. For their derivation, we refer to classical textbooks in eletrodynamics.

The electromagnetic potential $\mathcal{A}$ is a real-valued 1-form defined on Minkowski space. We now view the real numbers as the Lie algebra of the compact one-dimensional group U(1), in which the only quantity describing the electron which is relevant in Maxwell's theory is encoded—its electric charge. Passing now to the description of more complex elementary particles which carry not just electric charge, but in addition other characteristics (color, ...), we have to replace the one-dimensional group U(1) by a higher-dimensional compact Lie group G. If we also want to include gravitation in the model, as understood in the sense of general relativity, flat Minkowski space has to be replaced by a pseudo-Riemannian manifold. Combining all this, we are led to define a "generalized electromagnetic field with gravitation" as a 4-tuple $(M,\, g,\, G,\, \mathcal{A})$ consisting of a pseudo-Riemannian manifold, a Lie group G, and a 1-form $\mathcal{A} : T(M) \longrightarrow \mathfrak{g}$ with values in the Lie algebra $\mathfrak{g}$ of G. The non-commutativity of the group G will become essential in the sequel. The field strength $F_{\mathcal{A}}$ is described by a $\mathfrak{g}$-valued 2-form, and to the Maxwell equation in the vacuum there corresponds the so-called *Yang-Mills equation,*

$$F_{\mathcal{A}} \;:=\; d\mathcal{A} \;+\; \frac{1}{2} \cdot [\mathcal{A}, \mathcal{A}], \quad D^{\mathcal{A}}(\,\star\, F_{\mathcal{A}}) \;:=\; d \star F_{\mathcal{A}} \;+\; [\mathcal{A}, \star F_{\mathcal{A}}] \;=\; 0\,.$$

The possibly complicated topological structure of the base space M may lead to a situation in which the generalized electromagnetic potential $\mathcal{A}$ is not a globally defined 1-form on M. This leads to topics studied in the differential geometry of principal fiber bundles and the theory of connections.

With their help, the models discussed in the physics of elementary particles can be formulated and studied rigorously.[2]

9.5. The Lorentz Force

In this section, we discuss the equation of motion for the Lagrangian of a massive particle in an electromagnetic field as introduced in Example 11 of Chapter 7. We show that it is a natural generalization of the Lorentz equation, and discuss its solution for a few electromagnetic fields, i.e., a time-independent homogeneous field and a Dirac monopole. Frenet's formulation of curve theory will prove to be a suitable tool for the description of the particle's motion.

Let (M^m, g) be a pseudo-Riemannian manifold, A a 1-form on it, and consider the Lagrange function

$$L(v) \;=\; \frac{m}{2} g(v,v) - A(v)\,.$$

Theorem 9 (Generalized Lorentz equation). *A curve $\varrho(s) : \mathbb{R} \to M^m$ is a motion of the Lagrange system (M^m, L) if and only if it satisfies*

$$m\frac{\nabla \dot\varrho}{ds} \;=\; \mathcal{V}(\dot\varrho \,\lrcorner\, dA)\,,$$

where the right hand side is meant to denote the vector field associated with the 1-form $\dot\varrho \,\lrcorner\, dA$ relative to the Minkowski metric.

Proof. We use local coordinates $x^1, \ldots, x^m$ on M^m. The Lagrangian can then be rewritten as

$$L(x^1,\ldots,x^m,\dot x^1,\ldots,\dot x^m) \;=\; \frac{m}{2}\sum_{i,j=1}^{m} g_{ij}\dot x^i \dot x^j - \sum_{i=1}^{m} A_i \dot x^i .$$

By definition, $\varrho(s) : \mathbb{R} \to M^m$ is a motion of the Lagrange system (M^m, L) if and only if it fulfills for all k between 1 and m the Euler-Lagrange equation

$$\frac{d}{ds}\left(\frac{\partial L}{\partial \dot x^k}\big(\dot\varrho(s)\big)\right) \;=\; \frac{\partial L}{\partial x^k}\big(\dot\varrho(s)\big).$$

Since $\dot\varrho(s) = \big(dx^1/ds, \ldots, dx^m/ds\big)$, the left and right sides can be computed to be

$$\frac{\partial L}{\partial x^k}\big(\dot\varrho(s)\big) \;=\; \frac{m}{2}\sum_{i,j}\frac{\partial g_{ij}}{\partial x^k}\frac{dx^i}{ds}\frac{dx^j}{ds} - \sum_i \frac{\partial A_i}{\partial x^k}\frac{dx^i}{ds}$$

[2]See Thomas Friedrich, *Dirac Operators in Riemannian Geometry*, Grad. Studies in Math. vol. 25, AMS, Providence, 2000.

and

$$\frac{d}{ds}\left(\frac{\partial L}{\partial \dot{x}^k}(\dot{\varrho}(s))\right) = m\sum_{i,j}\frac{\partial g_{ik}}{\partial x^k}\frac{dx^i}{ds}\frac{dx^j}{ds} + m\sum_i g_{ik}\frac{d^2x^i}{ds^2} - \sum_i \frac{\partial A_k}{\partial x^i}\frac{dx^i}{ds}\,.$$

Equating them and rearranging some terms, we obtain the equation

$$(*) \quad \sum_i \left[\frac{\partial A_k}{\partial x^i} - \frac{\partial A_i}{\partial x^k}\right]\frac{dx^i}{ds} = m\sum_i g_{ik}\frac{d^2x^i}{ds^2} + m\sum_{i,j}\left[\frac{\partial g_{ik}}{\partial x^j} - \frac{1}{2}\frac{\partial g_{ij}}{\partial x^k}\right]\frac{dx^i}{ds}\frac{dx^j}{ds}\,.$$

Hence it remains to interpret both sides geometrically, as done in the formulation of the theorem. We begin by recalling the definition of the covariant derivative as discussed in §5.7, Definition 21. From there, we know that the covariant derivative may be expressed in terms of the Christoffel symbols associated to the chosen local coordinates as

$$m\frac{\nabla\dot{\varrho}}{ds} = m\sum_l \frac{d^2x^l}{ds^2}\frac{\partial}{\partial x^l} + m\sum_{i,j,l}\Gamma^l_{ij}\frac{dx^i}{ds}\frac{dx^j}{ds}\frac{\partial}{\partial x^l}\,.$$

But the Christoffel symbols can by computed from the metric (§5.7, Theorem 42 ff.):

$$\Gamma^k_{ij} = \frac{1}{2}\sum_m g^{km}\left(\frac{\partial g_{im}}{\partial x^k} + \frac{\partial g_{jm}}{\partial x^i} - \frac{\partial g_{ij}}{\partial x^m}\right);$$

hence one checks that the right-hand side of the equation of motion $(*)$ is equal to

$$g\left(m\frac{\nabla\dot{\varrho}}{ds}, \frac{\partial}{\partial x^k}\right).$$

Similarly,

$$dA = \sum_{i,j}\frac{\partial A_i}{\partial x^j}\,dx^i \wedge dx^j$$

implies that

$$dA\left(\dot{\varrho}, \frac{\partial}{\partial x^k}\right) = \sum_i \left[\frac{\partial A_k}{\partial x^i} - \frac{\partial A_i}{\partial x^k}\right]\frac{dx^i}{ds}\,,$$

which is just the left-hand side of $(*)$. We conclude that equation $(*)$ is equivalent to

$$g\left(m\frac{\nabla\dot{\varrho}}{ds}, \frac{\partial}{\partial x^k}\right) = dA\left(\dot{\varrho}, \frac{\partial}{\partial x^k}\right).$$

Since this has to hold for all k, the claim is proved. □

The most important situation where this result can be applied is the case that M^m is the flat Minkowski space $\mathbb{R}^{3,1}$ with its pseudo-Riemannian metric. As 1-form on $\mathbb{R}^{3,1}$, we choose the form $\mathcal{A}$ introduced in the previous section, which is composed of the magnetic 1-form A on the euclidean space $\mathbb{R}^3$ and the electric potential ϕ:

$$\mathcal{A} \;=\; -\frac{e}{c}A + e\phi\, dt \;=\; -\frac{e}{c}(A_1 dx^1 + A_2 dx^2 + A_3 dx^3) + e\phi\, dt\,,$$

and it satisfies $d\mathcal{A} = (-e/c)\cdot F$. We emphasize that time t is viewed as the fourth coordinate of space, whereas the curve parameter s of ϱ is interpreted as the proper time of the particle. The curve $\varrho(s) = \big(x^1(s), x^2(s), x^3(s), t(s)\big)$ is said to be parametrized in *proper time* if its tangent vector is normalized to length $-c^2$,

$$\left|\left|\frac{d\varrho}{ds}\right|\right|^2 \;:=\; -c^2.$$

In classical mechanics as discussed in Chapter 7 (the Newtonian limit), all velocities were so small compared to the velocity of light c that it was safe to identify s with time. We shall henceforth abstain from denoting derivatives with dots or primes, in order to avoid confusion. By velocities, we mean the time derivatives $v_i := dx^i/dt$ with euclidean length $v^2 := v_1^2 + v_2^2 + v_3^2$. The parameters s and t are related through the relativistic γ-factor. To compute it, we express the length of $d\varrho(s)/ds$ as

$$\left|\left|\frac{d\varrho}{ds}\right|\right|^2 = \sum_{i=1}^{3}\left[\frac{dx^i}{ds}\right]^2 - c^2\left[\frac{dt}{ds}\right]^2 = \sum_{i=1}^{3}\left[\frac{dx^i}{dt}\frac{dt}{ds}\right]^2 - c^2\left[\frac{dt}{ds}\right]^2 = (v^2 - c^2)\left[\frac{dt}{ds}\right]^2$$

and deduce from its length normalization that

$$\frac{dt}{ds} \;=\; \frac{1}{\sqrt{1 - v^2/c^2}} \;=:\; \gamma, \quad \text{and} \quad \frac{ds}{dt} \;=\; 1/\gamma\,.$$

Thus, it is comfortable to rewrite $d\varrho/ds$ as

$$\frac{d\varrho}{ds} \;=\; \gamma\left(v_1\frac{\partial}{\partial x^1} + v_2\frac{\partial}{\partial x^2} + v_3\frac{\partial}{\partial x^3} + \frac{\partial}{\partial t}\right).$$

The right-hand side of the generalized Lorentz equation now reads, using the definition of F,

$$\frac{d\varrho}{ds}\,\lrcorner\, d\mathcal{A} \;=\; -\frac{e}{c}\frac{d\varrho}{ds}\,\lrcorner\, F \;=\; +\frac{e}{c}\gamma\left[-(v\cdot E)c\, dt + \sum_i \big(cE_i + (v\times B)_i\big)dx^i\right],$$

where $v\cdot E$ denotes the euclidean scalar product and $v\times B$ the vector product in $\mathbb{R}^3$. Its dual vector is

$$\mathcal{V}\Big(\frac{d\varrho}{ds}\,\lrcorner\, d\mathcal{A}\Big) \;=\; \gamma\left[+\frac{e}{c^2}(v\cdot E)\frac{\partial}{\partial t} + \sum_i \Big(eE_i + \frac{e}{c}(v\times B)_i\Big)\frac{\partial}{\partial x^i}\right].$$

Since the left-hand side of the generalized Lorentz equation is

$$m\frac{d}{ds}\frac{d\varrho}{ds} \;=\; m\gamma\frac{d}{dt}\frac{d\varrho}{ds}\,,$$

we conclude that its spacelike components yield after simplification the *classical Lorentz equation*

$$(1)\qquad m\frac{d}{dt}(\gamma v) \;=\; e\cdot E + \frac{e}{c}v\times B\,,$$

and its time component leads to $md\gamma/dt = +(e/c^2)v\cdot E$. This equation describes in fact the change in time of the kinetic energy $\mathcal{E} = mc^2\gamma$ of the particle,

$$(2)\qquad \frac{d\mathcal{E}}{dt} \;=\; e(v\cdot E)\,.$$

This proves that only the electric field exercises power on the particle, as claimed in Example 13 of 7.5.

We shall now describe the motion of a charged particle in some special electromagnetic fields. Consider the case of a vanishing electric field ($E = 0$). Equation (2) then implies that the kinetic energy is constant; hence the γ-factor and, in particular, the total velocity v^2 are constant, too. The spatial part of the particle's motion $\tilde{\varrho}(s) = \big(x^1(s), x^2(s), x^3(s)\big)$ satisfies $||d\tilde{\varrho}/ds||^2 = v^2 = \text{const}$; hence the reparametrized curve $\eta(s) := \tilde{\varrho}(s/v)$ is given in its natural parametrization. The Lorentz equation for this curve can be written as

$$m\gamma v\frac{d^2\eta}{ds^2} \;=\; \frac{e}{c}\frac{d\eta}{ds}\times B\,,$$

and it is natural to use Frenet's frame to describe the curve $\eta(s)$ in euclidean space $\mathbb{R}^3$. It is safe again to denote derivatives with respect to the natural parameter s by dots.

Example 3 (Static and homogeneous magnetic field)**.** Suppose that the magnetic field B is static and homogeneous, i. e., constant in space and time. Recall that $\vec{t}, \vec{h}, \vec{b}$ denote the tangent, the principal normal and the binormal vector, respectively, and that κ and τ are the curvature and the torsion of η. The Lorentz equation is thus geometrically equivalent to

$$m\gamma v\cdot\kappa\vec{h} \;=\; \frac{e}{c}\vec{t}\times B\,.$$

We differentiate this equation with respect to s and use Frenet's formulas for the derivative of $\vec{h}$ and $\vec{t}$:

$$m\gamma v(\dot{\kappa}\vec{h} + \kappa(-\kappa\vec{t} + \tau\vec{b})) \;=\; \frac{e}{c}\kappa\,\vec{h}\times B\,.$$

Since the vectors $\vec{t}, \vec{h}, \vec{b}$ constitute, at every point, an orthonormal frame, we conclude that $\dot{\kappa} = 0$. We compute the value of the curvature:

$$\kappa = ||\ddot{\eta}|| = \frac{e}{m\gamma vc}||\dot{\eta} \times B|| = \frac{e||B||}{m\gamma vc}\sin(\dot{\eta}, B) = \frac{e||B||}{m\gamma vc}\sqrt{1 - \langle\dot{\eta}, B\rangle / ||B||}.$$

In fact, on can check directly that $\langle\dot{\eta}, B\rangle$ has to be constant, for its derivative is $2\langle\ddot{\eta}, B\rangle$; but by the Lorentz equation, $\ddot{\eta}$ is perpendicular to B. Alternatively, we could also have derived it from Noether's theorem. In order to determine the curve's torsion, we first compute its binormal vector,

$$\vec{b} = \vec{t} \times \vec{h} = \frac{1}{\kappa}\dot{\eta} \times \ddot{\eta},$$

as well as its derivative,

$$\dot{\vec{b}} = \frac{1}{\kappa}\dot{\eta} \times \dddot{\eta} = \frac{e}{\gamma mvc}\dot{\eta} \times (\vec{h} \times B) = \frac{e}{\gamma mvc}\left(\vec{h}\langle\vec{t}, B\rangle - B\langle\vec{h}, \vec{t}\rangle\right) = \frac{e\langle\vec{t}, B\rangle}{m\gamma vc}\vec{h}.$$

The torsion is then

$$\tau = -\langle\dot{\vec{b}}, \vec{h}\rangle = -\frac{e\langle\vec{t}, B\rangle}{\gamma mvc},$$

and again constant. From Chapter 5, Exercice 7 we know that a curve whose torsion and curvature are constant is necessarily a helix, the direction of which is given by B. In the special case that $\langle\vec{t}, B\rangle = 0$, the curve is a circle in the plane perpendicular to B with radius of curvature r and *synchrotron frequency* ω:

$$r = \frac{1}{\kappa} = \frac{m\gamma vc}{e||B||}, \qquad \omega = \frac{v}{r} = \frac{e||B||}{\gamma mc}.$$

Example 4 (Dirac monopole[3]). Although their physical existence could never be established, Dirac monopoles have proved important models in theoretical physics. A monopole at the origin of $\mathbb{R}^3$ induces the field

$$B = \frac{c}{2e||\eta||^3}\eta.$$

The Lorentz equation is thus

$$(*) \qquad \ddot{\eta} = -\frac{1}{2\gamma mv||\eta||^3}\eta \times \dot{\eta}.$$

We are going to prove that *the particle moves along a geodesic of a cone with vertex the origin.* As in the integration of the geodesic flow or the abstract discussion of Lagrange systems, the proof makes crucial use of a clever invariant of motion. But the invariant of the Dirac monopole differs from all invariants encountered so far in the fact that it is *not* linear in $\dot{\eta}$;

[3]This discussion has been published by Katharina Habermann (formerly Neitzke) in her article K. Neitkze, *Die Lorentz-Kraft auf pseudo-Riemannschen Mannigfaltigkeiten*, Math. Nachr. **149** (1990), 183-214.

hence its existence does not follow from Noether's theorem, but only from a direct calculation. First, we observe that

$$\frac{d}{ds}\langle\eta,\dot\eta\rangle \;=\; \langle\eta,\ddot\eta\rangle + \langle\dot\eta,\dot\eta\rangle \;=\; 0+1,$$

because η is perpendicular to $\ddot\eta$ by the Lorentz equation $(*)$, and so there exists a constant $a \in \mathbb{R}$ with $\langle\eta,\dot\eta\rangle = s+a$. It also implies that the following derivative vanishes:

$$\frac{d}{ds}\left[||\eta||^2 - \langle\eta,\dot\eta\rangle^2\right] \;=\; 2\langle\eta,\dot\eta\rangle - 2\langle\eta,\dot\eta\rangle\cdot 1 \;=\; 0\,,$$

which yields the invariant we were looking for. By the Cauchy-Schwarz inequality, it has to be non-negative; hence there exists a constant $k \geq 0$ such that

$$||\eta||^2 - \langle\eta,\dot\eta\rangle^2 \;=\; k^2 \quad\text{and}\quad ||\eta||^2 \;=\; k^2+(s+a)^2.$$

A short calculation for the curvature of the particle's curve of motion yields

$$\begin{aligned}\kappa = ||\ddot\eta|| &= \frac{||\eta\times\dot\eta||}{2\gamma m v||\eta||^3} = \frac{\sin(\eta,\dot\eta)}{2\gamma m v||\eta||^2} = \frac{\sqrt{1-\cos^2(\eta,\dot\eta)}}{2\gamma m v||\eta||^2}\\ &= \frac{\sqrt{1-\langle\eta,\dot\eta\rangle^2/||\eta||^2}}{2\gamma m v||\eta||^2} = \frac{\sqrt{||\eta||^2-\langle\eta,\dot\eta\rangle^2}}{2\gamma m v||\eta||^3} = \frac{k}{2\gamma m v||\eta||^3}\,.\end{aligned}$$

If $k=0$, the curvature vanishes and the curve is a straight line, i. e., there exist vectors $\vec v, \vec w \in \mathbb{R}^3$ such that $\eta(s) = s\cdot\vec v + \vec w$. The equation of motion $(*)$ then implies $\vec v\times\vec w = 0$, i. e., $\vec v$ and $\vec w$ are linearly dependent, so thet we finally obtain

$$\eta(s) \;=\; (s+\lambda)\vec v, \quad \lambda\in\mathbb{R}\,,$$

which is the equation of a line through the origin. Let us assume from now on that $k>0$. We shall derive the expression for the torsion of the curve. By Frenet's formulas, the principal normal vector $\vec h$ is

$$\vec h \;=\; \frac{\ddot\eta}{\kappa} \;=\; -\frac{1}{k}\eta\times\dot\eta\,.$$

The binormal vector then becomes

$$\vec b \;=\; \dot\eta\times\vec h \;=\; -\frac{1}{k}\dot\eta\times(\eta\times\dot\eta),$$

with derivative

$$\frac{d}{ds}\vec b \;=\; -\frac{1}{k}\dot\eta\times(\eta\times\ddot\eta) \;=\; \frac{\dot\eta\times(\eta\times(\eta\times\dot\eta))}{2kmv\gamma||\eta||^3}\,.$$

Using Grassmann's identity $\vec w\times(\vec u\times\vec v) = \vec u\,\langle\vec v,\vec w\rangle - \vec v\,\langle\vec u,\vec w\rangle$, one shows that $\vec v\times(\vec u\times(\vec u\times\vec v)) = -\langle\vec u,\vec v\rangle\,\vec u\times\vec v$, so that the former expression can further be simplified to

$$\frac{d}{ds}\vec{b} = \frac{\langle \eta, \dot{\eta} \rangle}{2mv\gamma ||\eta||^3}\left(-\frac{1}{k}\eta \times \dot{\eta}\right) = \frac{\langle \eta, \dot{\eta} \rangle}{2mv\gamma ||\eta||^3}\vec{h}\,.$$

The torsion is thus equal to

$$\tau = -\langle \dot{\vec{b}}, \vec{h} \rangle = -\frac{\langle \eta, \dot{\eta} \rangle}{2mv\gamma ||\eta||^3} = -\frac{s+a}{2mv\gamma \sqrt{k^2+(s+a)^2}^3}.$$

In particular, the quotient of torsion and curvature satisfies the simple identity $\tau/\kappa = -(s+a)/k$. We prove next that the curve lies on a cone, that is, that there exists a constant vector $\vec{u}$ whose angle φ with η is also constant.

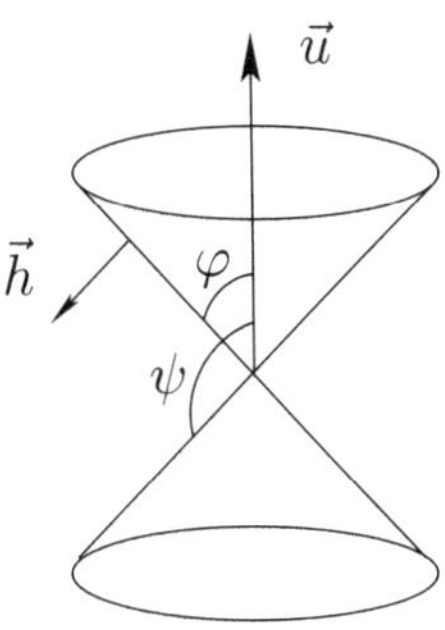

Define the functions

$$f(s) := \frac{s+a}{2km\gamma v\sqrt{k^2+(s+a)^2}}, \quad g(s) = -\frac{1}{2m\gamma v\sqrt{k^2+(s+a)^2}}\,.$$

They are primitives of κ and $-\tau$ and have the same quotient,

$$\frac{d\,f(s)}{ds} = \kappa, \quad \frac{d\,g(s)}{ds} = -\tau, \quad \frac{f(s)}{g(s)} = \frac{\kappa}{\tau}\,.$$

Frenet's formulas thus imply immediately that the vector

$$\vec{u} := \vec{h} + f(s)\cdot\vec{t} + g(s)\cdot\vec{b}$$

is constant. To compute its angle φ with η, we derive an exact expression for η by applying Grassmann's identity to the binormal,

$$\vec{b} = -\frac{1}{k}\dot{\eta}\times(\eta\times\dot{\eta}) = -\frac{1}{k}\left[\eta - \dot{\eta}(s+a)\right], \quad \eta = (s+a)\vec{t} - k\vec{b},$$

and remark that $\vec{u}$ can be rewritten $\vec{u} = \eta/2mk\gamma v||\eta|| + \vec{h}$. For the opening angle of the cone we obtain

$$\cos(\varphi) = \frac{\langle \eta, \vec{u} \rangle}{||\eta||\cdot||\vec{u}||} = \frac{1}{\sqrt{1+(2mk\gamma v)^2}}.$$

By definition, a curve is a geodesic on a given surface if and only if its principal normal vector is orthogonal to the tangent plane of the surface at any of its points. Denoting by ψ the angle between $\vec{h}$ and $\vec{u}$, we get

$$\cos(\psi) = \frac{\langle \vec{h}, \vec{u} \rangle}{||\vec{u}||} = \frac{1}{||\vec{u}||} = \frac{2mkv\gamma}{\sqrt{1+(2mkv\gamma)^2}},$$

and we see that this is equal to $\sqrt{1-\cos^2(\varphi)} = \sin(\varphi)$. Hence, ψ and φ are related through $\psi - \varphi = \pi/2$, and we conclude that $\vec{h}$ is indeed perpendicular to the tangent plane of the curve. The particle moves along a geodesic of the cone, as claimed.

Exercises

1 (Kirchhoff Formula). Let $u(x,t)$ be a function defined on $\mathbb{R}^3 \times \mathbb{R}$ and $\Omega \subset \mathbb{R}^3$ a compact domain with smooth boundary. Prove, for each point $(x_0, t_0) \in \Omega \times \mathbb{R}$, the *Kirchhoff formula:*

$$4\pi \cdot u(x_0, t_0) \;=\; -\int_\Omega \frac{\Box u(y, t_0 - r/c)}{r}\, dy \;+ \int\limits_{\partial\Omega} \left[\frac{\partial u\big(\alpha, t_0 - \frac{r}{c}\big)}{r \cdot \partial \mathcal{N}} \right.$$

$$\left. + \frac{\partial u\big(\alpha, t_0 - \frac{r}{c}\big)}{cr \cdot \partial t} \cdot \frac{\partial r\big(\alpha, t_0\big)}{\partial \mathcal{N}} - u\big(\alpha, t_0 - \tfrac{r}{c}\big) \cdot \frac{\partial\, r^{-1}(\alpha, t_0)}{\partial \mathcal{N}} \right] d(\partial\Omega)(\alpha),$$

where $r := ||x_0 - y||$ is the distance to the spatial point $x_0 \in \mathbb{R}^3$, $\mathcal{N}$ is the normal vector to the boundary $\partial\Omega$, and $\Box = \Delta_x - 1/c^2 \partial_{tt}$ is the wave operator. Deduce from this the solution formulas for the Cauchy problem of the wave equation by choosing for Ω a 3-dimensional ball.

2. Under the assumptions of Helmholtz' theorem, the electric and the magnetic field E and B, as well as the current density field J can be written as the sum of a divergence-free and a curl-free vector field,

$$E \;=\; E_{\text{div}} + E_{\text{curl}}, \quad B \;=\; B_{\text{div}} + B_{\text{curl}} \quad \text{and} \quad J \;=\; J_{\text{div}} + J_{\text{curl}}\,.$$

a) Prove that the Maxwell equations can then be written as follows:

$$\operatorname{curl}(E_{\text{div}}) \;=\; -\frac{1}{c}\frac{\partial B_{\text{div}}}{\partial t}, \quad \operatorname{div}(B_{\text{curl}}) \;=\; 0$$

$$\operatorname{curl}(B_{\text{div}}) \;=\; \frac{1}{c}\frac{\partial E_{\text{div}}}{\partial t} + \frac{4\pi}{c} J_{\text{div}}, \quad \operatorname{div}(E_{\text{curl}}) \;=\; 4\pi\varrho\,.$$

The continuity equation then becomes

$$\operatorname{div}(J_{\text{curl}}) + \frac{\partial \varrho}{\partial t} \;=\; 0\,.$$

Hint: In the proof of Theorem 2, we proved that, under the assumptions made here, any vector field which is at the same time divergence- and curl-free, has to vanish identically.

b) On the other hand, for the electric field E we already know the decomposition

$$E \;=\; -\frac{1}{c}\frac{\partial A}{\partial t} + \operatorname{grad}(-\phi)\,.$$

Prove that this is precisely the Helmholtz decomposition of E if the magnetic potential satisfies the condition called *Coulomb gauge*:

$$\operatorname{div}(A) \;=\; 0\,.$$

3. Deduce from Theorem 4 that if $E_0(x)$ and $B_0(x)$ are C^3-functions on $\mathbb{R}^3$ such that $\operatorname{div}E_0 = \operatorname{div}B_0 = 0$, then the Cauchy initial value problem for an electromagnetic wave

$$\operatorname{curl}(B) \;=\; \frac{1}{c}\frac{\partial E}{\partial t}\,, \quad \operatorname{curl}(E) \;=\; -\frac{1}{c}\frac{\partial B}{\partial t}\,,$$
$$\operatorname{div}(B) \;=\; 0\,, \quad \operatorname{div}(E) \;=\; 0\,, \quad E(x,0) \;=\; E_0(x), \quad B(x,0) \;=\; B_0(x)$$

has the unique solution

$$E(x,ct) \;=\; \frac{1}{4\pi ct}\int\limits_{S^2(x,ct)} \operatorname{curl} B_0(y)dy + \frac{1}{4\pi c}\frac{\partial}{\partial t}\left[\;\int\limits_{S^2(x,ct)} E_0(y)dy\right],$$

$$B(x,ct) \;=\; -\frac{1}{4\pi ct}\int\limits_{S^2(x,ct)} \operatorname{curl} E_0(y)dy + \frac{1}{4\pi c}\frac{\partial}{\partial t}\left[\;\int\limits_{S^2(x,ct)} B_0(y)dy\right].$$

4. Using Theorems 1 and 2, determine the electric and the magnetic field in $\mathbb{R}^3$ which is generated in the following situations:

a) a homogeneously charged ball of radius R with constant charge density ϱ;

b) a charged spherical shell of radius R with constant surface charge density σ;

c) an infinitely extended straight wire of radius R through which a current with constant current density j flows.

5. Describe the solution of the classical Lorentz equation for a homogeneous, static electric field ($E = \text{const}$) and vanishing magnetic field ($B = 0$). What happens in the non-relativistic limit of small velocities $v \ll c$?

6. The radiation rate of a Lorentz electron is proportional to the second derivative of the vector $\varrho(s) \in \mathbb{R}^{3,1}$:

$$\mathcal{R} \;=\; \frac{2e^2}{3c^2}\left|\left|\frac{d^2\varrho}{ds^2}\right|\right|.$$

Compute the general expression for $\mathcal{R}$ and discuss the case of planar circular synchrotron radiation.

7. Show that the Lie algebra $\mathfrak{o}(3,1)$ of the Lorentz group is not simple—more precisely, that it is the sum of two 3-dimensional ideals.

8 (Infinitely Extended String with Large Oscillations)**.** Consider the Cauchy problem for the one-dimensional wave equation determined by two functions $u_0(x) \in C^2(\mathbb{R})$, $u_1(x) \in C^1(\mathbb{R})$:

$$\begin{aligned} \partial_{tt} u &= c^2 \partial_{xx} u, \quad t \geq 0, \\ u(x,0) &= u_0(x), \quad \partial_t u(x,0) = u_1(x). \end{aligned}$$

Part 1. General Shape of the Solution. Prove that there exist two $\mathcal{C}^2$-functions f and g satisfying

$$u(x,t) = f(x+ct) + g(x-ct).$$

The solution is hence the superposition of a wave traveling to the left and one traveling to the right. Hint: Introduce the new coordinates $x_\pm = x \pm ct$ and show that the wave equation is equivalent to the differential equation

$$\frac{\partial^2 u}{\partial x_+ \partial x_-} = 0.$$

Part 2. Solution of the Cauchy Problem. With the above Ansatz for the solution, the following relations have to hold:

$$u_0(x) = f(x) + g(x), \quad u_1(x) = c(f'(x) - g'(x)).$$

By integration of the second equation over the interval $[0, x]$, prove that the general solution has to be given by

$$u(x,t) = \frac{1}{2}[u_0(x+ct) + u_0(x-ct)] + \frac{1}{2c}\int_{x-ct}^{x+ct} u_1(s)\, ds.$$

In particular, this argument shows the *uniqueness* of the solution. How can the result be understood qualitatively by means of the light-cone ?

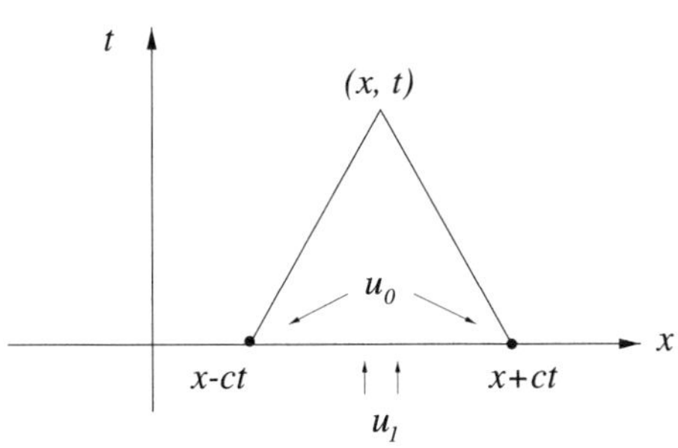

9 (One-sided Infinitely Extended Oscillating String)**.** An oscillating string extending infinitely in the positive x-direction is modeled by the one-dimensional wave equation,

$$\partial_{tt} u = c^2 \partial_{xx} u, \quad t \geq 0, \quad x \geq 0 .$$

In addition to the initial conditions,

$$u(x,0) = u_0(x), \quad \partial_t u(x,0) = u_1(x),$$

one has to impose boundary condition which describe the behavior of the "wall" at $x = 0$ for all times t:

$$u(0,t) = \varphi(t) .$$

Here $u_0, \varphi \in C^2(\mathbb{R}_+)$ and $u_1 \in C^1(\mathbb{R}_+)$ are required. Prove (using a similar Ansatz as in the preceding exercise) that the solution is

$$u(x,t) = \begin{cases} \frac{1}{2}[u_0(x+ct) + u_0(x-ct)] + \frac{1}{2c}\displaystyle\int_{x-ct}^{x+ct} u_1(s)\,ds, \\ \frac{1}{2}[u_0(x+ct) - u_0(ct-x)] + \varphi(t - x/c) + \frac{1}{2c}\displaystyle\int_{ct-x}^{ct+x} u_1(s)\,ds, \end{cases}$$

where the upper line is to be taken for points (x,t) in the region I, i. e. below the line $x = ct$, and the lower line for points (x,t) in the region II, above the line $x = ct$. What can be said concerning the behavior of the light-cone, in particular in II? Describe, moreover, the regularity properties of the solution on the line $x = ct$, and explain carefully under which additional conditions it is of class C^2 there.

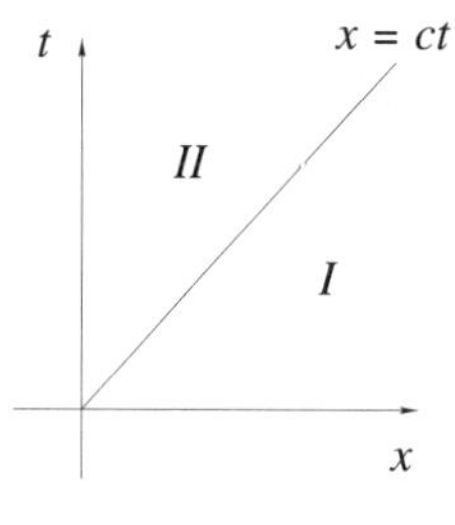

10 (Oscillating String Fixed on Either Side)**.** For the wave equation on bounded domains a separation Ansatz going back to Bernoulli proved to be successful, in that it reduces the problem to one for Fourier series. We are looking for a solution of the one-dimensional wave equation on the interval $[0,1]$,

$$\partial_{tt} u = c^2 \partial_{xx} u, \quad t \geq 0, \quad 0 \leq x \leq 1 .$$

The initial and boundary conditions are

$$u(x,0) = u_0(x), \quad \partial_t u(x,0) = u_1(x), \quad u(0,t) = 0, \quad u(1,t) = 0.$$

The *explanation* for the Fourier method and the related regularity properties to be required of u_0 and u_1 will be treated separately in the next exercise. Start from the Ansatz

$$u(x,t) \;=\; T(t)\cdot X(x)\,.$$

a) Prove that there exists a constant λ, such that

$$\frac{X''}{X} \;=\; \frac{T''}{c^2T} \;=\; \lambda\,.$$

b) Show that the cases $\lambda = 0$ and $\lambda > 0$ necessarily lead to the trivial solution $u = 0$; in the case $\lambda < 0$ there has to exist a natural number k such that $\lambda = -k^2\pi^2$ and

$$u_k(x,t) \;=\; X_k(x)T_k(t) \;=\; \sin k\pi x\cdot(a_k \sin k\pi ct + b_k\cos k\pi ct)\,.$$

The general solution is then obtained as the series

$$u(x,t) \;=\; \sum_{k=1}^{\infty} u_k(x,t) \;=\; \sum_{k=1}^{\infty} \sin k\pi x\cdot(a_k \sin k\pi ct + b_k\cos k\pi ct)\,.$$

c) Find an integral formula for the coefficients a_k, b_k depending on the initial conditions u_0 and u_1. Solution:

$$a_k \;=\; \frac{2}{k\pi c}\int_0^1 u_1(x)\sin k\pi x\,dx, \quad b_k \;=\; 2\int_0^1 u_0(x)\sin k\pi x\,dx\,.$$

11 (Validity of the Fourier Method)**.** By a theorem of Dirichlet, the Fourier series of a function $f \in C^1([0,1])$ satisfying $f(0) = f(1) = 0$ is uniformly convergent and tends pointwise to f. Prove the following lemma:

For a function $f \in C^k([0,1])$ such that $f^{(0)}, \ldots, f^{(k-1)}$ vanish at 0 and 1, there exists a constant A for which the Fourier coefficients of f satisfy the following inequality:

$$|c_n| \;=\; \left|\int_0^1 f(x)\sin n\pi x\,dx\right| \;\le\; \frac{A}{n^k}\,.$$

Formulate necessary conditions for u_0 and u_1, under which the solution $u(x,t)$ constructed in the preceding exercise is twice differentiable with respect to x and t and satisfies the initial conditions.

12 (Inhomogeneous Wave Equation)**.**

Part 1. The *inhomogeneous wave equation* is understood to be the equation

$$(*) \qquad \partial_{tt}u \;=\; c^2\partial_{xx}u + f(x,t)$$

with a given function f. We consider this equation on the interval $[0,1]$ with the initial conditions

$$u(x,0) = u_0(x), \quad \partial_t u(x,0) = u_1(x)$$

as well as, first, trivial boundary conditions,

$$u(0,t) = 0, \quad u(1,t) = 0 .$$

Show that the Ansatz

$$u(x,t) = \sum_{n=1}^{\infty} a_n(t)\sin(n\pi x)$$

satisfies the boundary conditions, and prove that the functions $a_n(t)$ are uniquely determined. To do so, prove that they solve an ordinary differential equation of second order, and that the initial values $a_n(0), a_n'(0)$ can be computed from the initial conditions. Hint: The Fourier coefficients of u_0, u_1 and f play an important role here.

Part 2. We are looking for a solution $u(x,t)$ of the inhomogeneous wave equation $(*)$ with arbitrarily given boundary conditions,

$$u(0,t) = \varphi_0(t), \quad u(1,t) = \varphi_1(t) .$$

Show that the solution of this equation can be reduced to the solution of an inhomogeneous wave equation with trivial boundary conditions (and a different inhomogeneity f!). Hint: Let $\Phi(x,t)$ be any function in two variables satisfying $\Phi(0,t) = \varphi_0(t)$ and $\Phi(1,t) = \varphi_1(t)$, e.g. $\Phi(x,t) = (1-x)\varphi_0(t) + x\varphi_1(t)$. Then consider the function $v = u - \Phi$.

13 (Hadamard Example)**.** This exercise intends to illustrate that the Cauchy problem is not posed correctly for *every* differential equation, i.e., the solution does not necessarily depend "continuously" on the initial conditions. Consider the following Cauchy problem for the Laplace equation: Determine the solution of the differential equation

$$\frac{\partial^2 u}{\partial t^2} + \frac{\partial^2 u}{\partial x^2} = 0,$$

which at $t = 0$ satisfy the conditions

$$u(0,x) = 0, \quad \partial_t u(0,x) = \frac{1}{n^k}\sin nx,$$

where k and n are positive integers. Suppose, moreover, that $x \in \mathbb{R}$ and $t \geq 0$.

a) Verify that this problem has the solution

$$u(t,x) \;=\; \frac{e^{nt} - e^{-nt}}{2\,n^{k+1}} \sin nx.$$

How do $|\partial_t u(0,x)|$ and $u(t,x)$ behave for arbitrarily small t, if n is sufficiently large?

b) Now we suppose that we have found the solution $\tilde{u}(t,x)$ for the initial conditions

$$u(0,x) \;=\; u_0(x), \quad \partial_t u(0,x) \;=\; u_1(x).$$

What is then the solution of the Cauchy problem with the initial conditions

$$u(0,x) \;=\; u_0(x), \quad \partial_t u(0,x) \;=\; u_1(x) + \frac{1}{n^k} \sin nx?$$

Conclude that the Cauchy problem for the two-dimensional Laplace equation is ill-posed.

Bibliography

Textbooks on Analysis on Manifolds

M. P. do Carmo, *Differential forms and applications*, Universitext, Springer, Berlin, 1994.

H. Flanders, *Differential forms. With applications to the physical sciences*, Academic Press, New York–London, 1963.

O. Forster, *Analysis 3. Integralrechnung im* $\mathbb{R}^n$ *mit Anwendungen*, Vieweg-Studium, Bd. 52, 2. Auflage, Vieweg-Verlag, Braunschweig–Wiesbaden, 1989.

K. Maurin, *Analysis II*, Reidel/PWN, Warsaw, 1980.

W. Rudin, *Principles of mathematical analysis*, 3rd ed., McGraw-Hill, New York, 1976.

M. Spivak, *Calculus on manifolds*, Addison-Wesley, Reading, MA, 1965.

Textbooks on Differential Geometry

W. Blaschke, H. Reichardt, *Einführung in die Differentialgeometrie*, Grundl. der math. Wiss., Bd. 58, Springer-Verlag, Berlin, 1960.

M. P. do Carmo, *Riemannian geometry*, Birkhäuser, Basel, 1992.

M. P. do Carmo, *Differential geometry of curves and surfaces*, Prentice-Hall, Englewood Cliffs, NJ, 1976.

C. Godbillon, *Géométrie différentielle et mécanique analytique*, Hermann, Paris, 1969.

A. Gray, *Modern differential geometry of curves and surfaces with Mathematica*, 2nd ed., CRC Press, Boca Raton, FL, 1998.

M. Spivak, *Differential geometry I–V*, Publish or Perish Inc., since 1975.

K. Strubecker, *Differentialgeometrie I–III*, Sammlung Göschen, W. de Gruyter Verlag, Berlin, 1969.

R. Sulanke, P. Wintgen, *Differentialgeometrie und Faserbündel*, VEB Dt. Verlag der Wiss., Berlin, 1972.

A. Svec, *Global differential geometry of surfaces*, VEB Dt. Verlag der Wiss., Berlin, 1981.

Textbooks on Lie Groups and Lie Algebras

J. F. Adams, *Lectures on Lie groups*, Benjamin, New York, and Univ. of Chicago Press, Chicago, IL, 1969.

Th. Bröcker, T. tom Dieck, *Representations of compact Lie groups*, GTM 98, Springer-Verlag, Berlin, 1985.

J. J. Duistermaat, J. A. C. Kolk, *Lie groups*, Universitext, Springer, Berlin, 2000.

W. Fulton, J. Harris, *Representation theory – a first course*, GTM 129, Springer, Berlin, 1991.

Textbooks on Symplectic Geometry and Mechanics

V. I. Arnold, *Mathematical methods of classical mechanics*, GTM 60, Springer, Berlin, 1989.

A. T. Fomenko, *Symplectic geometry*, Advanced Studies in Contemporary Mathematics 5, Gordon and Breach Publ., Amsterdam, 1995.

L. D. Landau, E. M. Lifschitz, *Course of theoretical physics, vol. I: Mechanics*, Pergamon Press, Oxford, 3rd corr. ed., 1994.

A. Sommerfeld, *Lectures on theoretical physics, vol. I: Mechanics*, Academic Press, New York, 1952.

W. Thirring, *A course in mathematical physics, vol. 1: Classical dynamical systems*, Springer, Berlin, 1978.

Textbooks on Statistical Mechanics and Thermodynamics

R. Becker, *Theorie der Wärme*, Heidelb. Taschenb. Bd. 10, Springer-Verlag, Berlin, 1985.

L. D. Landau, E. M. Lifschitz, *Course of theoretical physics, vol V: Statistical physics*, Pergamon Press, Oxford, 3rd rev. ed., 1996.

A. Sommerfeld, *Lectures on theoretical physics, vol. V: Thermodynamics and statistical mechanics*, Academic Press, New York, 1956.

D. N. Zubarev, V. Morozov, G. Roepke, *Statistical mechanics of non-equilibrium processes*, vols. 1 and 2, Akademie-Verlag, Berlin, 1996, 1997.

Textbooks on Electrodynamics

L. D. Landau, E. M. Lifschitz, *Course of theoretical physics, vol. II: The classical theory of fields*, 4th rev. ed., Butterworth-Heinemann, Oxford, 1996.

A. Sommerfeld, *Lectures on theoretical physics, vol. III: Electrodynamics*, Academic Press, New York, 1952.

A. Sommerfeld, *Partial differential equations in physics*, Academic Press, New York, 1949.

W. Thirring, *A course in mathematical physics, vol. 2: Classical field theory*, Springer, Berlin, 1986.

Symbols

Index

Titles in This Series

52 **Ilka Agricola and Thomas Friedrich,** Global analysis: Differential forms in analysis, geometry and physics, 2002
51 **Y. A. Abramovich and C. D. Aliprantis,** Problems in operator theory, 2002
50 **Y. A. Abramovich and C. D. Aliprantis,** An invitation to operator theory, 2002
49 **John R. Harper,** Secondary cohomology operations, 2002
48 **Y. Eliashberg and N. Mishachev,** Introduction to the h-principle, 2002
47 **A. Yu. Kitaev, A. H. Shen, and M. N. Vyalyi,** Classical and quantum computation, 2002
46 **Joseph L. Taylor,** Several complex variables with connections to algebraic geometry and Lie groups, 2002
45 **Inder K. Rana,** An introduction to measure and integration, second edition, 2002
44 **Jim Agler and John E. McCarthy,** Pick interpolation and Hilbert function spaces, 2002
43 **N. V. Krylov,** Introduction to the theory of random processes, 2002
42 **Jin Hong and Seok-Jin Kang,** Introduction to quantum groups and crystal bases, 2002
41 **Georgi V. Smirnov,** Introduction to the theory of differential inclusions, 2002
40 **Robert E. Greene and Steven G. Krantz,** Function theory of one complex variable, 2002
39 **Larry C. Grove,** Classical groups and geometric algebra, 2002
38 **Elton P. Hsu,** Stochastic analysis on manifolds, 2002
37 **Hershel M. Farkas and Irwin Kra,** Theta constants, Riemann surfaces and the modular group, 2001
36 **Martin Schechter,** Principles of functional analysis, second edition, 2002
35 **James F. Davis and Paul Kirk,** Lecture notes in algebraic topology, 2001
34 **Sigurdur Helgason,** Differential geometry, Lie groups, and symmetric spaces, 2001
33 **Dmitri Burago, Yuri Burago, and Sergei Ivanov,** A course in metric geometry, 2001
32 **Robert G. Bartle,** A modern theory of integration, 2001
31 **Ralf Korn and Elke Korn,** Option pricing and portfolio optimization: Modern methods of financial mathematics, 2001
30 **J. C. McConnell and J. C. Robson,** Noncommutative Noetherian rings, 2001
29 **Javier Duoandikoetxea,** Fourier analysis, 2001
28 **Liviu I. Nicolaescu,** Notes on Seiberg-Witten theory, 2000
27 **Thierry Aubin,** A course in differential geometry, 2001
26 **Rolf Berndt,** An introduction to symplectic geometry, 2001
25 **Thomas Friedrich,** Dirac operators in Riemannian geometry, 2000
24 **Helmut Koch,** Number theory: Algebraic numbers and functions, 2000
23 **Alberto Candel and Lawrence Conlon,** Foliations I, 2000
22 **Günter R. Krause and Thomas H. Lenagan,** Growth of algebras and Gelfand-Kirillov dimension, 2000
21 **John B. Conway,** A course in operator theory, 2000
20 **Robert E. Gompf and András I. Stipsicz,** 4-manifolds and Kirby calculus, 1999
19 **Lawrence C. Evans,** Partial differential equations, 1998
18 **Winfried Just and Martin Weese,** Discovering modern set theory. II: Set-theoretic tools for every mathematician, 1997
17 **Henryk Iwaniec,** Topics in classical automorphic forms, 1997
16 **Richard V. Kadison and John R. Ringrose,** Fundamentals of the theory of operator algebras. Volume II: Advanced theory, 1997
15 **Richard V. Kadison and John R. Ringrose,** Fundamentals of the theory of operator algebras. Volume I: Elementary theory, 1997

TITLES IN THIS SERIES

14 **Elliott H. Lieb and Michael Loss,** Analysis, 1997
13 **Paul C. Shields,** The ergodic theory of discrete sample paths, 1996
12 **N. V. Krylov,** Lectures on elliptic and parabolic equations in Hölder spaces, 1996
11 **Jacques Dixmier,** Enveloping algebras, 1996 Printing
10 **Barry Simon,** Representations of finite and compact groups, 1996
9 **Dino Lorenzini,** An invitation to arithmetic geometry, 1996
8 **Winfried Just and Martin Weese,** Discovering modern set theory. I: The basics, 1996
7 **Gerald J. Janusz,** Algebraic number fields, second edition, 1996
6 **Jens Carsten Jantzen,** Lectures on quantum groups, 1996
5 **Rick Miranda,** Algebraic curves and Riemann surfaces, 1995
4 **Russell A. Gordon,** The integrals of Lebesgue, Denjoy, Perron, and Henstock, 1994
3 **William W. Adams and Philippe Loustaunau,** An introduction to Gröbner bases, 1994
2 **Jack Graver, Brigitte Servatius, and Herman Servatius,** Combinatorial rigidity, 1993
1 **Ethan Akin,** The general topology of dynamical systems, 1993